MRE

Materials Research and Engineering
Edited by B. Ilschner and N. J. Grant

E. Dörre · H. Hübner

Alumina

Processing, Properties, and Applications

With 178 Figures

Springer-Verlag
Berlin Heidelberg New York Tokyo
1984

Dr. ERHARD DÖRRE

Feldmühle Aktiengesellschaft, Plochingen

Prof. Dr. HEINZ HÜBNER

Arbeitsbereich Werkstoffphysik und -technologie,
Technische Universität Hamburg-Harburg

Dr. rer. nat. BERNHARD ILSCHNER

o. Professor, Laboratoire de Métallurgie Mécanique,
Département des Matériaux, EPFL, Lausanne

Prof. NICHOLAS J. GRANT

Department of Materials Science and Engineering, Cambridge

Library of Congress Cataloging in Publication Data.

Dörre, E. (Erhard)
Alumina: processing, properties, and applications.
(Materials resarch and engineering;)
Bibliography: p.
Includes index.
1. Aluminum oxide. 2. Ceramic materials.
I. Hübner, H. (Heinz). II. Title. III. Series.
TP810.5.D67 1984 666'.028 84-14090

ISBN-13:978-3-642-82306-0   e-ISBN-13:978-3-642-82304-6
DOI: 10.1007/978-3-642-82304-6

2061/3020-543210

*Dedicated to
Walther Dawihl*

Editor's Preface

This is the third book in the new series "Material Research and Engineering", devoted to the science and technology of materials. "MRE" evolves from a previous series on "Reine und Angewandte Metallkunde", which was edited by Werner Köster until his eightieth birthday in 1976. For the new series, the presentation as well as the scope had to be modified. In particular, the scientific and technological links between volumes on metallic, non-metallic, and composite materials should reflect the successful development of materials science and engineering within the last two decades.

Thus, the material provided by Dörre and Hübner for the present volume is particularly welcome. Alumina as a ceramic material has received very large attention as an object of scientific investigation in all of its aspects. Additionally, it plays a leading role as a nonmetallic material in many fields of technical application.

This book deals with both aspects: in Chapter 2 (physical properties) and 3 (mechanical properties), H. Hübner presents an outstanding documentation of what one might call the science of alumina, based on 560 literature references and 15 years of personal experience gained from experimental and theoretical work in university laboratories in Erlangen, Rio de Janeiro, and Hamburg. In Chapter 4 (fabrication) and 5 (applications), E. Dörre is the responsible author, and again the reader may take advantage of the author's great experience from his longstanding affiliation with industry and from his own contributions to the development of high quality alumina ceramics for application in electronics, mechanical engineering and other fields.

In conclusion: The present book by Dörre and Hübner places comprehensive and up-to-date information on both the scientific and the technological aspects of alumina at the disposition of the reader. It may be expected to have a further impact on research and development in this field.

Lausanne, Cambridge, Mass., USA   B. Ilschner, N.J. Grant
September 1984

Author's Preface

Activity in the fields of processing, research, and applications of structural ceramics has increased considerably in recent years. More and more laboratories deal with ceramic materials and contribute to a growing number of publications. Similarly, increasing numbers of companies produce larger quantities and new types of ceramic components.

In these times of ever scarcer raw materials and increasing energy costs, the potential of ceramic materials has been recognized more fully. The interest which ceramics enjoy today is based on a few essential features: (1) Ceramic materials are not strategic materials, but can be obtained in unlimited quantities. (2) In energy conversion equipment, ceramic materials permit distinctly higher operating temperatures and, therefore, greater efficiencies. For this reason the use of ceramics in automobile engines and in stationary as well as aircraft turbines is intensively studied today. (3) The low specific weight of ceramic materials as compared to metallic superalloys leads to further cost and energy savings. Lighter moving parts result in smaller mechanical loads and thus allow additional reduction of section thickness. In particular, reduction of the total weight in aerospace applications leads to a considerable fuel saving. Consequently, attempts are being made to substitute refractory metals with ceramics. Additionally, new applications are continually becoming available by using ceramics of improved strength.

Aluminum oxide is that structural ceramic which has been successfully employed for the longest time. This material is so well established, that it has gradually lost attention as the subject of advanced research. At the same time, other ceramics, such as the new silicon ceramics or ZrO_2-based materials, are currently attracting the attention of workers. This seems to be an opportune time to take account of the accumulated information on alumina by compiling the available knowledge and presenting it in a monograph.

However, the present book is not simply a data collection on alumina. On the contrary, it was our intention to find a compromise between a reference book and a text-book in how this knowledge was presented. The properties and phenomena remain ill-understood if they cannot be related to the fundamental mechanisms and theoretical framework of materials science. Therefore, in this book the designer and potential user of alumina will find not only a description of the current applications and needed data for the material, but also a treatment of the mechanisms controlling its behavior. The material scientist interested in ceramic research will find an updated interpretation of material behavior in terms of current models. The access for the graduate student who wishes to familiarize himself with the behavior of a specific structural ceramic is facilitated through introductory passages in each section.

The authors would like to thank Professor Bernhard Ilschner for stimulating this monograph. Furthermore, our thanks are due to Feldmühle Aktiengesellschaft, Plochingen, which encouraged the production of this book by providing extensive illustrative material as well as financial support. Grateful acknowledgement is made to the following colleagues of Feldmühle for valuable discussions: Robert Abel, Dr. Ulf Dworak, Heinz Eichas, Karl Dieter Fuchs, Volker Gomoll, Hans Jud, Hans-Gerd Rittel, and Gert Wloka. The use of the technical services granted by the Instituto Militar de Engenharia, Rio de Janeiro, where the greater part of the literature study on the physical and mechanical properties was carried out, is gratefully acknowledged. Thanks are also due to Dr. Jules R. Routbort for a critical revision of the chapters on the material properties. We also thank Dr. Jean K. Gregory for her extensive and competent linguistic revision of the English text. Finally, the work of Christine Patzschke who carefully typed the manuscript is gratefully appreciated.

Plochingen and Hamburg, August 1984　　　　　　　　　　　　E. Dörre

　　　　　　　　　　　　　　　　　　　　　　　　　　　　　H. Hübner

Contents

1 Introduction

1.1 Scope of the Book

This book is concerned with the ceramic material alumina (α-Al_2O_3). Like many other ceramics, alumina is not only fabricated and used in its pure, single-phase configuration Al_2O_3, but there exists a wide series of material compositions which start from Al_2O_3 as a base material and contain one or more other ceramic or even metallic phases. This book will deal with material compositions which consist of at least 95% alumina. For this range of composition the term high alumina ceramics has come into use in the literature. With increasing content of alumina towards pure aluminum oxide materials containing at least 99% alumina will be called high-purity alumina ceramics, having additives in the range of some tenths of a per cent in order to control the sintering process. When dealing with any material from the range above 95% alumina without referring specificly to its exact composition it will, for brevity, be denoted alumina, aluminum oxide, or Al_2O_3.

In some cases the behavior of single-crystal α-Al_2O_3 (sapphire) will also be discussed. A detailed treatment of sapphires and rubies, however, is beyond the scope of the book. Materials from the series of β-aluminas will not be reviewed either, because they are not considered to be a modification of Al_2O_3 but a compound consisting of Al_2O_3 and alkaline oxides. Alumina-based ceramics of high porosity and γ-alumina will be excluded as well.

It is intended in this book to give a review and a summary of the present state of knowledge of high alumina ceramics and high-purity alumina ceramics. The physical and mechanical material properties will be discussed in Sections 2 and 3. Optical and magnetic properties will not be treated because they do not play an important role in the applications of alumina. Section 4 will deal with fabrication processes and the applications will be decribed in Section 5.

Before starting with a detailed review of alumina, some fundamental aspects of the behavior of ceramic materials will be addressed and the history of the development and applications of alumina will be traced briefly.

1.2 General Remarks on the Use and the Behavior of Ceramic Materials

The field of ceramic materials is extremely vast and varied. This holds for compositions and forms as well as for properties and applications. On the one hand, one finds the traditional ceramic materials based on silicate products which in some cases have been known and used for many centuries or even for millenia. The manifold types of clayware, pottery, structural clay products, sanitary ware, cement, porcelains, and silicate glasses belong to this group of ceramics. Beyond this traditional field, on the other hand, an extraordinary variety of special ceramics has arisen within the last decades which has found applications in many areas of industry and research. Depending on the intended application, these special ceramics may be classified into four principal groups, i.e., ceramics for electronic, optical, nuclear, and structural applications. Special ceramics often show very specialized properties because they have been designed for exactly defined applications.

Alumina can be considered a typical representative of the group of structural ceramics. These ceramics are intended particularly to serve as structural parts subjected to mechanical loads, in many cases at high temperatures. The main function of these materials is to sustain mechanical loading. Thus, the common feature of structural ceramics is good mechanical behavior and, therefore, efforts in developing, fabricating, and optimizing these materials are mainly concentrated on the aim of a high strength. In many cases the development of high-strength ceramics is carried out with the objective of substituting metallic materials. As examples of the group of structural ceramics we cite, in addition to alumina, some other pure oxide ceramics such as yttria (Y_2O_3), titania (TiO_2), zirconia (ZrO_2), magnesia (MgO), and alumina-magnesia spinel ($MgAl_2O_4$), as well as the two ceramics which are currently being developed for gas turbine applications, silicon nitride (Si_3N_4) and silicon carbide (SiC).

In many other practical applications alumina is used as an electrical insulating material. Thus, it is clear that the electrical properties are more pertinent in such cases. From these brief considerations it can be concluded that reviewing the properties of alumina is to pay particular attention to the mechanical and electrical behavior.

2

A definition of what is a ceramic material has necessarily to be very general if it is to cover the whole spectrum of compositions and properties of ceramics. The description given by Kingery et al. [1.1] which is generally accepted today, defines ceramics as materials "which have as their essential component, and are composed in large part of, inorganic nonmetallic materials."

This definition characterizes an entire class of materials which in general differs very strongly from metallic materials. In comparison to metals, ceramics have a series of prejudicial properties which, for a long time, impeded widespread use and led to only limited applications. This holds particularly for the mechanical behavior because ceramic materials are characterized by low strength when subjected to tensile loading; by brittleness, i.e., by a complete absence of plasticity at the usual service temperatures; by a considerable statistical dispersion of strength values; by susceptibility to thermal and mechanical shock loading; and by the phenomenon of time-dependent strength degradation, a property which is found particularly with alumina grades of lower purity. On the other hand, ceramics show a unique combination of favorable properties, especially at high temperatures. Among these are good creep strength, high hardness and high wear resistance, chemical inertness and resistance against high temperature corrosion even when operated in air, as well as high electrical resistivity. Without doubt it is due to these good material properties that a trend toward an increasing utilization of ceramic materials has been initiated, whether in novel technical applications or as a substitute for metallic materials. This development will be shown in detail in the case of alumina.

The driving force behind this progress in ceramics, which also applies to alumina, may be attributed to three different developments: The appearance of new and more rigorous demands with respect to novel materials and properties; the improvement of fabrication facilities; and a deeper and expanded understanding of the behavior and properties of ceramics. As to alumina and its compositions, demands from very different industrial areas have caused the intensified occupation with this material. To mention some of them, applications in mechanical and electrical engineering, the paper, textile, machine tool, and chemical industries, the field of laboratory equipment, electronics, the computer industry, and medicine are cited. All these applications will be discussed extensively in Section 5. Other special ceramics have been designed, for example, for electronic purposes, semiconductor devices, computer memories, optical filters, and nuclear fuels.

The improvement of fabrication facilities and techniques has strongly contributed to an improvement of the quality of ceramic products. This observation holds for the whole family of new ceramic materials, including alumina, as is shown in Section 4. Most ceramics are fabricated by sintering. Many properties of sintered bodies can be improved by the use of higher sintering temperatures as well as by a more precise control of the temperature during the sintering process. The freedom in the choice of the furnace atmosphere (oxidizing, reducing, neutral, or vacuum) even at very high temperatures signifies great progress in the manufacturing technology. The manufacture of many ceramics of suitable quality turns out to be possible only if the temperature and the atmosphere can be controlled accurately. A further aspect, which also accounts for the improvement in the quality of sintered ceramic parts, is the purity of the raw material. A high degree of purity of the starting powder contributes essentially to an improvement in the properties of the products. The increase in purity results in a diminution of the impurity level and hence in the reduction of second phase precipitates, of grain boundary segregations, and of lattice defects. Controlling the impurity level makes it possible to considerably improve the electrical, chemical, optical, and mechanical properties of a ceramic material.

The problem of impurities is closely related to the origin of the starting material. Essentially higher degrees of purity can be achieved when synthetic starting materials, rather than natural, i.e., mineral raw materials, are used. Modern special ceramics, for example, the whole series of pure oxide ceramics, ceramic semiconductors, ferroelectrics, and optical ceramics, are characterized by the fact that they are produced exclusively from synthetic raw materials. For the manufacture of the traditional silicate ceramics, however, raw materials of natural occurrence are normally used.

Ceramic research has contributed increasingly during the last decades to expanding the understanding of the behavior of ceramic materials. Starting from elementary foundations of solid state physics and materials science, many properties and effects of ceramic materials have been studied and, at least in principle, explained. The utilization of the knowledge acquired on ceramic behavior and effects may be viewed under two different aspects, i.e., with respect to the optimization of the sintering process, and with respect to the design of ceramic materials of desired properties. As to the first aspect, fabrication has been facilitated and improved by an understanding of transport phenomena in ceramics. The properties of the finished product can be optimized by a suitable choice of the sintering conditions, for example, atmosphere, temperature, and characteristics of the starting powder. The addition of sintering aids results in significant improvements in density and grain size control and, therefore, in all those properties that depend on these two basic values. Sintering aids also result in a reduction of sintering temperature and time, thus reducing the price of the product.

4

As to the second aspect, designing ceramic materials has become possible due to the awareness of the fact that the properties of a ceramic material are determined, not so much by its composition as by its microstructure. Important microstructural parameters controlling the behavior of a ceramic material are grain size and porosity, the size and the spatial distribution of pores, the presence and distribution of second phases, and segregation effects. Also of importance are internal stresses due to elastic and thermal anisotropies. Properties such as mechanical strength, electrical conductivity, and diffusion rate are extremely dependent on the microstructural characteristics. These effects will be discussed extensively in Section 2 and 3. The control of the microstructure is an important means of obtaining modifications of physical and mechanical properties in a given and desired way and turns out to be the crucial factor in the design of ceramic materials. Recent successes in designing ceramics are the development of the new high-strength turbine ceramics and the toughening of brittle materials by controlled microcracking and transformation [1.2].

1.3 The History of Alumina

Scientific investigations on alumina date back to the last century. The first knowledge of a commercial use, however, goes back to 1907, when a patent was applied for, describing the production of a high alumina ceramic material [1.3]. Commercial production and application on a larger scale started in the late 1920s and the early 1930s [1.4,1.5], after a series of basic requirements for material properties could be satisfied as a result of the development of the suitable high-temperature furnace technology. The fundamental way of sintering high-purity alumina ceramics of the composition still in use at present, i.e., the addition of a small amount of MgO as a sintering aid, was reported for the first time in 1936 by a patent [1.6] that described the use of magnesium compounds which form MgO during sintering.

The first practical uses of high alumina ceramics were as spark-plug insulators and laboratory equipment, followed by further applications in the field of electronics and mechanical engineering after World War II. Since spark-plug insulators normally contain less than 95% Al_2O_3, they will be excluded from the following statistical considerations. Laboratory equipment will be treated in Section 5.5 under "Other Applications."

The general utilization of alumina spread quickly into many areas of modern industry. Figure 1.1 gives the progress of utilization of high alumina ceramics in the various fields of application during the last five decades, but without considering the actual importance or the production volume of the respective areas.

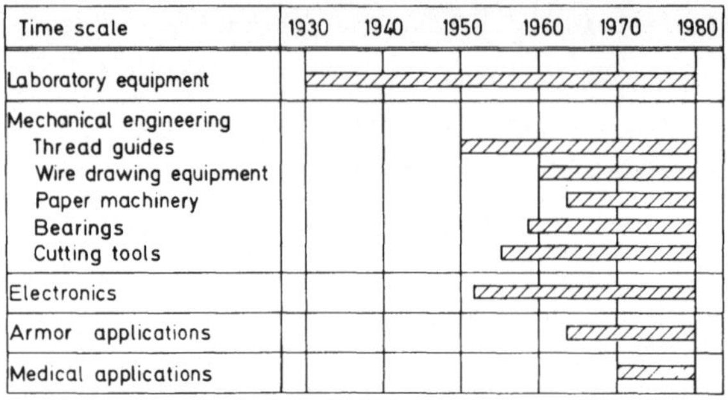

Time scale	1930	1940	1950	1960	1970	1980
Laboratory equipment	▨	▨	▨	▨	▨	
Mechanical engineering						
Thread guides			▨	▨	▨	
Wire drawing equipment				▨	▨	
Paper machinery				▨	▨	
Bearings			▨	▨	▨	
Cutting tools			▨	▨	▨	
Electronics			▨	▨	▨	
Armor applications				▨	▨	
Medical applications					▨	

Fig. 1.1. Advance of utilization of high alumina ceramics in various fields of application

As already outlined in Section 1.2, the growth of alumina applications can be explained by three different facts: new demands, improvements in fabrication facilities, and increase in knowledge. The history of increasing utilization of high alumina ceramics is strongly related to advances in high-temperature furnace technology. The improvement of furnace materials and the increase in sintering temperature during the period between 1930 and 1950 resulted in improved quality of the sintered products, particularly with respect to the purity and density of the material

The technical and economic importance of high alumina ceramics is illustrated in Fig. 1.2, which shows the growth of the annual production volume of the material with respect to producing countries or geographic regions. The production costs decreased with increasing quantities, compensating approximately for the inflation rate. Based on an evaluation of a large amount of miscellaneous information, Fig. 1.2 illustrates trends in world production. Figure 1.3 includes a distribution of the production volume related to major fields of application. Section 5 returns to this subject and the various applications of the material are described in detail.

Further information on the continuously increasing importance of alumina can be obtained by tracing the number of publications which deal with scientific or technological activities on the subject. The activities increased exponentially during the last three decades. This fact documents the strong interest which still exists among scientists and manufacturers in increasing their understanding of the material behavior and improving its performance.

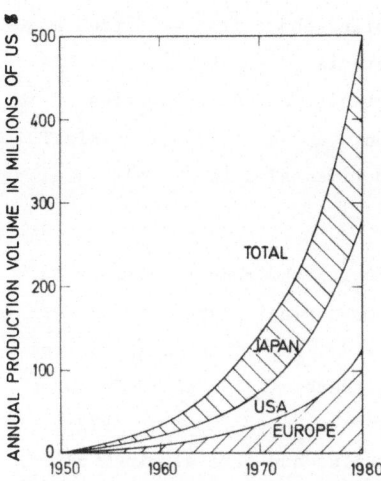

Fig. 1.2. Development of world production volume of high alumina ceramics

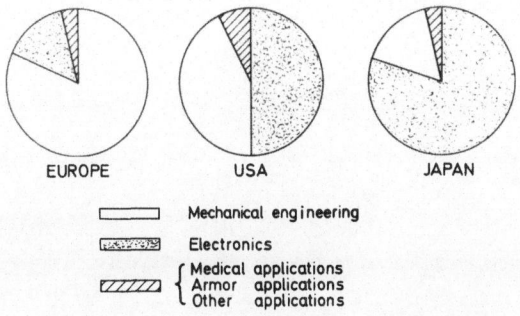

Fig. 1.3. Present distribution of production volume of high alumina ceramics to various fields of application

1.4 Preceding Summarizing Literature on Alumina

The first comprehensive description of the fabrication and properties of alumina appeared in the work of Ryshkewitch entitled "Oxide Ceramics." The first edition was published in German in 1948 [1.7], followed by an English version in 1960 [1.8]. Due to the scope of Ryshkewitch's book, which was to give a summary of the current state-of-the-art of metal oxides, only a small amount of space was devoted to the treatment of aluminum oxide.

The next review work to appear on alumina was compiled and edited by Gitzen and published under the title "Alumina as a Ceramic Material," in 1970 [1.9]. This work is mainly concerned with a treatise of characteristics and properties of the starting materials used for alumina fabrication; yet, a series of physical and mechanical properties is reviewed and many experimental data on these are compiled.

There are two further review works on ceramics to which the reader is referred. Evans and Langdon published a detailed summary on mechanical properties of ceramic materials in 1976 entitled "Structural Ceramics" [1.2]. Much information on the brittle and plastic behavior of Al_2O_3 can be found there. This also holds for Davidge's recent book "Mechanical Behaviour of Ceramics," published in 1979 [1.10] which is concerned to some extent with alumina ceramics and pure aluminum oxide.

Finally, for an introductory study of the general behavior of ceramics, the reader is referred to a standard work on ceramics, "Introduction to Ceramics" by Kingery et al. [1.1]. The book also cites much original research work on Al_2O_3 and describes a range of properties of this material.

2 Physical Properties

In this chapter we review those physical properties of alumina which are of particular significance for technical and structural applications. Properties of both single-crystal and polycrystalline material will be treated.

After dealing with some basic facts on the crystal structure a series of thermal properties will be described, e.g., heat capacity, thermal expansion, and thermal conductivity. Then the defect structure and its relation to the physical properties like electrical conductivity, diffusion, and sintering behavior will be discussed. Finally, some results on the segregation behavior of the material will be compiled.

When possible and necessary, special emphasis will be put on the effects of technically important variables on the material properties such as porosity, grain size, impurities and/or dopants, and ambient atmosphere.

2.1 Structure

The extensive historic development of the discovery and research of the individual crystallographic phases of alumina bas been traced by Gitzen [2.1] in his ample work on aluminum oxide ceramic technology. He reported that seven crystallographic phases of calcined, water-free alumina have been found. Structural applications of aluminum oxide, however, are limited almost entirely to the α phase (α-Al_2O_3), also called corundum or, in its single-crystalline form, sapphire. After Gitzen [2.1], the first approximate determination of the crystallographic structure of corundum dates from the work of Bragg and Bragg in 1915, and the first exact attribution to the rhombohedral structure ("corundum structure") was made by Pauling and Hendricks in 1925. Further crystallographic studies were carried out by Winchell [2.2]. He proposed a numerical procedure

and a particular stereographic projection map of corundum for the determination
of the orientation of principal crystallographic planes and directions from Laue
back-reflection patterns. A very detailed description of the crystallography of
sapphire single crystals was given by Kronberg [2.3]. The structure of aluminum
oxide consists of close packed planes of the large oxygen ions stacked in the
sequence A-B-A-B, thus forming a hexagonal close packed array of anions. The
cations are placed on the octohedral sites of this basic array and form another
type of close packed planes which are inserted between the oxygen layers. To
maintain charge neutrality, however, only two thirds of the octohedral sites
available are filled with cations. Figure 2.1, which is taken from Ref. [2.3],
illustrates the packing of Al and O in the basal plane. Since the vacant octohe-

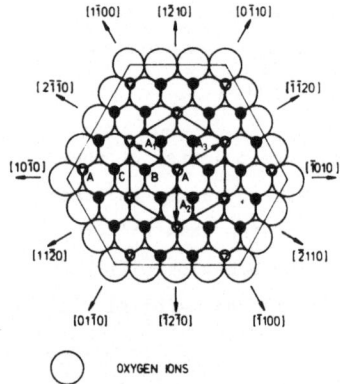

OXYGEN IONS
ALUMINUM IONS
EMPTY OCTAHEDRAL SITES

Fig. 2.1. Packing of Al and O ions in the basal plane. The
upper layer of O ions is not shown. Basal hexagonal cell
vectors and directions are indicated. - From Kronberg [2.3]

dral sites also form a regular hexagonal array, three different types of cation
layers can be defined, depending on the position of the vacant cation site
within the layer, which may be named a, b, and c, and which are stacked in the
sequence a-b-c-a-b-c. This gives the complete stacking sequence of anion and
cation layers of the form A-a-B-b-A-c-B-a-A-b-B-c-A. It is only reproduced
after the sixths oxygen layer, or after the sequence a-b-c is repeated twice.
The unit cell defined in this way is called the crystallopraphic, or structural
unit cell, in contrast to the morphological unit cell, where the cation sequence
is repeated only once and the height is half the height of the structural cell.
This important difference between the two cells which has given rise to some
confusion in the literature dealing with crystallographic indices of sapphire,
was pointed out clearly by Kronberg [2.3].

The structure of α-Al_2O_3 results in coordination numbers of 6 and 4 for the
cation and the anion, respectively. The ionic radii for this coordination are
0.053 nm for Al^{3+} and 0.138 nm for O^{2-} [2.4].

10

Lattice parameters a_o and c_o are given in Table 1. The data of Swanson et al. [2.5] were considered by Gitzen [2.1] as the best values in 1970, and were also cited by Lynch [2.6] in 1975. The values published by Jan et al. [2.7], however, seem to be somewhat more precise. The latter values were also used by Rasmussen

Table 2.1. Lattice parameters and theoretical density at room temperature

Author	Ref.	Year	a_o, nm	c_o, nm	ρ_{th}, g/cm^3
Kronberg	2.3	1957	0.475	1.297	4.01
Swanson et al.	2.5	1960	0.4758	1.2991	3.987
Jan et al.	2.7	1960	0.47591	1.29894	3.9862
Rossi and Lawrence	2.9	1970	0.4758	1.2990	3.987
Phillips et al.	2.10	1980	0.475923	1.299208	3.9851

and Kingery [2.8] as reference data in the evaluation of high precision density measurements of doped Al_2O_3 single crystals. Additional and more recent data were given by Rossi and Lawrence [2.9] and by Phillips et al. [2.10].

The theoretical density, ρ_{th}, can be calculated from the lattice parameters using the relation (2.1)

$$\rho_{th} = \frac{4 \cdot 3^{1/2} M}{N \cdot a_o^2 \cdot c_o} , \qquad (2.1)$$

where M is the molecular weight, N is Avogadro's number, and the numerical factor is determined by the particular structure of α-Al_2O_3. Table 2.1 also contains some ρ_{th} data.

2.2 Thermal Properties

In this section we shall discuss some selected thermal properties of Al_2O_3, the knowledge of which is of some use in practice. For a more detailed treatment of the subject the reader is referred to special compilations of thermal properties existing in the literature, for example, to the handbooks of Goldsmith et al. [2.11], Gitzen [2.1], Weast [2.12], and Lynch [2.6].

The melting point of aluminum oxide has been determined in many investigations. The first experimental determination after the adoption of the International Temperature Scale in 1948 was carried out by Diamond and Schneider [2.13]. They found a melting point of 2025°C in a solar furnace. Gitzen [2.1] cites four different determinations from 1966, with values between 2037 and 2051°C, giving preference to a value of $2051.0 \pm 9.7^{\circ}$C. The melting temperature cited by Lynch [2.6] is $2047.5 \pm 8^{\circ}$C.

The surface energy per unit area or thermodynamic surface energy, γ_o, is an important parameter in sintering because it controls both sintering kinetics and the formation of surface roughness of as-sintered surfaces. It is also of some interest in the understanding of fracture processes, especially of single crystals. A theoretical assessment of γ_o was given recently by Davidge [2.14] using a simple model originally developed by Orowan which assumes a sinusoidal function of stress versus displacement during the separation of two lattice planes. The model gives $\gamma_o \simeq Er_o/125$, where E is the elastic modulus of the material and r_o is the equilibrium distance between atoms. For Al_2O_3, the formula gives $\gamma_o \simeq 1.2$ J/m^2 [2.14], which agrees reasonably well with experimental data presented below.

Experimental determinations of the specific surface energy of alumina have employed essentially two different techniques, namely, the measurement of the contact angle of a liquid metal on the alumina surface [2.15, 2.16] and the pendant-drop technique [2.17,2.18,2.19]. The experimental value determined by Norton et al. [2.15] at 1850°C is 0.905 J/m^2 and the value estimated for 1000°C is 1.0 J/m^2. The work of Rhee [2.16] gave, for the (0001) plane of single crystals, values from 0.772 J/m^2 to 0.646 J/m^2 for temperatures from 1000°C up to the melting point. The surface energy at the melting point was determined by Kingery [2.17] and, more recently, by Rasmussen and Nelson [2.18]. The values found were 0.69 J/m^2 and 0.638 ± 0.100 J/m^2, respectively. Rasmussen [2.19] reported that there was no difference between the surface energy of molten polycrystalline aluminum oxide containing 0.25% MgO and molten single-crystal material at the melting point.

The value of the dihedral angle of grain-boundary grooves on the surface of polycrystalline alumina was determined by Kingery [2.20] to be 152 degrees, which resulted in a ratio of grain-boundary energy, γ_{gb}, to surface energy, $\gamma_{gb}/\gamma_o = 0.48$.

The heat capacity as a function of temperature is shown in Fig. 2.2. The three

different curves taken from the works of Engel [2.21], Gitzen [2.1], and Kingery
et al. [2.4] agree within ± 6%. At high temperatures, the curves vary only
slightly with temperature or even become constant at a level close to
$3R = 24.94$ J/g-atom·K, where R is the universal gas constant (R = 8.314 J/mol·K),
as postulated by the classical kinetic theory of heat.

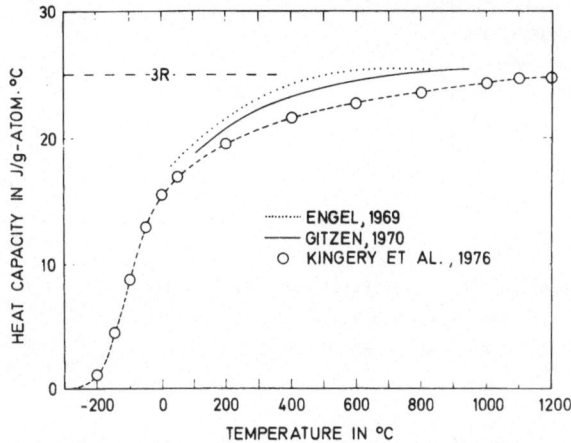

Fig. 2.2. Heat capacity of Al_2O_3 at various temperatures [2.1,2.4,2.21]

Figure 2.3 shows some results on the thermal expansion at various temperatures.
Coble and Kingery [2.22] measured the linear thermal expansion coefficient between room temperature and 1200°C on samples having an average grain size of
23 μm and an artificially introduced porosity between 4% and 49%. No influence
of the porosity on the expansion coefficient was detected. Wachtman et al. [2.23]
determined the relative linear thermal expansion of single-crystal Al_2O_3 parallel
and perpendicular to the c-axis and of a polycrystalline material between 100 and
1100 K. In Fig. 2.3, the slope of their curves which is the expansion coefficient
is plotted. The anisotropy of the hexagonal structure is obvious, the expansion
parallel to the c-axis exceeding the expansion in the basal plane by about 10%
between 100 and 800°C. The values for the polycrystalline aluminum oxide are
intermediate between those for the two single-crystal orientations. Expansion-
coefficient data up to 1800°C were reported by Nielsen and Leipold [2.24], and
are also included in Fig. 2.3. No difference in the results was found between
pure Al_2O_3 and on Al_2O_3 containing 0.1% MgO.

The data of Fig. 2.3 show the typical levelling-off of the thermal expansion
coefficient at elevated temperatures, and the analogy to the change in heat
capacity with temperature, Fig. 2.2, is evident. This observation is in accordance
with basic predictions of solid state physics [2.25]. Due to the fact that both

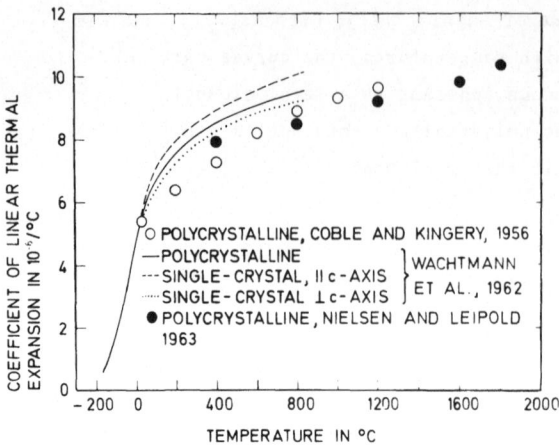

Fig. 2.3. Coefficient of linear thermal expansion of single-crystal and polycrystalline
Al_2O_3 as a function of temperature. Data points from Coble and Kingery [2.22] and Nielsen
and Leipold [2.24]. Curves evaluated from data of Wachtman et al. [2.23]

changes in volume and changes in heat content are provoked by increased lattice
vibrations, a parallel change in the thermal expansion coefficient and in heat capacity is expected. The validity of this statement was verified for Al_2O_3 and a
series of other refractory materials by Hoch and Vernardakis [2.26].

The transport of heat through a ceramic material is achieved by essentially two
parallel transport mechanisms, namely, phonon conduction and electromagnetic
radiation [2.4]. At low temperatures the transport process is controlled mainly
by the mean free path of phonons between two collisions, which varies with
the inverse of temperature, $1/T$. At elevated temperatures, however, electromagnetic radiation becomes increasingly predominant, which compensates for the
decrease of thermal conductivity, k, caused by the increasing scattering of
phonons, and leads to a variation of $k \propto T^3$. Thus, a minimum in k as a function
of temperature is expected; this is found to lie in the temperature range
of about 1000 to $1500^{\circ}C$ for many ceramic materials [2.4].

The thermal conductivity of Al_2O_3 has been determined by many investigators.
Figure 2.4 demonstrates that the agreement between the results obtained by
several authors is very good and that the material shows all the general features
stated above, the minimum in k occurring at about $1300^{\circ}C$. Charvat and Kingery
[2.27] reported values for both single crystals and polycrystalline aggregates.
Since thermal conductivity is strongly affected by porosity, P, the results
on polycrystalline samples were corrected for zero porosity using the expression

14

$k = k_p/(1 - P)$, where k_p is the experimental conductivity. No effect of grain size was detected, but the single-crystal values were found to be increasingly higher with increasing temperature, with the minimum lying at a lower temperature, which was interpreted in terms of the greater transparency of the single crystals compared to the polycrystalline samples. The curve from Gitzen [2.1] represents the mean of nine different references cited there. Nishijima et al. [2.28] fitted the expression $k = A/(B + T) + CT^3$, with A, B, and C being three different constants, to their experimental data, which is also plotted in Fig. 2.4. The data points of Fitzer and Weisenburger [2.29] agree very well with Charvat and Kingery's

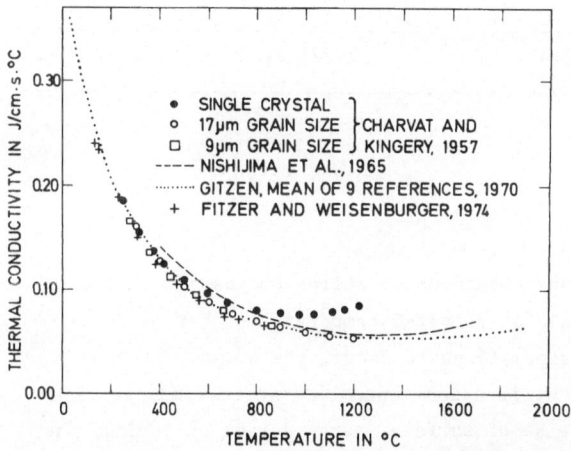

Fig. 2.4. Thermal conductivity of Al$_2$O$_3$ at various temperatures. Data points of single-crystal and polycrystalline material from [2.27,2.29], curves from [2.1,2.28]

values. Slight deviations between different results may be due to the variation of even small amounts of impurities, which can considerably affect the experimental conductivity, as pointed out in [2.27]. The effect can be understood in terms of the loss of periodicity of the regular ionic array caused by the incorporation of both substitutional or interstitial impurity ions.

In Table 2.2 numerical values of the thermal properties treated in the foregoing paragraphs are summarized for various temperatures.

Table 2.2. Summary of selected thermal properties of aluminum oxide at various temperatures

Temperature	Specific surface energy, γ_o J/m^2	Heat capacity, C_p $J/mol \cdot K$	Linear thermal expansion $10^{-6}/K$		Thermal conducti- vity, k $J/cm \cdot s \cdot K$
Room temp.	–	15.5	5.43	–	0.360
400°C	–	21.7	7.33	8.00	0.128
800°C	–	23.7	8.97	8.53	0.070
1200°C	–	24.8	9.73	9.27	0.055
1600°C	–	–	–	9.87	0.056
Melting point	0.638	–	–	–	–
References	2.19	2.4	2.22	2.24	2.1

2.3 Diffusion

Diffusion is one of the most important phenomena in solids because it controls nearly all the properties of materials at elevated temperatures. For micro-structural changes or chemical transformations to occur, the atoms must have some mobility. The driving force of a diffusion-controlled process may be of very different kind, e.g., the decrease of surface energy during sintering, the change of the free energy in a phase transformation, or the gain of mechanical work in creep. The kinetics of the process, however, is determined in most cases by the mobility of the diffusing atoms. (Exceptions exist when the rate of the process ist fixed by the rate of the transformation reaction). All following sections of this chapter (electrical conductivity, sintering, grain growth, hot-pressing) as well as some sections of the next chapter (especially creep) ultimately lead to different aspects of diffusion.

Our approach to the description of diffusion and diffusion-controlled phenomena in Al_2O_3 will be gradual. For an understanding of the characteristics of diffusion in Al_2O_3, which are sometimes complex and confusing, it is necessary to recall some fundamental knowledge about defect structures and disorder mechanisms. At first, some general remarks on diffusion phenomena in ceramics and in Al_2O_3 will be given (Section 2.3.1), followed by a discussion of intrinsic and ex-trinsic material behavior (Sections 2.3.2 and 2.3.3) and of the problems asso-ciated with the measurement of diffusivity (Section 2.3.4), and finally published diffusion data on Al_2O_3 will be reviewed (Section 2.3.5).

2.3.1 General Remarks on Diffusion Phenomena in Ceramics

A series of important processes and phenomena in Al_2O_3 which are affected by diffusion are compiled in Table 2.3. The ionic component of the electrical

Table 2.3. Diffusion-controlled phenomena in Al_2O_3

Phenomenon	Elements of an explicative theory	Influencing variables theory
Ionic conductivity		
Sintering	Predominant type of disorder	Temperature
Grain growth	(Schottky, Frenkel)	Oxygen partial pressure
Creep	Defect model	Impurity content
Subcritical crack growth at elevated temperatures	$D \propto \sum_i C_i M_i$ $D = D_o \exp(-Q/kT)$	Dopants Grain size
Thermal surface finishing		Segregation
Color boundary migration		
Annihilation of irradiation damage		

conductivity is directly proportional to the diffusivity of the faster diffusing ionic species. Sintering, grain growth, and diffusional creep are caused by the diffusional mass transport of the two ionic species in the stoichiometric ratio. Both volume and surface diffusion are involved in subcritical crack growth at elevated temperatures and in thermal surface finishing processes, such as crack healing, thermal grooving, scratch smoothing, and flame polishing. Color changes are caused by the oxidation/reduction of color centers and hence by the incorporation and diffusion of oxygen. The annihilation of irradiation damage depends on the mobilities of the created defects.

The rate of all these processes can be altered by a series of influencing variables, which are listed in the third column of Table 2.3. The diffusion of ions in ceramic materials is known to be strongly affected by temperature, by the oxygen partial pressure, p_{O2}, by the impurity and/or dopant content and their segregation tendency, and by the type of the diffusion path [2.4].

Since all these variables may act simultaneously, the results of diffusivity measurements are frequently very dissimilar and contradictory, thus creating the difficult task of interpreting them by means of a suitable model.

The way to arrive at an explicative theory which is able to explain the effects of the experimental variables on the measured phenomena is to establish the defect model of the material in question. For that purpose it is necessary to indicate the predominant type of discorder. This problem is still not resolved unequivocally for Al_2O_3 [2.30,2.31].

The defect model is based on two fundamental facts. On the one hand, the neutrality condition is established which states which type of native defect is created in order to make up the charge balance with other intrinsic or extrinsic defects. On the other hand, it is presumed that for the charge compensation of foreign, aliovalent ions the same native defect is used which is also the majority defect when intrinsic disorder dominates [2.32]. In the best case, the defect model predicts the defect structure of the material in question, the type and concentration of the diffusing defect (vacancy, interstitial), and the transition from extrinsic to intrinsic behavior. It should also explain the variation of the diffusion coefficient with temperature, p_{O_2}, and impurity content under varying experimental conditions. In practice, however, the experimental data are limited, and precise values for all the necessary equilibrium constants are not available for any oxide system [2.4].

As shown in Table 2.3, the diffusion coefficient D may be written as proportional to the sum of the products of concentrations, C_i, and mobilities, M_i, of all defects which contribute to the motion of one ionic species (in general vacancies and interstitials) [2.4]:

$$D \propto \sum_i C_i M_i \quad . \tag{2.2}$$

Since both ionic species, Al and O, may diffuse independently of each other, D must be determined for Al and O seperately. The mobility is a strong function of temperature [2.4],

$$M = M_o \exp(-H_M/kT) \tag{2.3}$$

(where H_M is the activation energy of motion), is supposed to be independent of concentration [2.32], and depends strongly on the diffusion path, i.e. is different for volume, grain boundary, and surface diffusion. For example, rapid oxygen diffusion along grain boundaries is well known to occur in Al_2O_3 [2.33].

18

As to the concentration of the diffusing defect, C_i, it is distinguished between intrinsic (temperature-controlled) defects and extrinsic (impurity-controlled) defects. These two cases will be treated in the next sections. In most cases only one defect is predominant in Eq. 2.2, so that just the different M_i's must be considered.

The relation

$$D = D_o \exp(-Q/kT) \qquad (2.4)$$

of Table 2.3 is a phenomenological description of the temperature dependence of D, where Q is called the experimental activation energy.

2.3.2 Intrinsic Diffusion and Disorder Mechanism

Intrinsic diffusion, or more generally, intrinsic properties of a ceramic material, means that the point defects through which the atomic mobility is accomplished are created exclusively by the temperature. The atomic disorder of the crystal is caused, then without the participation of foreign atoms.

Pure Al_2O_3 is a very stable oxide, having a cation with a preference for only a single valence state, i.e., Al^{3+}. Therefore, in the absence of foreign atoms, the range of stoichiometry is supposed to be very limited [2.4], and nonstoichiometric effects are related almost entirely to the presence of impurities of variable valence.

The various types of defects that frequently occur in ionic crystals are described most conveniently by a system of notation which is known as the Kröger-Vink notation [2.34] and which we shall use in this book, too. In this notation, the charge of an ion is noted only when it deviates from the charge normally encountered on the lattice site in question. Positive and negative effective charges are marked by the superscripts $\cdot$ and $'$, respectively, whereas for the indication of effective charge equality the symbol x is used. The position of the ion is given by the subscript which may denote the regular lattice sites of the anion or cation by their respective symbols or an interstitial site by the symbol i. Thus, $Al_i^{\cdots}$ is an interstitial Al ion, O_O^x represents an oxygen ion on its regular lattice site, and Mg_{Al}', Fe_{Al}^x, and $Ti_{Al}^{\cdot}$ are examples for various impurities which substitute Al on its regular site and whose valences are 2+, 3+, and 4+, respectively.

The charged native defects that are most likely to occur in Al_2O_3 are V_{Al}''', $Al_i^{\cdots}$, $V_O^{\cdot\cdot}$ and O_i''. These defects are the most important ones observed in Al_2O_3 and are the only ones that need to be considered here. However, a series of other defects has been identified and compiled by Mitchell et. al. [2.31]. Many of them have been produced by irradiation and detected by spectroscopic techniques. Among these are V_O^x, $V_O^\cdot$, and several $(V_{Al}O_O)$ associates, O_i' and $O_i^\cdot$, Al_i with various charge states, and also $Al_O^\cdot$ and the $(AlO)_{OO}^\cdot$ associate [2.31].

Several different types of intrinsic disorder can occur in ionic crystals. The specific disorder mechanism that predominates in a particular ceramic system and that is responsible for the formation of the majority point defects is an important matter, because it also determines the type of defects formed when the material is doped by aliovalent additives [2.32]. If Schottky disorder is dominant in Al_2O_3, doping with tetravalent ions (donors with respect to Al^{3+}, as for example Ti^{4+}) causes Al vacancies to form, whereas doping with divalent ions (acceptors with respect to Al^{3+}, as for example Ca^{2+}) results in the formation of O vacancies [2.35]. On the other hand, if Frenkel disorder of Al predominates, the addition of a donor results in the formation of Al vacancies and the addition of an acceptor causes Al interstitials to form. In the same way, oxygen interstitials or oxygen vacancies are created by doping with a donor or an acceptor, respectively, if Frenkel disorder of O is the dominant type of disorder. Thus, knowledge of the disorder mechanism indicates the type of native defects formed by doping, and knowledge of the native defects formed on doping leads to knowledge of the major disorder mechanism [2.35].

In Table 2.4, the reaction equations for the formation of several types of native defects and the relations between defect concentrations and the equilibrium constants are summarized for the case of Al_2O_3. The equilibrium constant of defect formation, $K_j^{1/n}$, is given by [2.4]

$$K_j^{1/n} = N_o \exp(-G_j/nkT) = N_o \exp(S_j/nk) \exp(-H_j/nkT) \ \mathrm{cm}^{-3} , \qquad (2.5)$$

where n = 2 when j means Frenkel disorder of Al or O (symbol F(Al) or F (O), respectively) and intrinsic electronic disorder (symbol i), n = 5 when j means Schottky disorder (symbol S), $N_o = 2.36 \cdot 10^{22} \mathrm{cm}^{-3}$ is the number of Al_2O_3 molecules per unit volume, G_j/n is the free energy necessary to create one defect, S_j/n and H_j/n are the entropy and enthalpy envolved to create one defect, and k and T have their usual meaning.

Table 2.4 also gives values of the formation energies and the preexponential terms, $\exp(S_j/nk)$, of Schottky and Frenkel defects as far as they are know today.

The data of Dienes et al. [2.36] were achieved by theoretical calculations based on a polarizable point-ion shell model for α-Al_2O_3. The results predict that Schottky defects are energetically more favorable than Frenkel pairs.

Table 2.4. Possible disorder mechanisms in Al_2O_3

Type of disorder	Equation of formation and equilibrium constant	Formation energy H_j/n, in eV		Preexponential term, $\exp(S_j/nk)$	
Frenkel disorder of Al	$Al_{Al}^x \rightleftharpoons Al_i^{\cdots} + V_{Al}^{'''}$	10.0	2.36[*])	$\approx 10^{-4}$	2.35 [*])
		8.3	2.31		
	$[V_{Al}^{'''}] = [Al_i^{\cdots}] = K_{F(Al)}^{1/2}$	4.45	2.35		
	$K_{F(Al)}^{1/2} = 1.32 \cdot 10^{18} \exp(-4.45/kT)$ cm^{-3}				2.35
Frenkel disorder of O	$O_O^x \rightleftharpoons O_i^{''} + V_O^{\cdot\cdot}$	7.0	2.36		
	$[V_O^{\cdot\cdot}] = [O_i^{''}] = K_{F(O)}^{1/2}$	7.1	2.31		
Schottky disorder	null $\rightleftharpoons 2V_{Al}^{'''} + 3V_O^{\cdot\cdot}$	5.7	2.36	$\approx 10^{-4}$	2.35
		5.2	2.31		
	$V_{Al}^{'''} = \frac{2}{3}[V_O^{\cdot\cdot}] = \left(\frac{2}{3}\right)^{3/5} K_S^{1/5}$	4.1	2.33		
		3.83	2.35		
	$K_S^{1/5} = 1.36 \cdot 10^{19} \exp(-3.83/kT)$ cm^{-3}				2.35
Intrinsic electronic disorder	null $\rightleftharpoons e' + h^{\cdot}$	5.18	2.35	150	2.35
	$[h^{\cdot}] = [e'] = K_i^{1/2}$				
	$K_i^{1/2} = 3.14 \cdot 10^{19}\ T^{3/2} \exp(-5.18/kT)$ cm^{-3}				2.35

[*]) Numbers in this column are references

The equilibrium constants and energy values of Mohapatra and Kröger [2.35] are based on experimental results and were calculated by combining equilibrium constants obtained with donor and acceptor doped crystals. Although considerably smaller than the formation energies of [2.36], they show the same trend, indicating that Schottky disorder is the main atomic disorder mechanism in Al_2O_3. The equilibrium constants obtained in [2.35] are plotted in Fig. 2.5 as a function of temperature. As indicated by the three curves, the concentration of Frenkel pairs is expected to be smaller than the concentration of Schottky defects by two to three orders of magnitude at temperatures ranging from $800^{\circ}C$ to the melting point. Intrinsic electronic disorder may also occur, but can only become dominant with impurity levels of aliovalent ions smaller than 10^{12} cm^{-3}, i.e., with relative impurity concentrations in the order of 10^{-10}. Therefore, Mohapatra and Kröger [2.35] concluded that it may prove impossible to prepare a crystal pure enough to have its defect structure dominated by electronic defects.

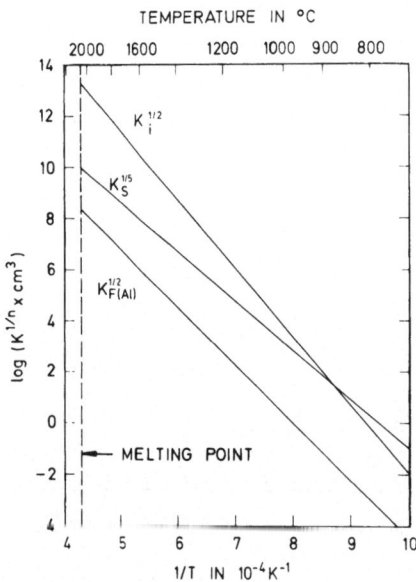

Fig. 2.5. Equilibrium constants for possible disorder mechanisms in Al_2O_3. $K_i^{1/2}$ is equilibrium constant for intrinsic electronic disorder, $K_S^{1/5}$ for Schottky disorder, and $K_{F(Al)}^{1/2}$ for Frenkel disorder of Al. – After Mohapatra and Kröger [2.35]

We have seen by means of Table 2.4 and Fig. 2.5 that there is much evidence for Schottky disorder to dominate in Al_2O_3. However, many experimental results on diffusion-controlled properties have been interpreted satisfactorily in terms of Frenkel disorder of Al, and the question of which disorder mechanism determines the behavior of pure and doped Al_2O_3 is still open. Table 2.5 summarizes experimental work which has dealt with the subject. Various properties have been used to study the problem, mainly electrical conductivity, optical absorption, sintering, and creep. The argument used in favor of the particular type of disorder is indicated. In many cases, Schottky disorder or Frenkel disorder of Al has been proposed. Some papers report that the question could not be decided unequivocally, either because of lack of accuracy in the experimental data or because of the fact that different types of disorder may explain the experimental findings equally well.

The only reference known to us that reports on Frenkel disorder of O in Al_2O_3 (last line of Table 2.5) is a paper by Phillips et al. [2.10]. These authors studied the formation of TiO_2 precipitates in supersaturated Al_2O_3:Ti solid solutions, which was shown to be accompanied by the formation of prismatic interstitial dislocation loops around the precipitates. The observation of interstitial loops which act as a sink for extra material was taken as evidence for the fact that the solid solution had a higher density than the two-phase mixture which, in terms of defect chemistry, can only be achieved when the charge-compensating defect for Ti^{4+} in Al_2O_3 is the oxygen interstitial O_i'' and not the aluminum vacancy V_{Al}''' [2.10]. This result contradicts, however, the findings of Rasmussen and Kingery [2.8], who observed a density decrease on doping Al_2O_3 with Ti, as well as the spectroscopic studies of Cox [2.52], in single-crystal Al_2O_3:Ti who found that the charge-compensating native defect was V_{Al}'''.

Most of the authors who studied the defect structure of Al_2O_3 agree that more experimental work is needed in order to elucidate the disorder mechanism. A broader range of dopant levels and of oxygen partial pressure, materials of higher purity, and increased experimental accuracy in determining the effects of temperature and p_{O2} are suggested as means to clarify the situation.

As a result of the foregoing considerations, we can assume that in the intrinsic regime of diffusion, whatever the specific disorder mechanism may be, the concentration term of Eq. (2.2) depends on the temperature in the manner indicated in Table 2.4. The activation energy is equal to the formation energy per defect, H_j/n, and this gives, according to Eq. (2.4), a total activation energy of diffusion $Q = H_j/n + H_M$. In fact, the stronger variation of the diffusion coefficient with temperature in the intrinsic region is used to distinguish between intrinsic and extrinsic behavior.

However, numerical values of the equilibrium constants of defect formation and of the respective formation energies indicate that intrinsic defect concentrations are very small in oxides [2.4]. Typical high-temperature intrinsic defect concentrations in oxides are in the range 10^{-9} or lower [2.31]. From [2.35] it follows (see Fig. 2.5) that the concentration of Schottky defects in Al_2O_3 near

Table 2.5. Type of disorder found in pure and doped Al_2O_3

Authors	Ref. Year	Effect used	Sample cond.[1]	Dopants	Disorder	Argument
Pappis & Kingery	2.37 1961	Electrical conductivity	sc pc	Fe	Intrinsic electronic	Type of charge carrier, P_{O2} dependence, ae^2
Rasmussen & Kingery	2.8 1970	Density charges due to aliovalent additives	sc	Si, Ti, Ca, Mg	Schottky	Decrease of density in all cases
Brook, Yee & Kröger	2.38 1971	Electrical conductivity, tn^3	sc	pure, Mg, Co	Frenkel of Al	Type of charge carrier, coincidence with Al self-diffusion data
Brook	2.39 1972	Initial sintering rate	pc	Ti, Mg	Frenkel of Al	Variation of diffusion coefficient with Ti
Yee & Kröger	2.40 1973	Electrical conductivity, tn	sc pc	Mg, Co Ti	Not decided	Unexplained contribution from hole conduction
Hollenberg & Gordon	2.41 1973	Creep	pc	Cr, Fe Ti	Frenkel of Al	Increase of Al diffusion due to Fe addition
Raja Rao & Cutler	2.42 1973	Sintering, shrinkage rate	pc	Fe, Ti	Frenkel of Al	Variation of Al diffusion with Fe content
Raja Rao & Cutler	2.43 1973	Heating, quenching, subs. sintering of particles	pc	Fe	Frenkel of Al	Increase in diffusion coeff. due to $Al_i^{\cdots}$ conc.
Kitazawa & Coble	2.44 1974	Electrical conductivity, tn	sc pc	Mg	Frenkel of Al	Coincidence with Al self-diffusion data
Dutt, Hurrell & Kröger	2.45 1975	Electrical conductivity, tn, optical absorption	sc	Co	Frenkel of Al	Coincidence with Al self-diffusion data
Dutt & Kröger	2.46 1975	Interpretation of data of Pappis & Kingery	sc pc	Fe	Frenkel of Al	Oxygen pressure exponent
Frederikse & Hosler	2.47 1975	Electrical conductivity	sc pc	pure	Schottky	Coincidence with Al self-diffusion data

Table 2.5. (continued)

Author	Ref./Year	Method	type[1]	Dopant	Disorder	Remarks
Mohapatra & Kröger	2.48 1977	Electrical conductivity, tn, optical absorption	sc	Mg	Not decided	Uncertainty as to the P_{O_2} exponent
Mohapatra & Kröger	2.49 1977	Electrical conductivity, tn, optical absorption	sc	Ti	Not decided	O_i'' not ruled out unequivocally
Mohapatra & Kröger	2.35 1977	Interpretation of literature data	sc	Fe, Ti	Schottky	Comparison of equilibrium constants
Yen & Coble	2.50 1979	Optical absorption after induced irradiation damage	sc	Fe	Not decided	Spectrum explicable by both Frenkel and Schottky defects
Mohapatra, Tiku & Kröger	2.32 1979	Electrical conductivity, tn	sc	pure	Schottky	Best agreement of data from several dopings
Hou, Tiku, Wang & Kröger	2.30 1979	Electrical conductivity, tn, creep	pc	Fe, Ti	Not decided	Exp. data equally well explained by Frenkel and Schottky disorder
Harmer, Roberts & Brook	2.51 1979	Density change in sintering	pc	Mg, Ti	Frenkel of Al	Formation of Al_i''' by Mg doping
Phillips et al.	2.10 1980	Formation of interstitial loops due to density changes	sc	Ti	Frenkel of O	Increase in density with Ti concentration

1 sc single-crystal, pc polycrystalline
2 ae activation energy
3 tn transference number

the melting point is of the order 10^{-12}. That means that in Al_2O_3 intrinsic defects should not become important, even close to the melting point, because their
number is smaller by several orders of magnitude than the number of native defects
formed to compensate for aliovalent impurities. For intrinsic behavior to be
observed in Al_2O_3 at $2000^{\circ}C$, the impurity concentration must be smaller than
10^{-12}, which is a purity level far beyond present techniques of crystal growing
[2.31]. Impurity contents of a fraction of a ppm are sufficient to control the
defect concentration, and it is therefore unlikely that intrinsic diffusion has
ever been observed in refractory oxides [2.4].

Nevertheless, we would like to return to the arguments pointed out above which
justify the study of the defect structure: even if intrinsic defects do not
dominate the diffusion-controlled properties of Al_2O_3 directly, knowledge of
the dominant disorder mechanism enables us to indicate which native defects are
formed by the presence of aliovalent impurities [2.35] and provides a rational
basis for the interpretation of material behavior.

2.3.3 Extrinsic Diffusion

Since intrinsic point defect concentrations are very small in Al_2O_3, diffusion-
controlled properties are supposed to depend strongly on aliovalent impurity or
dopant concentration. There are two different forms of impurity-controlled behavior that shall be described briefly. In the first one, the impurity ion has a
fixed valence. An example may be Mg, which is thought to have a fixed valence
state and to occur in Al_2O_3 as Mg'_{Al}. (There are indications, however, that at
higher oxygen pressure the Mg'_{Al} concentration decreases due to the formation of
Mg^x_{Al} centers, which are conceived to consist of a hole trapped near the Mg'_{Al}
[2.48,2.52]). Then the neutrality condition is given by $[Mg'_{Al}] = 3[Al^{\cdots}_i]$ for
Frenkel disorder of Al to be dominant, and $[Mg'_{Al}] = 2[V^{\cdots}_O]$ for Schottky disorder to be dominant. For even a very small dopant content in the ppm range, the
concentration of native defects that results from the requirement of charge balance is much greater than the number of defects formed by any intrinsic disorder mechanism, and is independent of temperature and p_{O2}. This situation was
encountered by Hou et al. [2.30], who found that the conduction and creep behavior of polycrystalline Al_2O_3 doped with Fe and Ti was completely dominated
by an unintentional impurity content of 65 ppm Mg. The ionic part of the electrical conductivity, σ_i, which was shown to be proportional to $[V^{\cdots}_O]$, was independent
of p_{O2} and had an activation energy related to the mobility of O in Al_2O_3, $H_M(O)$,

only. The creep rate, $\dot{\varepsilon}$, turned out to be proportional to $[Al_i^{...}]$, was independent of p_{O2} and had an activation energy equal to the activation energy of motion of $Al_i^{...}$ in Al_2O_3, $H_{M(Al)}$.

Let us consider the variation of Al_i and the diffusion coefficient with temperature, admitting that intrinsic disorder also becomes involved in the formation of native defects at very high temperatures. If Frenkel disorder dominates (which was considered possible in [2.30]), the coefficient of Al self-diffusion in the temperature range investigated in [2.30] can be written as

$$D_{Al} \propto [Al_i^{...}]\exp(-H_{M(Al)}/kT) \, , \tag{2.6}$$

with $[Al_i^{...}] = 1/3 \, [Mg'_{Al}] = \text{const.}$ At increasing temperatures, the number of intrinsic Frenkel defects may exceed the extrinsic $Al_i^{...}$ concentration, and a behavior as depicted in Fig. 2.6 is expected. At high temperatures, $[Al_i^{...}]$ will also depend on temperature, and D_{Al} will be

$$D_{Al} \propto \exp(-H_{F(Al)}/2kT)\exp(-H_{M(Al)}/kT) \quad . \tag{2.7}$$

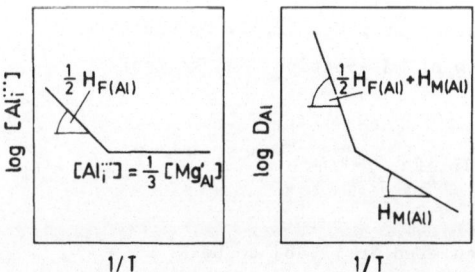

Fig. 2.6. Variation of the concentration of the diffusing defect and of the diffusion coefficient with temperature in the extrinsic and intrinsic region. In the extrinsic region, it is assumed that the $Al_i^{...}$ concentration is fixed by the content of a constant valence dopant, Mg'_{Al}, through the neutrality condition $[Mg'_{Al}] = 3[Al_i^{...}]$

Both the curve $\log [Al_i^{...}]$ and the curve $\log D_{Al}$ vs. $1/T$ will show a bend which is a characteristic for the transition from extrinsic to intrinsic behavior. There are indications in the literature that this bend has been observed in Al_2O_3 [2.44, 2.53], but there are other authors [2.4] who doubt those interpretations.

The second form of impurity-controlled behavior is found when the foreign ion has a variable valence. This is the case with transition metal ions such as Fe, Co, or Ti. Let us consider the example of Fe as the dopant in Al_2O_3, following the work of Hollenberg and Gordon [2.41], who studied the effect of Fe additions and p_{O2} on the creep rate of polycrystalline Al_2O_3, and the paper of Dutt and Kröger [2.46] on the defect structure of Fe-doped Al_2O_3. If Fe is present in its divalent state, positive native defects must form to maintain electrical neutrality of the crystal. Assuming again Frenkel disorder of Al [2.41], the neutrality condition is

$$[Fe'_{Al}] = 3[Al_i^{\cdots}] .$$ (2.8)

Since Fe^{3+} is an acceptor in Al_2O_3, its reduction to Fe^{2+} is given by the reaction.

$$Fe^x_{Al} \rightleftharpoons Fe'_{Al} + h^{\cdot} ,$$ (2.9)

with the equilibrium constant K_{Fe}. The oxidation of Fe^{2+} in Al_2O_3 occurs by the incorporation of gaseous oxygen according to the reaction

$$(3/4)O_2(g) + Al_i^{\cdots} + 3Fe'_{Al} \rightleftharpoons (3/2)O_O^x + Al^x_{Al} + 3Fe^x_{Al}$$ (2.10)

with the equilibrium constant $K^{Fe}_{ox,i}$. The oxidation-reduction of Fe gives rise to a p_{O2} dependence of $[Al_i^{\cdots}]$, which is obtained from Eq. (2.10) by substituting $[Fe'_{Al}]$ from Eq. (2.8), applying the law of mass-action, and setting $[Fe^x_{Al}] \simeq [Fe_{tot}]$, which gives

$$[Al_i^{\cdots}] \propto [Fe_{tot}]^{3/4} p_{O2}^{-3/16} (K^{Fe}_{ox,i})^{-1/4} ,$$ (2.11)

as already derived in [2.41]. $K^{Fe}_{ox,i}$ was calculated in [2.46] to be $K^{Fe}_{ox,i} = 1.16 \cdot 10^{-4} exp(7.32eV/kT) atm^{-3/4} molfr^{-1}$. Therefore, the temperature dependence of $[Al_i^{\cdots}]$ should be controlled by an activation energy of $H^{Fe}_{ox,i}/4 = 1.83$ eV. Inserting Eq. (2.11) into Eq. (2.6) yields

$$D_{Al} \propto [Fe_{tot}]^{3/4} p_{O2}^{-3/16} exp(-H^{Fe}_{ox,i}/4kT) exp(-H_{M(Al)}/kT) .$$ (2.12)

Equation (2.12) means that in the presence of Fe (or, in general, of a variable-valence dopant), the diffusion coefficient has an increased activation energy and depends on p_{O2}, due to the incorporation reaction of the ambient oxygen and the oxidation-reduction reaction of the dopant. Inserting Eq. (2.9) into Eq. (2.10) gives the reaction which controls the variation of stoichiometry of pure Al_2O_3 if Frenkel disorder of Al dominates,

$$(3/4)O_2(g) + Al_i^{\cdots} \rightleftharpoons (3/2)O_O^x + Al^x_{Al} + 3h^{\cdot} ,$$ (2.13)

with the equilibrium constant K_x. The three equilibrium constants are related by $K_x = K_{Fe}^3 \cdot K_{ox,i}^{Fe}$. From Eq. (2.13) we easily obtain

$$[h^\cdot] \propto p_{O2}^{3/16} [Fe_{tot}]^{1/4} K_x^{1/4} \quad . \qquad (2.14)$$

It has been shown in [2.46] that the electrical conductivity data of Pappis and Kingery [2.37] can be explained very well in terms of the temperature and p_{O2} dependence predicted by Eqs. (2.12) and (2.14). The ionic conductivity which dominates at low p_{O2} is proportional to D_{Al} of Eq. (2.12), whereas the electronic part which dominates at high p_{O2} is proportional to $[h^\cdot]$ of Eq. (2.14). Good agreement was also obtained between creep data and Eq. (2.12) in [2.41], and the same model was applied in [2.42] to interpret sintering data. Thus, assuming a defect model in which the dopant in its divalent form is compensated by interstitial Al ions, a series of material properties could be interpreted satisfactorily by the idea of extrinsic, impurity-controlled diffusion.

Figure 2.7 shows schematically the variation of $[Al_i^{\cdots}]$ from Eq. (2.11) and D_{Al} from Eq. (2.12), with temperature and p_{O2} for the case of impurity-induced nonstoichiometry as discussed above. Again, transition to intrinsic behavior is assumed to occur at high temperatures. The distinction from Fig. 2.6 is obvious.

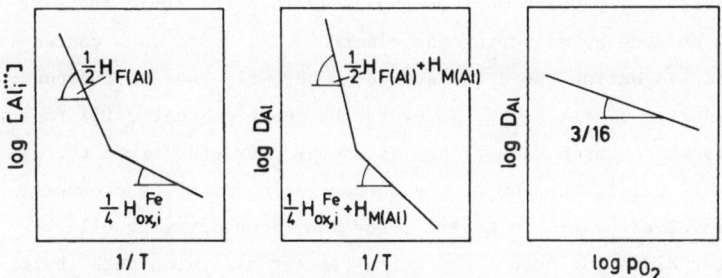

Fig. 2.7. Variation of the concentration of the diffusing defect and of the diffusion coefficient with temperature and oxygen partial pressure in the extrinsic and intrinsic region. In the extrinsic region, it is assumed that the native defect concentration is fixed by the amount of a, variable valence dopant in its divalent state, e.g., Fe'_{Al}, through $[Fe'_{Al}] = 3[Al_i^{\cdots}]$. This leads also to a p_{O2} dependence of D_{Al}

It has been pointed out [2.54] that anomalously high activation energies found in sintering and creep of refractory oxides can be misinterpreted as indications of intrinsic diffusion. They may, instead be caused by the incorporation reaction

of oxygen due to the valence change of a transition metal impurity, as shown
here, or by the dissolution of precipitates, or by both.

2.3.4 Problems in Determining the Diffusion Coefficient

For the measurement of the diffusion coefficient all phenomena, in principle, may
be used that depend on it (see Table 2.3). If there exists a model which describes
the dependence of the process rate on the diffusion coefficient, the relation may
be inverted to determine the diffusion coefficient from the kinetics of the process. Thus, diffusion data for Al_2O_3 have been determined from electrical conductivity measurements [2.38,2.44,2.48.2.55], from sintering experiments [2.39,2.42,
2.56 to 2.58], and from creep. Creep diffusion data have been reviewed by Cannon
and Coble [2.59]. The use of radioactive tracers is of minor importance in Al_2O_3
because of the lack of suitable isotopes, but has been reported as well [2.33,
2.60]. The kinetics of color boundary migration on oxidizing doped polycrystalline Al_2O_3 has been used to measure O grain-boundary diffusivities [2.61].

The interpretation of diffusion data is unambiguous when determined by the tracer
technique. Yet, misinterpretation may occur when a diffusion coefficient is to be
calculated from electrical conductivity or creep data. In measuring the current
passing through a sample at elevated temperature, the experiment yields a value
for the total conductivity,σ_{tot}, which must be separated into its ionic and electronic parts. This can be done by measuring the electromotive force in a concentration cell [2.38] and evaluating the transference number, t_i. The ionic conductivity, σ_i, is obtained from $\sigma_i = t_i \cdot \sigma_{tot}$. However, the question that still remains to be answered is which ionic defects act as charge carriers, since the
ionic current is proportional to the sum of the concentration-mobility products
of all ionic species involved [2.4]. In general, the ionic conductivity will be
determined by the faster diffusing defect along its fastest diffusion path (bulk,
grain boundaries, or surface). Even if one is able to determine the sign of the
charge carrier it remains to be clarified whether, say, $V_O^{\cdot\cdot}$ or $Al_i^{\cdot\cdot\cdot}$ is the moving
species.

In sintering or diffusional creep both the cations and anions must diffuse and
at steady state the total fluxes of the two components must be in the stoichiometric ratio. The rate of the overall process is controlled by the slower diffusing ion along its fastest diffusion path. This path may be through the bulk
or along the grain boundaries. In general, the two ionic species may have diffusivities that differ from each other and that are different along different

paths. This gives rise to four different material fluxes which are proportional to the four quantities

$$D_{Al}^{\ell} \qquad\qquad D_{O}^{\ell}$$

$$\pi\delta_{Al}D_{Al}^{b}/GS \qquad\qquad \pi\delta_{O}D_{O}^{b}/GS,$$

where D_{Al}^{ℓ} and D_{Al}^{b} are the Al lattice and grain-boundary diffusivities, D_{O}^{ℓ} and D_{O}^{b} are the O lattice and grain-boundary diffusivities, δ_{Al} and δ_{O} are the effective grain-boundary widths for Al and O, respectively, and GS is the grain size.

Considering the priority condition of diffusional mass transport, which states that the process rate is controlled by the slower diffusing species along its fastest path, leads to a matter transport model in which the four individual fluxes are arranged partly in series and partly in parallel to give the overall flux of matter which may be expressed by an apparent diffusion coefficient, $D_{complex}$. Sintering models have been developed which permit the independent determination of grain-boundary or volume diffusivities of the rate-controlling ion [2.62 to 2.64]. The expression for $D_{complex}$, as derived by Gordon[2.65, 2.66] and Evans and Langdon [2.67], is given by

$$D_{complex} = \frac{1}{\alpha} \frac{D_{Al}^{\ell} + \pi\delta_{Al}D_{Al}^{b}/GS}{1 + \dfrac{\beta}{\alpha}\dfrac{D_{Al}^{\ell} + \pi\delta_{Al}D_{Al}^{b}/GS}{D_{O}^{\ell} + \pi\delta_{O}D_{O}^{b}/GS}} \qquad (2.15)$$

For Al_2O_3, $\alpha = 2$, and $\beta = 3$.

Equation (2.15) demonstrates that the apparent diffusion coefficient obtained from a diffusional mass transport experiment is a complex function of the individual volume and grain-boundary diffusivities of the two ionic species and is also a function of the grain size.

As will be shown further, the effective grain-boundary diffusivity of oxygen, $\pi\delta_{O}D_{O}^{b}/GS$, by far exceeds all other diffusivities in a wide range of grain sizes in alumina. This means that $\pi\delta_{O}D_{O}^{b}/GS \gg D_{Al}^{\ell}$ and $\delta_{O}D_{O}^{b} \gg \delta_{Al}D_{Al}^{b}$, and therefore permits a considerable simplification of Eq. (2.15), giving

$$D_{complex} = \frac{1}{\alpha} (D_{Al}^{\ell} + \pi\delta_{Al}D_{Al}^{b}/GS) \quad . \qquad (2.16)$$

Equation (2.16) shows that, for the assumptions made above, the process rate is independent of the O diffusivity and the contributions of Al lattice and grain-boundary diffusivities to the apparent diffusion coefficient are additive, the relative amount depending on the grain size.

To summarize, the determination of diffusion coefficients from electrical conductivity, diffusional creep, or sintering contains a series of uncertainties. The type of the diffusing ion which dominates the observed effect is not known a priori, nor the diffusion path used. If, in diffusional mass transport, the diffusion of one species occurs by the movement of a neutral defect, as has been argued for fast oxygen grain-boundary diffusion [2.30], the movement of the other species must be accompanied by the ambipolar diffusion of further native defects, which may be electrons or holes in the case of $Al_i^{\cdots}$ or $V_{Al}^{\prime\prime\prime}$ diffusion, respectively. Small amounts of impurities may control the concentration of the diffusing defects and can lead to dissolution/precipitation and oxidation/reduction processes, thus contributing to the experimental activation energy which may be misinterpreted as being intrinsic when the chemical composition is not exactly known.

2.3.5 Diffusion Data

Diffusion data of Al_2O_3 will be summarized in the sequence: lattice diffusion, grain-boundary diffusion, surface diffusion, and impurity diffusion. Diffusion data of oxides have been reviewed by Harrop in 1968 [2.68] and, very recently, by Freer in 1980 [2.53].

2.3.5.1 Lattice Diffusion

Lattice diffusivities of Al and O as functions of temperature are shown in Fig. 2.8, and some of these data are given numerically in part A of Table 2.6. There are only a few data obtained by the direct determination of self-diffusion coefficients in Al_2O_3 via radioactive tracers. Among these are the basic and much cited studies of Kingery and co-workers on the self-diffusion coefficient of O [2.33] and Al [2.60], (curves 1, 2, 4 and 5 of Fig. 2.8). Most data have been inferred, however, from diffusional creep and sintering experiments. In their review paper, Cannon and Coble [2.59] have reevaluated the results of a total of nine diffusional creep studies on MgO-doped Al_2O_3 and found that the Al diffusivities do not differ by more than a factor of two from the expression $D = 1.36 \cdot 10^5 \exp(-577kJ/RT)$

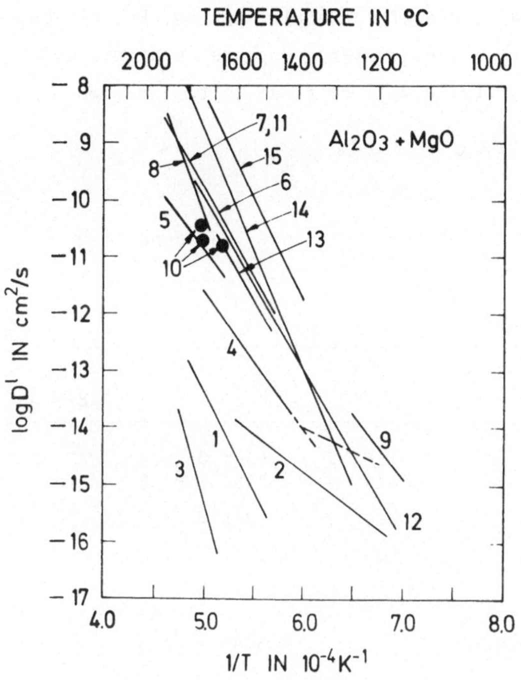

Fig. 2.8. Aluminum and oxygen lattice diffusivities in MgO-doped Al_2O_3. 1 to 3: D_O^{ℓ} in single-crystal Al_2O_3, 1 and 2: [2.33], 3: [2.69], 4: D_O in polycrystalline Al_2O_3 [2.33], 5 to 14: D_{Al}^{ℓ} in polycrystalline Al_2O_3. 5: [2.60], 6 to 13 from creep experiments, 6: [2.70], 7: [2.71], 8: [2.72], 9: [2.73], 10: [2.74], 11: [2.75], 12: [2.76], 13: [2.77], 14 and 15 from sintering experiments, 14: [2.78], 15: [2.56]. Curves 6 to 12 after the compilation of [2.59]

which was found by Cannon et al. [2.76] by evaluating diffusional creep experiments on fine-grained alumina. Some of these reevaluated diffusion data are included in Fig. 2.8 as curves 6 to 12.

The lattice diffusivity of Al is greater than the lattice diffusivity of O by about three orders of magnitude in the observed temperature range, $D_{Al}^{\ell} \gg D_O^{\ell}$, as it is also found to be for many other oxides [2.4]. The activation energies show some spread which does not seem to be unexpected, bearing in mind the various ways of how very small impurity concentrations contribute to the experimental activation energy, as has been discussed in the previous section. There exists agreement today that all experimental work on diffusion in Al_2O_3 has been performed in the extrinsic range. The observation that roughly coincident diffusion coefficients are obtained from creep experiments on various nominally pure and MgO-saturated materials [2.59,2.76] is attributed to the fact that even "nominally pure" samples contain aliovalent impurities in the range of some 10 to 100 ppm, thus already causing composition-controlled behavior [2.30].

It has been observed by many investigators that aliovalent additives, both donors and acceptors, increase the diffusivity of Al and affect the activation energy of diffusion. Additives which have been studied are, for example, Mg, Mn, Fe, Ni, Co, and Ti. The p_{O2} dependence in the case of variable valence dopants is attributed to impurity-controlled nonstoichiometry. Details will be given in the sections on sintering and creep.

Table 2.6. Some selected self-diffusion data of nominally pure and doped Al_2O_3

(A) Lattice diffusion, $D^{\ell} = (D^{\ell})_o \exp(-Q/RT)$

Authors	Ref. Year	Method	Ion	Sample cond.*)	$(D_2^{\ell})_o$ cm^2/s	Q kJ/mol	Curve no. in Fig.2.8
Oishi & Kingery	2.33 1960	Tracer diffusion	O O	sc sc	$1.9 \cdot 10^3$ $6.3 \cdot 10^{-8}$	636 ± 105 238	1 2
Reed & Wuensch	2.69 1980	Tracer diffusion	O	sc	$6.4 \cdot 10^5$	787 ± 29	3
Paladino & Kingery	2.60 1962	Tracer diffusion	Al	pc	28	477 ± 63	5
Folweiler	2.70 1961	Diffusional creep	Al	pc		544	6
Warshaw & Norton	2.71 1962	Diffusional creep	Al	pc		544	7
Cannon et al.	2.76 1980	Diffusional creep	Al	pc	$1.36 \cdot 10^5$	577	12
Engelhardt & Thümmler	2.77 1970	Diffusional creep	Al	pc		585 ± 71	13
Coble	2.78 1958	Initial-stage sintering	Al	pc		763	14
Coble	2.56 1961	Interm.-stage sintering	Al	pc		627	15

(B) Grain-boundary diffusion, $D^b = (D^b)_o \exp(-Q/RT)$

Authors	Ref. Year	Method	Ion	Sample cond.*)	$(D^b)_o$ cm^3/s	Q kJ/mol	Curve no. in Fig.2.9
Oishi & Kingery	2.33 1960	Tracer diffusion	O	pc	$1.6 \cdot 10^{-3}$	460 ± 63	4
Cannon et al.	2.76 1980	Diffusional creep	Al	pc	$8.60 \cdot 10^{-4}$	418	12
Heuer et al.	2.79 1970	Diffusional creep	Al	pc		515	14

*)sc single-crystal, pc polycrystalline

In many cases it is not quite clear whether the diffusivity determined experimentally is due to the motion of V'''_{Al} or Al'''_i [2.59]. This is also true for the activation energies of defect mobilities, $H_M(V'''_{Al})$ and $H_M(Al'''_i)$. Table 2.7 summarizes the available data. The results of Jones et al. [2.80], Kitazawa and Coble [2.44],

Table 2.7. Activation energies (in eV) of defect mobilities in Al_2O_3

Authors	Ref. Year	$H_M(V'''_{Al})$	$H_M(Al'''_i)$	$H_M(V''_O)$	Method
Dienes et al.	2.36 1975	3.8	4.8	2.9	Theoretical calculations
Jones et al.	2.80 1969		3.5		Color boundary migration
Kitazawa & Coble	2.44 1974		2.5		Electrical conductivity
Cannon & Coble	2.59 1975		4.0		Diffusional creep
Mohapatra & Kröger	2.49 1977	3.78			Electrical conductivity
Brook et al.	2.38 1971		2.9		Electrical conductivity
Mohapatra & Kröger	2.48 1975		4.39 4.84		$\perp$ c) Electrical $\parallel$ c) conductivity
Dutt et al.	2.45 1975		3.97		Electrical conductivity
Oishi & Kingery	2.33 1960			2.5	Extrinsic diffusion of O
Mohapatra et al.	2.32 1979			3.60	Electrical conductivity

and Cannon and Coble [2.59] could not be ascribed unambiguously to either vacancy or interstitial mobility of Al. The activation energy of defect mobility given in [2.59] was obtained from creep studies, applying the argument that the experimental activation energy of diffusional creep, Q, in MgO-saturated Al_2O_3 is the sum of the activation energy of the mobility of the diffusing defect, H_M, and the solution enthalpy of MgO in Al_2O_3, H_S. Applying defect chemistry correctly yields $Q = H_M + (3/4)H_S$ and not $Q = H_M + H_S$ as in [2.59]. Since H_S is known from [2.81] to be 61 kcal/mol, using Q = 138 kcal/mol from [2.59], H_M = 92 kcal/mol and not 77 kcal/mol as in [2.59]. This gives the value of 4.0 eV shown in Table 2.7, which is even closer to the theoretical numbers of [2.36].

Many of the activation energy data obtained by Kröger and co-workers from electrical conductivity measurements [2.32,2.38,2.45,2.48,2.49], Table 2.7, agree well with the results of theoretical calculations made by Dienes et al. [2.36], thus facilitating the establishement of the dominant disorder mechanism in Al_2O_3 (see also Table 2.5).

2.3.5.2 Grain-Boundary Diffusion

Grain-boundary diffusion of both Al and O constitutes a fast diffusion path in alumina not only for the anion, as was found first by Oishi and Kingery [2.33], but also for the cation if the grain size is small enough so that the second bracket term in Eq. (2.16) dominates the apparent diffusion coefficient.

There does not exist any direct determination of Al grain-boundary diffusivities in the literature. All data have been inferred from creep or sintering experiments. Figure 2.9 shows some apparent grain-boundary diffusivities, $\delta_{Al}D_{Al}^b$, as a function of temperature. Curves 6 to 12 again are from the review work of Cannon and Coble [2.59] and have been obtained from diffusional creep data, considering independent contributions of lattice and grain-boundary diffusion to the total creep rate using Eq. (2.16). Curves 6 to 12 of Fig. 2.9 have been obtained, therefore, from the same original data as the identically numbered curves of Fig. 2.8. Curve 14 shows $\delta_{Al}D_{Al}^b$ data from the work of Heuer et al. [2.79]. Some numerical grain-boundary diffusion data are listed in part B of Table 2.6.

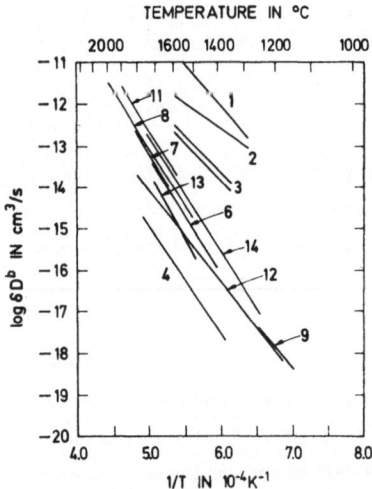

Fig. 2.9. Grain-boundary diffusivities in pure and doped Al_2O_3. 1 to 4: $\delta_O D_O^b$. 1: [2.61], 3 w/o Fe; 2: [2.61], 0.7 w/o Ti; 3: [2.85], 2%Fe, upper curve reducing atmosphere, lower curve oxidizing atmosphere; 4: [2.33], using $D_O^\ell = \pi \delta_O D_O^b/GS$ and GS = 25 μm. 6 to 14: $\delta_{Al}D_{Al}^b$. 6 to 13: same numbers and references as in Fig. 2.8; 14: [2.79], Mg doped

Only a few data are available in the literature on O grain-boundary diffusivities. The lattice diffusion coefficients D_0^{ℓ} determined by Oishi and Kingery in polycrystalline alumina with 25 μm grain size (curve 4 of Fig. 2.8) were transformed by Mistler and Coble [2.82] into grain-boundary diffusivities by means of the expression $D_0^{\ell} = \pi \delta_0 D_0^b / GS$ because it was thought that the oxygen motion was mainly due to grain-boundary diffusion. This gives curve 4 of Fig. 2.9, which lies below the $\delta_{Al} D_{Al}^b$ data by some orders of magnitude. However, it is generally accepted today that grain-boundary diffusion of O is very rapid in Al_2O_3 [2.30,2.33,2.59,2.65 to 2.67,2.76,2.79,2.83 to 2.85]. Lessing and Gordon [2.85] evaluated O grain-boundary diffusivities from creep measurements on Fe-doped alumina (curve 3 of Fig. 2.9). The only paper that reports on a direct measurement of $\delta_0 D_0^b$ is the work of Wang and Kröger [2.61]. From the kinetics of the movement of a color front on oxidation of reduced samples they determined oxygen diffusion data for Fe- and Ti-doped alumina as shown in Fig. 2.9.

The two last-mentioned references indeed show that O grain-boundary diffusion is very fast compared with both Al grain-boundary and Al lattice diffusion. The same result has been inferred repeatedly from creep studies, where it has always been observed that the creep rate depends only on the motion of Al. In [2.59] it was shown that the ratio $\delta_0 D_0^b / \delta_{Al} D_{Al}^b$ must be at least 100 if Al diffusion kinetics can be observed, and in [2.85] the same ratio was determined to be about 200. For the ratio of the chemical diffusion coefficient of O along the grain boundaries to D_{Al}^{ℓ} a value of 200 was assessed in [2.61]. An interesting consequence of these observations is that O diffusion kinetics should never be rate controlling in polycrystalline Al_2O_3 unless at very large grain size (in the range 1000 μm, as indicated in [2.84]), or if Al lattice diffusion can be accelerated markedly due to aliovalent doping, as was shown in [2.85].

There is evidence that aliovalent dopings enhance the grain-boundary diffusivity of O, but not the grain-boundary diffusivity of Al. An increase of $\delta_0 D_0^b$ due to doping with Mg was found in [2.55] and [2.75], due to doping with Fe in [2.61,2.65. 2.85,2.86], and due to doping with Ti in [2.61,2.65]. Diffusion kinetics controlled by $\delta_{Al} D_{Al}^b$, however, were reported in [2.76] not to be altered by Mg additives. For the case of enhanced O diffusivity it was pointed out in [2.55] that an increased concentration of aliovalent ions in the boundary may result in an excess charge which may lower the energy barrier of motion, thus increasing the mobility but not the concentration of the diffusing charged defect.

Fast oxygen migration along the grain boundary in alumina, however, is now supposed to occur by neutral intersticial oxygen, O_i^x. This interpretation was primarily

inferred by Hou et al. [2.30] and by El-Aiat et al. [2.86] from the observation
that the ionic part of the electrical conductivity in polycrystalline alumina did
not show a dependence on grain size, thus indicating that there is no fast migration of charged ionic species along grain boundaries. Later, Wang and Kröger [2.61]
were able to prove this conclusion experimentally by determining the dependence
of O grain-boundary diffusivity on oxygen pressure and comparing the observed
pressure exponent with that following from the equation of defect formation.

The grain-boundary thickness in alumina, as found recently by Carter et al. [2.87]
in a TEM study, was reported to be 6 to 9 nm.

2.3.5.3 Surface Diffusion

The state of the literature on surface diffusion in Al_2O_3 has been reviewed recently by Gupta [2.88]. Table 2.8 and Fig. 2.10 are taken from this reference. Table

Table 2.8. Surface diffusion in Al_2O_3 (after Gupta [2.88])

Author	Ref. Year	Technique	D_{O_2} cm^2/s	Q kJ/mol	Sample cond.*)	Curve no. in Fig.2.10
Robertson & Chang	2.89 1966	Thermal grooving	$7.0 \cdot 10^2$	314	pc	1
Shackelford & Scott	2.90 1968	Thermal grooving	$5.0 \cdot 10^5$	460	sc	2
Robertson & Ekstrom	2.91 1969	Thermal grooving Thermal grooving Thermal grooving	$1.0 \cdot 10^8$ $8.0 \cdot 10^8$ $9.0 \cdot 10^2$	535 556 314	pc pc pc	3 3 3
Prochazka & Coble	2.92 1970	Initial sintering Initial sintering	$4.8 \cdot 10^{-2}$ $1.15 \cdot 10^1$	234 280	pc pc	4 4
Yen & Coble	2.93 1972	Void breakup during crack healing	$2.38 \cdot 10^5$	397	sc	5
Raja Rao & Cutler	2.58 1972	Initial sintering	$D(1200°C) \approx 10^{-15} - 10^{-16}$			–
Moriyoshi & Komatsu	2.94 1973	Initial sintering	$1.06 \cdot 10^2$	267	pc	6
Maruyama & Komatsu	2.95 1975	Scratch smoothing	$1.5 \cdot 10^{10}$	544	sc	7
Gupta	2.88 1978	Void breakup during crack healing	$4.2 \cdot 10^6$	493	pc	8

*) sc single-crystal, pc polycrystalline

2.8 contains values of D_o and Q in $D = D_o exp(-Q/RT)$ together with the measuring technique used. Most of the experimental work has been performed in thermal grooving and sintering, and there is agreement among the investigators that these techniques do not give information on the rate-controlling species, i.e., whether Al or O is the slower moving ion in surface diffusion [2.88,2.95].

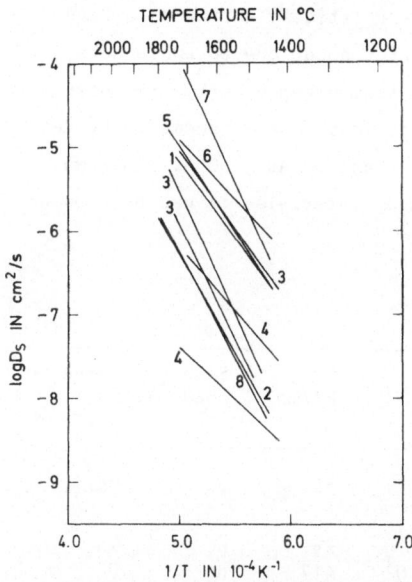

Fig. 2.10. Surface diffusion coefficients of Al_2O_3. 1: [2.89], 2: [2.90], 3: [2.91], 4: [2.92], 5: [2.93], 6: [2.94], 7: [2.95], 8: [2.88]. − After Gupta [2.88]

The data of Table 2.8 are plotted in Fig. 2.10 as a function of temperature. There is considerable scatter of both diffusivities and activation energies obtained by several investigators depending on the experimental method and on the type of material used. Practically nothing is known about the effect of composition on the surface diffusivity in Al_2O_3. Without doubt, more experimental work is needed to establish the details of surface diffusional behavior of Al_2O_3.

2.3.5.4 Impurity Diffusion

A careful reexamination of the literature resulted in information on the diffusion of 17 different impurities in Al_2O_3. In the 1960s there was some interest in using Al_2O_3 as a cladding material for nuclear fuels [2.96]. This interest gave rise to studies of the diffusion of fissile materials (all three fissile elements Th, U, and Pu are included in our compilation) and of a series of fission products. Later on, in the 1970s, Al_2O_3 was taken into consideration as a candidate material for nuclear fusion devices [2.97], which resulted in an increasing interest in the diffusion of hydrogen isotopes.

Data of impurity diffusion are summarized in Table 2.9, and diffusion coefficients are plotted in Fig. 2.11 as a function of temperature. As can be seen from the figure, most diffusion coefficients fall within a relatively narrow data band which, when compared with Fig. 2.8, is at a position about one to two orders of magnitude above the Al diffusivities obtained from creep experiments. However, there are also some exceptions. Na diffusion (curve Na-1) seems to be very fast in polycrystalline Al_2O_3 even at temperatures as low as $1200^\circ C$. Since Na is often present as a common impurity in many Al_2O_3 materials, it has been poin-

Table 2.9. Impurity diffusion in Al_2O_3

Authors	Ref. Year	Diffusing element	Temperature range ($^\circ$C)	D_0 cm^2/s	Q kJ/mol	Sample cond. [*]	Curve no. in Fig. 2.11
Bobleter et al.	2.98 1965	Pm 147	1350–1540	10^8	657	pc	Pm-1
Fiedler & Bobleter	2.99 1967	Th 229 U 233	1200–1400 1350–1500	10^{-2} $1.7 \cdot 10^1$	267 417	pc pc	Th U
Fiedler & Grass	2.100 1967	Nb 95 Ce 141 Pm 147 Sm 153	1200–1500 1200–1500 1200–1500 1200–1500		437 464 474 757	pc pc pc pc	Nb Ce Pm-2 Sm
Matzke	2.101 1967	Ra, Xe	400–1400	$3 \cdot 10^{1\perp1}$	355	pc	Ra, Xe
Frischat	2.102 1971	Na 22 Na 22 Na 22 Ca 45	1200–1750 1505 1568 1367	$2 \cdot 10^{-2}$ $D = 1.58 \cdot 10^{-11}$ $D = 2.29 \cdot 10^{-11}$ $D = 3.1 \cdot 10^{-9}$	209	pc sc sc pc	Na-1 Na-2 Na-2 Ca
Hirota & Komatsu	2.103 1977	Ni, D^ℓ Ni, D_{gb}	1220–1370 1220–1370	$1.26 \cdot 10^4$ $2.14 \cdot 10^{-1}$	435 264	pc pc	Ni-1 Ni-2
Fowler et al.	2.104 1977	T	300–1200	$1.3 \cdot 10^{-6}$	129	pc	T-1
Roberts et al.	2.105 1979	T	1200–1450		201	pc	T-2
Freer	2.53 1980	Am 242 Pu 239 Fe 59 S 35 S 35	1200–1430 1200–1450 900–1100 1000–1200 1300	$4.94 \cdot 10^6$ $1.54 \cdot 10^8$ $9.18 \cdot 10^{-8}$ $4.80 \cdot 10^{-7}$ $D \approx 10^{-15}$	564 594 112 194	pc pc pc pc sc	Am Pu Fe S-1 S-2

[*]) sc single-crystal, pc polycrystalline

ted out in [2.102] that the ionic conductivity of polycrystalline Al$_2$O$_3$ may be mainly due to the mobility of Na.

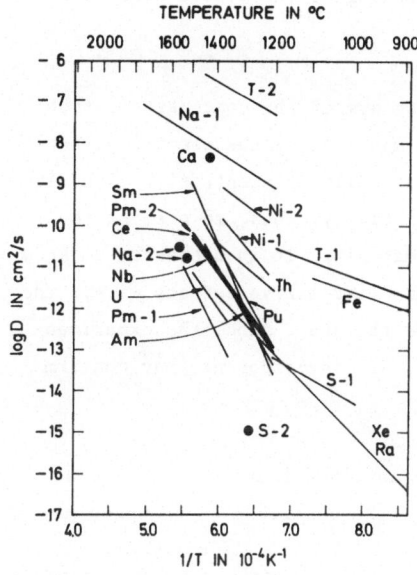

Fig. 2.11. Impurity diffusion coefficients in Al$_2$O$_3$. Pm-1: [2.98]; Th,U: [2.99]; Ce,Nb, Pm-2,Sm: [2.100]; Ra,Xe: [2.101]; Ca,Na-1,Na-2: [2.102]; Ni-2,Ni-2: [2.103]; T-1: [2.104]; T-2: [2.105]; Am,Fe,S-1,S-2,Pu: [2.53].

To resume the compilation of the diffusional behavior of Al$_2$O$_3$, it can be stated that much work has been performed in the acquisition of experimental data, but a series of uncertainties and unresolved problems still remains. Among these, there is the question of the type of mobile defect that causes cation mobility, V'''_{Al} or $Al_i^{...}$, a problem which is intimately related to the disorder mechanism, the reason for the high O mobility along the grain boundaries, and the effect of impurities and dopants on both lattice and grain-boundary diffusivities of Al and O, the investigation of which is just beginning.

2.4 Electrical Conductivity

Aluminum oxide is a very good electrical insulator which retains its high electrical resistivity up to very high temperatures. It is widely used, therefore, as an insulating material. Its applications in electronics and computer industry and as insulating parts in general will be given in Section 5.1. The electrical resistivity of aluminum oxide is one of its technically most important properties. Therefore, the investigation of the electrical conductivity and of conduction mechanisms has gained much attention among researchers. Moreover, the study of the electrical conductivity represents a key to the understanding of the defect structure of the material.

The electrical conductivity is proportional to the sum of the concentration-mobility products of the charge carriers present. In many cases, the electrical behavior is dominated by the majority defect, but the smaller concentration of a minority defect may be compensated for by a greater mobility. Thus, to predict the electrical conductivity of a particular ceramic material, it is necessary to know the type and concentration of the majority and minority native defects, i.e., the defect structure, as well as their mobilities. On the other hand, the experimental determination of the electrical conductivity and its components may confirm or reject a defect model tentatively established.

The elctrical conductivity of aluminum oxide, σ, is often found to be composed of an ionic part, σ_i, and an electronic part, σ_e, the contribution of which may vary with both temperature and P_{O2},

$$\sigma = \sigma_i + \sigma_e \quad . \tag{2.17}$$

The relative contributions of the ionic charge carriers, $t_i = \sigma_i/\sigma$, and the electronic charge carriers, $t_e = \sigma_e/\sigma$, to the over-all conductivity are called the ionic and electronic transference numbers, respectively; their sum is equal to unity:

$$t_i + t_e = 1 \quad . \tag{2.18}$$

The ionic part of the electrical conductivity is proportional to the diffusivity of the charge-carrying ionic defect,

$$\sigma_i = n_i z_i^2 e^2 D_i/kT \quad , \tag{2.19}$$

where n_i is the number of lattice sites per unit volume, z_i is the valence of the migrating defect, e is the elementary charge, D_i the diffusion coefficient, and k the Boltzmann constant. Equation (2.19) is known as the Nernst-Einstein relation. The ionic transference number can be obtained from emf measurements on an oxygen concentration cell with P_{O2} = constant at one side (I) and variable P_{O2} at the other side (II). The emf of such a cell is given by

$$E = (RT/4F) \cdot \int_I^{II} t_i \, d\ell n P_{O2} \quad , \tag{2.20}$$

with R the gas constant and F the Faraday constant. The slope of the curve $E = f(p_{O2}^{II})$ gives the ionic transference number at p_{O2}^{II}.

A problem in the measurement of the electrical conductivity of a highly iso-lating material such as aluminum oxide at temperatures above $1000^{\circ}C$ consists in the fact that the absolute value to be determined is of the same order of magni-tude as the surface conduction and the conduction through the gas phase [2.106]. This makes experimental results doubtful unless special precautions are taken to suppress surface and gas phase conduction, as has been pointed out by Peters et al. [2.106]. Improvements in the experimental techniques have been intro-duced by Özkan and Moulson [2.108], Brook et al. [2.38], and Yee and Kröger [2.40] using several safe guarding techniques to avoid surface and gas phase conduction. A detailed review of the literature on electrical conduction in aluminum oxide before 1971 may be found in the paper of Brook et al. [2.38], and there is strong evidence that experimental data before this date are not very reliable.

The electrical conductivity of aluminum oxide is found to depend on temperature, oxygen partial pressure, dopant or impurity content, and grain size. The subject will be treated in that sequence.

A summary of experimental data on the temperature dependence of the electrical conductivity is given in Fig. 2.12. Part (A) contains data of single-crystal Al_2O_3, and Part (B) of polycrystalline materials. For reasons discussed above only rela-tively recent measurements have been considered. Conductivity curves as a function of temperature calculated on the assumption of purely ionic conductivity by both the cation and the anion which follow from Eq. (2.19) are also included in Fig. 2.12 for comparison.

Table 2.10 contains values of the apparent activation energies of the curves of Fig. 2.12, (A) and (B), as reported in the references or assessed by us. The main impurity species or dopant is also given in Table 2.10 as well as the atmosphere, which in most of the cases was oxygen or air of one bar, and the grain size of the polycrystalline materials. When looking at Fig. 2.12 and Table 2.10, one observes a large variation of the conductivity values and activation energies obtained by several authors which, without doubt, may be attributed to different experimental techniques and varying material composition. Both single-crystal and polycrystal-line conductivities apparently lie closer to the curve which results from pure ionic conduction via Al diffusion rather than O diffusion. On the average, the electrical conductivity of polycrystalline aluminum oxide is greater and the ac-tivation energy is smaller than in single-crystals.

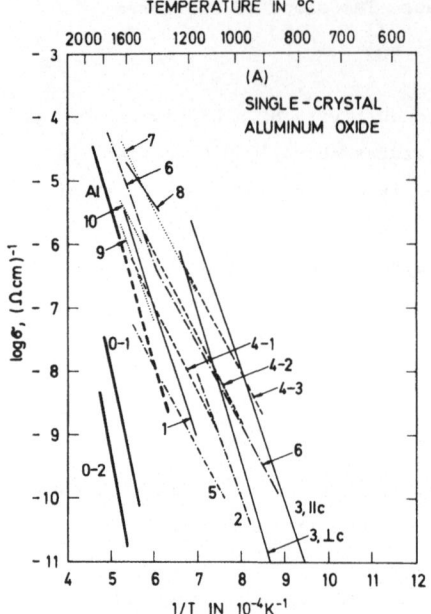
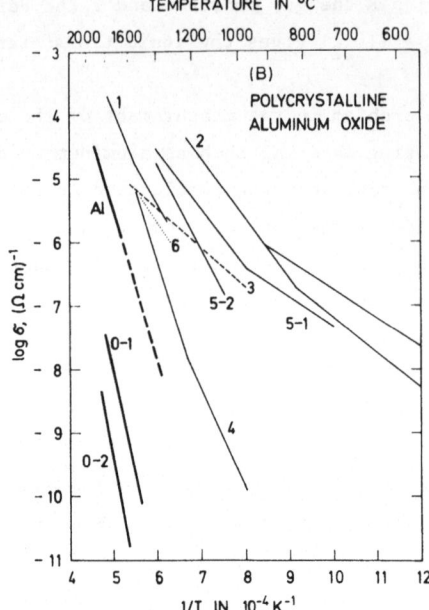

Fig. 2.12. Electrical conductivity of aluminum oxide as a function of temperature. (A) Single
crystal aluminum oxide. 1: [2.106]; 2: [2.107]; 3: [2.108]; 4: [2.38], various dopants;
5: [2.109]; 6: [2.44]; 7: [2.45]; 8: [2.48], Mg-doped; 9: [2.49], Ti-doped; 10: [2.32];
Al: σ calculated from Al self-diffusion data of [2.60]; 0-1: σ calculated from 0 self-
diffusion data of [2.33]; 0-2: 0 self-diffusion data of [2.69]
(B) Polycrystalline aluminum oxide, 1: [2.37]; 2: [2.108]; 3: [2.40]; 4: [2.110]; 5: [2.44],
undoped and MgO-doped; 6: [2.30]; Al, 0-1, and 0-2 same as in part (A)

When the electrical conductivity is determined as a function of p_{O_2} it is often
found that it increases with both small and large oxygen partial pressure, showing a minimum located at an intermediate pressure (10^{-6} to 10^{-4} bar) [2.30,2.32,
2.37,2.38,2.44 to 2.46,2.48,2.49]. This behavior seems to be independent of the
type of dopant (acceptors or donors) and to hold for both singe-crystal and polycrystalline materials. As examples, Fig. 2.13 shows the p_{O_2} dependence of the conductivity of two singe-crystal, acceptor-doped samples and a polycrystalline, donor-doped material.

A common method to find an interpretation of the observed behavior consists in
the determination of the ionic transference number and its dependence on tempera-

44

Table 2.10. Activation energy of electrical conduction at p_{O2} = 1 bar

(A) Single-crystal alumina

Authors	Ref. Year	Activ. energy in eV	Main impurity or dopant in ppm
Peters et al.	2.106 1965	4.2[1]	100 Fe
Moulson & Popper	2.107 1968	3.9[2]	pure
Özkan & Moulson	2.108 1970	4.0[3]	pure
Brook et al.	2.38 1971	2.9[2] 2.9[2] 2.9[2]	pure 670 Mg 85 Co
Lackey	2.109 1971	2.59	200 total
Kitazawa & Coble	2.44 1974	3.5 2.5	100 Si
Dutt et al.	2.45 1975	2.65	31 Co
Mohapatra & Kröger	2.48 1977	2.03	10 Mg
Mohapatra & Kröger	2.49 1977	3.48	430 Ti
Mohapatra et al.	2.32 1979	2.55	unknown acceptor

(B) Polycrystalline alumina

Authors	Ref. Year	Activ. energy in eV	Main impurity or dopant in ppm	Grain size in μm
Pappis & Kingery	2.37 1961	3.32 2.46	not specified	28
Özkan & Moulson	2.108 1970	1.9[3]	pure	12 20
Yee & Kröger	2.40 1973	1.3[2]	Mg-doped	20
Dilger	2.110 1974	4.46[4]	1000 Si	7
Kitazawa & Coble	2.44 1974	1.8 2.7	20 Na 5000 Mg	4 30
Hou et al.	2.30 1979	1.85	20 Mg	3

[1] in helium
[2] p_{O2} = 0.2 bar
[3] atmosphere not reported
[4] $3 \cdot 10^{-5}$ bar

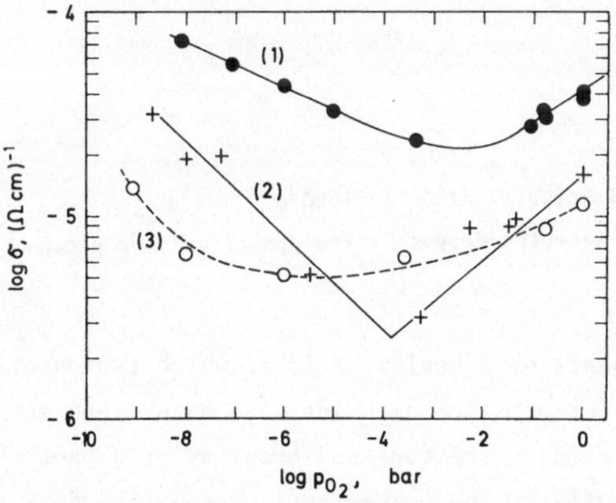

Fig. 2.13. Electrical conductivity of aluminum oxide as a function of p_{O2}.
1: single-crystal, 31 ppm Co, 1620°C [2.45]; 2: single-crystal, 10 ppm Co, 1650°C [2.44];
3: polycrystalline, GS = 6 μm, 80 ppm Ti, 1600°C [2.30]

ture and p_{O2}. Earlier experimental work has generally shown that t_i is near unity at low temperatures and decreases to values near zero at elevated temperatures. For that reason, the interpretation of many authors has been that aluminum oxide is an ionic conductor at low temperatures but becomes an electronic conductor at elevated temperatures [2.37,2.44,2.111 to 2.114]. It has been emphasized early, however, that the decrease of t_i at high temperatures is more representative of the increase in the gas phase conduction than of a real increase in t_e [2.38].

Later on, when improved experimental techniques became more and more available it could be demonstrated that t_i can be quite high at high temperatures and that the way in which t_i depends on p_{O2} is strongly affected by the dominant type of dopant. Figure 2.14 contains three recent measurements of the ionic transference number of single-crystal Al_2O_3, both donor and acceptor-dominated, as a function of p_{O2} at $1600°C$. With increasing p_{O2}, t_i of the two acceptor-dominated materials falls off from values near unity to zero, and the opposite is true for the donor-dominated crystal where t_i increase to unity when p_{O2} approaches 10^{-2} bar.

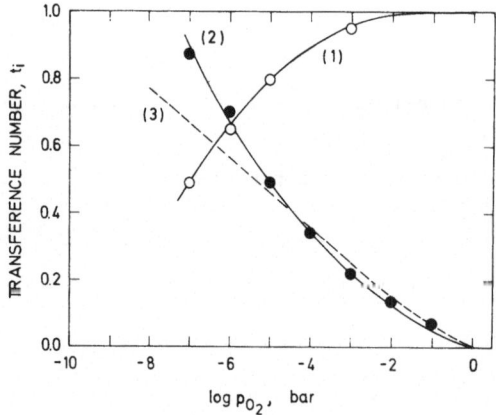

Fig. 2.14. Ionic transference number of single-crystal Al_2O_3 at $1600°C$.
1: donor-doped (430 ppm Ti) [2.49]; 2 and 3: acceptor-doped; 2: few ppm of unknown acceptor [2.32]; 3: 10 ppm Mg [2.48]

These findings can be explained on the basis of a series of ideas which represent the present view of the electrical behavior of aluminum oxide. The nature and magnitude of the electrical conductivity depend on the type and quantity of dopant and on p_{O2}. Doping with aliovalent ions affects the concentration of native defects and thereby the diffusivities of Al and O. The variation of the oxygen partial pressure changes the fraction of the dopant which is present in the aliovalent state.

46

In the case of acceptor doping (e.g., Fe, Co, and Mg), the foreign ion is present at low p_{O2} in its divalent state, which gives rise to the formation of positively charged native defects, $Al_i^{...}$ or $V_O^{..}$, so that the electrical conductivity is mainly ionic. With increasing p_{O2} the acceptor becomes increasingly oxidized whereby the concentration of ionic defects is decreased and the concentration of electron holes is increased due to the increased concentration of the acceptor in its isovalent state. Thus, as a function of increasing p_{O2}, the ionic conductivity decreases whereas the hole conductivity increases. This behavior can also be described quantitatively. For the case of Fe-doped Al_2O_3, it has been shown by Dutt and Kröger [2.46] that, starting from Eqs. (2.8) to (2.10), Eq. (2.12) gives an expression for the ionic part and Eq. (2.14) for the electronic part of the electrical conductivity, and that the sum of these two expressions meets the experimental data very ell. The same set of equations and calculation method for σ has been used successfully for Co-doped Al_2O_3 [2.45], just substituting Co for Fe; in a series of other papers on acceptor-dominated Al_2O_3 it was shown that theoretical conductivity curves based on the principles described above agree very well with experimental data [2.30,2.32,2.48].

A similar consideration may be applied to donor doping. At high p_{O2}, the donor is present in its oxidized, aliovalent form causing $V_{Al}^{'''}$ defects and hence predominantly ionic conductivity, whereas at low p_{O2} the dopant becomes isovalent so that the ionic defect concentration is very small, and electronic conductivity predominates. The corresponding set of equations for Ti-doped Al_2O_3, according to Mohapatra and Kröger [2.49], is

$$[Ti_{Al}^{.}] = 3[V_{Al}^{'''}] \quad , \tag{2.21}$$

$$(3/4)O_2(g) + 3Ti_{Al}^x \rightleftharpoons 3Ti_{Al}^{.} + V_{Al}^{'''} + (3/2)O_O^x \quad , \tag{2.22}$$

$$Ti_{Al}^x \rightleftharpoons Ti_{Al}^{.} + e' \quad . \tag{2.23}$$

Again, defect concentrations and their p_{O2} and temperature dependence are obtained by applying the mass-action law and solving Eqs. (2.21) to (2.23), provided the several equilibrium constants are known. In this way, conductivity curves have been constructed for some donor-doped aluminas which meet the experimental data well [2.32,2.49].

It should be emphasized, however, that the elementary treatment just shown above may give an over-simplified view of what happens in a real crystal. The unequivocal effect of a single acceptor may be masked and partly compensated by the simultaneous presence of a donor of either known or unknown concentration, as has been shown in [2.30,2.32,2.45]. This makes the theoretical analysis much more difficult, and the defect model that accounts for the electrical behavior of that material may be quite complicated. The reader who is interested in knowing more details about the subject is referred to the papers [2.30,2.32,2.86] where an extensive analysis of the effect of various dopants is given and more sophisticated defect models are presented.

To resume, the electrical behavior of aluminum oxide turns out to be strongly dependent on the type and amount of doping, on the ionization constant of the dominant dopant, K_i, and on the equilibrium constant which controls the incorporation of oxygen in pure Al_2O_3, K_{ox}. The temperature dependence of the electrical conductivity results from the temperature dependence of both the mobility and the concentration of the major charge-carrying defect which, in turn, depends on the temperature variation of K_{ox} and K_i. Therefore, the experimental activation energy of conduction is expected to be different for different dopants, and indeed a wide range of values is observed experimentally (Fig. 2.12). In some cases, however, it has been possible to interpret the experimental activation energy in terms of the mobility of the charge-carrying ionic defect [2.30,2.32, 2.45,2.46,2.48,2.49] which may be $V_O^{..}$, $V_{Al}^{'''}$ or $Al_i^{...}$, and of the energy level of the donor or acceptor in the band gap. Energy levels of several donor and acceptor dopants in Al_2O_3 as determined experimentally or predicted theoretically have been summarized by Tiku and Kröger [2.115].

The effect of foreign additives in the range up to several per cent can also be understood in terms of the model depicted above. Whereas the results of Vernetti and Cook [2.116] on the effect of additions of Co_2O_3, Fe_2O_3, Cr_2O_3, CuO, MnO_2, and NiO have been contradictory, it was observed by Kemp and Moulson [2.113] that Fe_2O_3 additions up to 1.0 w/o increase the conductivity if Fe is present in the divalent form. This result was confirmed recently up to 3 cation % Fe in a conductivity study made by El-Aiat et al. [2.86]. A conductivity increase of over four orders of magnitude was reported by Hennicke and Sturhahn [2.114] for Cr_2O_3 additions up to 100%. Dilger [2.110] also observed an increase in σ for NiO and ZnO additions up to 1%.

Finally, the electrical behavior of polycrystalline Al_2O_3 shall be considered. There is agreement among several experimental studies which compared polycrystalline and single-crystal materials [2.37,2.40,2.44,2.108] as to a series of observations: in polycrystalline alumina, the absolute value of σ is found to be

48

considerably greater (up to a factor of 10^4) than in single crystals, and the P_{O2} and temperature dependence (activation energy) as well as the value of the ionic transference number is found to be smaller than in single crystals. Since it is also observed that σ increases with decreasing grain size [2.108,2.110] it has been concluded by several authors [2.40,2.44,2.110] that the grain boundary is a region of enhanced conductivity and that hole conduction along the boundary is probably the prevailing conduction mechanism. Two particularly detailed studies of the electrical conductivity of polycrystalline Al_2O_3 are the papers of Hou et al.[2.30] and of El-Aiat et al. [2.86]. On determining the electronic and ionic conductivity as a function of temperature, P_{O2}, and grain size in both pure and Fe- or Ti-doped materials, no increase in the ionic part of the conductivity with decreasing grain size was observed, as would have been expected if there were any charged ion migration along the grain boundary. Therefore, it was concluded that neither O nor Al migrate via grain boundaries in charged form which led to the suggestion of fast O grain-boundary diffusion of neutral oxygen, O_i^x, as already discussed in Section 2.3.5.2. What was observed to increase, however, with decreasing grain size was the electronic conductivity, and this was attributed to an increased hole transport via grain boundaries [2.30,2.86].

2.5 Sintering and Grain Growth

In this section the fundamental characteristics of sintering phenomena in alumina will be described. Practical and technological aspects of sintering will be reported in Sections 4.2 and 4.3. The questions to be answered in the following paragraphs are: what happens in the material when alumina is sintered, how has it been possible to sinter alumina to high or even theoretical density, and what has been done to identify the controlling variables and mechanisms? Hence, the basic sintering mechanisms in Al_2O_3 will be dealt with first, and then some modifying effects caused by the occurrence of grain growth and the use of different additives and sintering atmospheres will be considered. Finally, the particular effects of MgO will be treated, and details on grain growth phenomena presented.

The fact that Al_2O_3 can be sintered to theoretical density was demonstrated for the first time by Cahoon and Christensen [2.117] in 1956 and a few years later by Coble [2.56]. These authors showed that the essential prerequisite for the complete removal of pores is the suppression of discontinuous grain growth, which was achieved by a small MgO addition. Research activities on the sintering of Al_2O_3 proceeded in the 1960's and 1970's and were advanced a great deal by the work of Coble [2.78,2.118 to 2.120] and Cutler and co-workers [2.42,2.43,2.57,2.58,2.121

to 2.125]. A detailed review of the literature until 1970 has been given by Gitzen [2.1].

2.5.1 Fundamental Sintering Mechanisms in Al_2O_3 During the Initial Stage

The basic problem of studies on sintering phenomena in a particular material consists in the identification of the dominant mechanism by which matter transport occurs [2.120]. The driving force of sintering in the solid state has early been recognized to be the surface tension [2.126]. The kinetics of the process, however, may be controlled during the initial stage of sintering by a series of alternate diffusion processes, which define different paths for matter transport. Figure 2.15 shows the typical sintering geometry between two powder particles

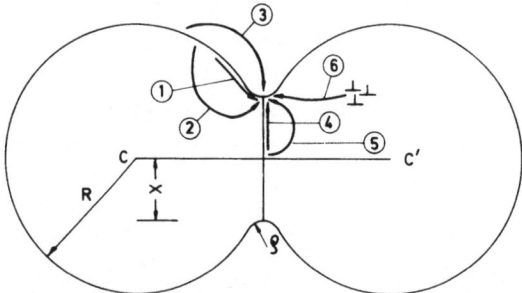

Fig. 2.15. Schematic of the geometry of sintering of two spheres, showing six alternate paths for matter transport. For numbers see text. – After Kingery and Berg [2.126] and Ashby [2.127]

according to Kingery and Berg [2.126] as well as six alternate paths for matter transport according to Ashby [2.127]. These are (1) surface diffusion, (2) lattice diffusion from the surface region, (3) vapor transport, (4) boundary diffusion, (5) lattice diffusion from the boundary region, and (6) lattice diffusion from dislocations or other internal sources to the neck surface. Plastic flow is considered not to play an important role in densification of aluminum oxide [2.120] except during hot pressing. It is usual to classify these transport mechanisms into two groups [2.127,2.128], i.e., those which can produce shrinkage (mechanisms 4 to 6) and those which cannot (mechanisms 1 to 3).

The first quantitative study on the initial sintering of alumina was published by Coble [2.78]. He determined the growth of the neck which is formed between a sphere and a plate as a function of time and sphere radius, and measured continuously the shrinkage of powder compacts as a function of time at various temperatures. These shrinkage measurements are reproduced in Fig. 2.16. Coble was able to

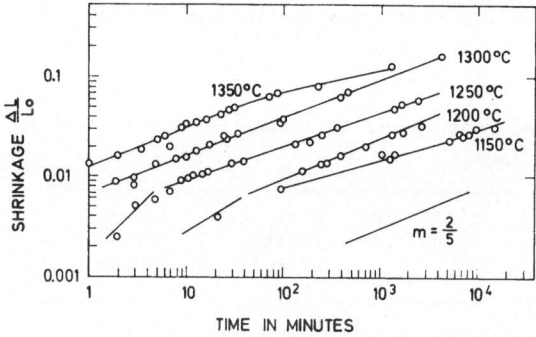

Fig. 2.16. Shrinkage isotherms of alumina powder compacts during initial stage sintering.
- After Coble [2.78]

show that the kinetics and size dependence of the sintering of alumina follow exactly the predictions of a quantitative sintering model presented by Kingery and
Berg [2.126]. This model is based on the assumptions that the driving force for
solid state sintering is the surface tension, the dominant matter transport mechanism is lattice diffusion, and the grain boundaries operate as vacancy sinks.
The lattice diffusion coefficient which follows from a comparison of Coble's experimental sintering data [2.78] with this model is plotted as curve 14 of Fig.
2.8 and shows excellent agreement with other diffusion data. In another early
work on the subject, Kuczynski et al. [2.129] confirmed Cobles results, repor-
ting that the sintering of single-crystal sapphire spheres is controlled by a
lattice diffusion mechanism.

Since then, more refined sintering models have been developed [2.62,2.63,2.131]
which describe neck growth and densification during the initial stage of sintering
by both lattice and grain-boundary diffusion, and a series of other experimental
studies on the initial sintering of both pure and doped alumina has been published,
which is summarized in Table 2.11. In some cases it has been found that the kinetics of initial sintering of alumina is controlled by grain-boundary diffusion,
but most of the authors agree that the process is controlled by lattice diffusion
and, mainly in more recent papers, it has been recognized that the rate controlling (slower moving) defect is V_{Al}''' or $Al_i^{...}$, depending on the dominant dopant, the
oxygen moving rapidly along the boundaries. Values of the activation energy of
sintering lie in the same range as the activation energy of lattice diffusion of
Al, Table 2.6 (A). For the cases of Ti doping [2.57,2.39] and Fe doping [2.42]
mentioned in Table 2.11 it has been shown that the magnitude of the activation
energy of sintering is composed of the mobility term of the rate-controlling moving defect, V_{Al}''' or $Al_i^{...}$, and a term which contains the activation enthalpy of

oxidation of the dopant, in much the same way as already outlined in Section 2.3.3 and described by Eq. (2.12).

Table 2.11. Sintering mechanisms and activation energies during initial sintering of pure and doped Al_2O_3

Authors	Ref. Year	Main objective of the work was to determine	Diffusion mechanism	Activation energy, kJ/mol	Rate-controlling defect
Coble	2.78 1958	kinetics of initial sintering	lattice	691	
Kuczynski et al.	2.129 1959	kinetics of sphere: sphere sintering	lattice		
Johnson & Cutler	2.121 1963	kinetics of different powders	boundary	628	$V_O^{\cdot\cdot}$
Keski & Cutler	2.122 1965	effect of MnO additions	lattice	586	$V_O^{\cdot\cdot}$
Wilcox & Cutler	2.123 1966	kinetics of initial sintering	boundary		
Keski & Cutler	2.124 1968	effect of MnO additions	lattice	649[1] 775[2]	$V_O^{\cdot\cdot}$
Bagley et al.	2.57 1970	effect of TiO_2 additions	lattice, boundary[4]	620 to 636[3]	$V_{Al}^{'''}$
Young & Cutler	2.125 1970	feasibility of constant rate of heating-technique	boundary	481	
Greskovich & Lay	2.130 1972	grain growth during initial sintering by SEM	lattice		
Raja Rao & Cutler	2.58 1972	matter transport by surface or lattice diffusion	lattice		$V_{Al}^{'''}$
Raja Rao & Cutler	2.42 1973	effect of Fe and Ti additions	lattice	628[1] 691[2]	$Al_i^{\cdot\cdot\cdot}$[5] $V_{Al}^{'''}$[5]

[1] pure; [2] doped; [3] for lattice diffusion; [4] depending on particle size; [5] depending on dopant

Whereas the interpretation of densification kinetics seems to be unequivocal, there exists some ambiguity as to the role of surface diffusion in neck growth. It has been stated by some investigators [2.92,2.130,2.132] that neck growth in alumina during the early stages of sintering is caused by surface diffusion. On the other hand, in two different papers Raja Rao and Cutler [2.42,2.58] raised doubts about these findings as well as the validity of reported surface diffusion coef-

ficients and claimed that neck growth is controlled preponderantly by lattice
diffusion, the contribution of surface diffusion being negligibly small. This
claim probably may be substantiated by an observation made by McAllister and Coble
[2.133], who reported that thermal grooving of Al_2O_3 (a "negative" neck growth
effect) could best be explained in terms of volume diffusion rather than surface
diffusion.

2.5.2 Intermediate and Final-Stage Sintering

The processes that occur in a powder compact during the later stages of sintering
are decisive for the final density of the product. The classification of sintering
into three stages was originally proposed by Coble [2.134] and is generally accepted today. The initial stage is characterized by the neck growth between the original powder particles and a slight increase in density of about 10%. The beginning
of the intermediate stage coincides with the beginning of grain growth. During
this stage, particles grow to a grain-like structure, the pore phase forming an
array of interconnected cylindrical channels lying on three-grain edges. The final
stage starts at about 95% porosity, when cylindrical pores are transformed into
spherical voids by a pinch-off process. The pore phase is now present as discontinuous pores lying at four-grain corners [2.134].

Intermediate and final-stage sintering kinetics in alumina were first investigated by Coble [2.56]. He found that during sintering the porosity decreased linearly with the logarithm of time, $P \propto \ln t$. This decrease agrees with that predicted
by his late-stage lattice diffusion sintering model [2.134] when the time dependence of grain growth during sintering is taken into account. Densification curves from Ref. [2.56] are shown in Fig. 2.17, which also demonstrates the beneficial effect of a 0.25 w/o MgO addition on both the densification rate and final
density. As can be seen from Fig. 2.17, density in the MgO-free material does never exceed 99%; in the MgO-containing material, however, theoretical density is
achieved at $1675^{\circ}C$ after about 300 minutes. Values of the lattice diffusion coefficient which follow from a comparison between experimental densification data
and the sintering model [2.134] are within a factor of 10 of the standard curve of
Cannon and Coble [2.59] and are shown as curve 15 of Fig. 2.8.

It has been pointed out by Coble and Burke [2.120] that the necessary condition
to achieve theoretical density in solid state sintering consists in eliminating
or suppressing the occurrence of discontinuous grain growth. Then, grain boundaries will remain attached to the pores, and the normal grain growth will be
sufficiently slow that the pores can follow the movement of the grain boundaries

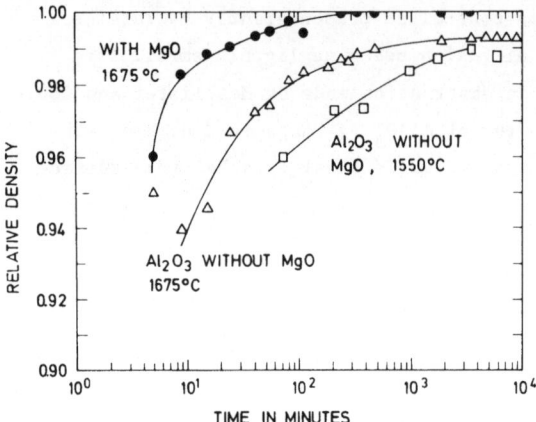

Fig. 2.17. Densification of alumina powder compacts during intermediate and final stage sintering. – After Coble [2.56]

and do not become trapped inside the grains. Since the pore phase remains inter-sected by the boundaries, the diffusion path is short and facilitates the comple-te removal of porosity. Discontinuous grain growth in alumina is obviously eli-minated by a small MgO addition [2.56,2.117]. This subject will be returned to in Section 2.5.4 when the effects of MgO are discussed in detail.

Another careful and detailed study on the sintering kinetics of alumina in the late stages was made by Bruch [2.135]. He also obtained nearly theoretical density by using an addition of 0.25 w/o of MgO. He further showed that the temperature dependence of both porosity and grain size could be fitted empirically to an Arrhenius term, $\exp(-Q/RT)$, with an equal activation energy Q for both densifi-cation and grain growth. Thus he arrived at an empirical relation between grain size and porosity which is independent of time and temperature. This relation, which is reproduced graphically in Fig. 2.18 predicts that, for constant initial porosity, the grain size is a unique function of the achieved porosity, and that a given combination of grain size and porosity can be obtained by innumerable com-binations of time and temperature. It is clear that the permutability of time and temperature observed by Bruch [2.135] reflects the fact that sintering is a ther-mally activated process.

Further examples of sintering of aluminum oxide in the intermediate and final stage may be found in the papers of Coble [2.119], Vink [2.136], and Peelen [2.137], and, more recently, of Johnson and Coble [2.138] and Harmer et al. [2.51]. The authors of the papers cited in this section agree about the follo-wing series of conclusions: (1) It is possible to sinter alumina to theoreti-

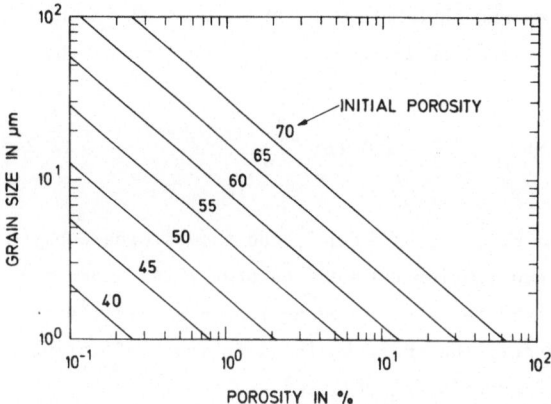

Fig. 2.18 Sintering diagram of MgO-doped alumina. - After Bruch [2.135]

cal or nearly theoretical density if discontinuous grain growth can be eliminated; (2) sintering kinetics of alumina in the intermediate and final stage is
controlled by lattice diffusion; (3) the slower moving ion is Al, the O diffusion occurring rapidly along grain boundaries; and (4) sintering data agree, at
least within an order-of-magnitude approximation, with predictions of late-stage
sintering models, as given p.e. by Coble [2.134] and Johnson [2.64].

2.5.3 Influence of Additives

In this section an overview of sintering additives other than MgO which have been
used successfully to accelerate the sintering rate of alumina will be given. The
effect of MgO will be discussed in the following section.

Since the first publication which reported that alumina can be sintered to theoretical density [2.117] it has been known that additives may have a beneficial influence on the sintering kinetics of alumina. In this early comprehensive study,
a total of 23 different additives was tested for their effect on density, grain
size, and compressive strength. It was found that mainly MgO and NiO are able to
inhibit discontinuous grain growth and to promote sintering kinetics which leads
to full density [2.117].

Until now, a series of other papers dealing with the same subject has appeared in
the literature. The effect of MnO additions in the range between 1000 and 4000 ppm
was studied by Keski and Cutler [2.122,2.124]. They found an increase in sintering
rate with increasing MnO content and decreasing p_{O2}. The proposed interpretation

55

was that Schottky disorder predominates and that Mn which substitutes Al in the Al_2O_3 lattice can be present in its divalent or trivalent form according to the reaction

$$O_O^x + 2Mn_{Al}^x \rightleftharpoons 2Mn_{Al}' + V_O^{\cdot\cdot} + \frac{1}{2} O_2(g) \quad . \tag{2.24}$$

Thus, at small p_{O2}, the additive is mainly present as Mn_{Al}^{2+} and should create oxygen vacancies to maintain the charge equilibrium through the neutrality condition $[Mn_{Al}'] = 2[V_O^{\cdot\cdot}]$. Since O was thought to be the slower moving ion in self-diffusion of alumina, this model predicted correctly the increase in diffusion coefficient with both increasing MnO content and decreasing p_{O2} [2.124].

Warman and Budworth [2.139], in searching for additives which serve as sintering aids in alumina, reported that full density could be achieved using NiO, ZnO, CoO, and SnO_2 in a quantity of 0.25 w/o. They stated that the common feature of the additives used, and what is perhaps the reason for their beneficial effect, is their volatility and insolubility in alumina.

Small additions of TiO_2 to alumina were studied by Bagley et al. [2.57]. Measuring the initial sintering rate, they found that increasing the TiO_2 content increased the volume-diffusion coefficient exponentially until it tapered off and reached a saturation value at a TiO_2 content which was interpreted to be the saturation limit of TiO_2 solid solution in Al_2O_3. The substitutional incorporation of Ti^{4+} in Al_2O_3 was thought to produce V_{Al}''' defects, and from that it was concluded that Al is the slower moving ion in sintering. The high activation energy of sintering (see Table 2.11) was attributed to the enthalpy of the incorporation reaction of oxygen and to the dissolution enthalpy of the $TiAl_2O_5$ precipitates.

In analyzing the foregoing results of Bagley et al. [2.57], Brook[2.39] proposed a defect model for Al_2O_3:Ti in which the excess charge of the Ti^{4+} ions is compensated by divalent impurities such as Mg_{Al}' and Ca_{Al}' so that the charge-neutrality condition is

$$[Ti_{Al}^{\cdot}] = [Mg_{Al}'] \quad , \tag{2.25}$$

where Mg represents all types of divalent impurities present. The oxidation-reduction of Ti in Al_2O_3 was assumed to be given by the reaction of Eq. (2.22). Starting from Eqs. (2.25) and (2.22), Brook showed that $[V_{Al}'''] = const. \cdot [Ti_{Al}^x]^3$, which is in excellent agreement with the third-power concentration dependence of the experimental diffusion coefficient data when plotted according to a power law.

56

Brook's defect model for Al_2O_3:Ti is shown in Fig. 2.19 together with the corresponding log-log plot of the volume diffusion coefficient of Al in Al_2O_3:Ti as a function of the Ti content.

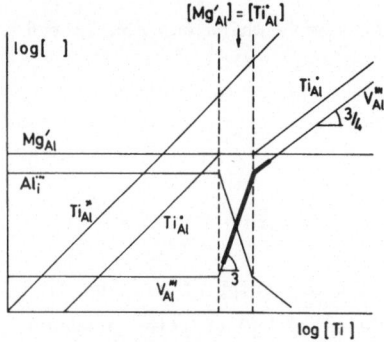 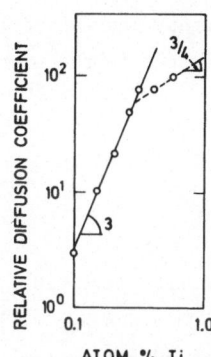

Fig. 2.19. Proposed defect model for Al_2O_3:Ti containing divalent impurities (A), and power-law dependence of diffusion coefficients on Ti concentration. - After Bagley et al. [2.57] and Brook [2.39]

Another careful study of the effect of Fe and Ti additions on the sintering kinetics of alumina was carried out by Raja Rao and Cutler [2.42]. It was found that both Fe^{2+} and Ti^{4+} additions in the range of some tenths of a per cent enhanced the diffusion rate by several orders of magnitude, and that the effect of the simultaneous presence of Fe^{2+} and Ti^{4+} was to partly compensate each other. Decreasing p_{O2} also increased the diffusion rate. The dependence of the diffusion coefficient on p_{O2} and the Fe^{2+} concentration was interpreted in terms of the defect model described in Section 2.3.3 which assumes that Frenkel disorder predominates and that Al interstitials are the major diffusing cationic defects. Equation (2.12) predicts the increase of the diffusion coefficient with decreasing p_{O2} as was observed experimentally; it also predicts a power law dependence of the diffusion coefficient on the total Fe concentration, $D_{Al} \propto [Fe_{tot}]^m$, with m = 3/4. Experimentally a value of 0.67 < m < 0.73 was found, which is also close to the theoretical value of m = 2/3 for Schottky disorder [2.42]. Nevertheless, the authors gave priority to Frenkel disorder because the results of the Ti-doped material could be explained likewise in terms of the same model.

A series of different additives to alumina were studied by Rossi and Burke [2.140] with the objective of finding sintering aids which give equally good results as MgO. The quality criteria applied by them were a homogeneous microstructure and a small final porosity. They found that the only successful sintering aid was NiO, whereas the other additions investigated (CaO, SrO, BaO, Y_2O_3, and ZrO_2) yielded unsatisfactory results.

The last study on sintering additives others than MgO that is reported here was published by Harmer et al. [2.51]. It was found that a 200 ppm TiO_2 addition increased the densification rate and the final density. The authors explained their results by assuming a defect model based on Eqs. (2.21) and (2.22), i.e. Ti^{4+} was thought to create V_{Al}''', thereby increasing the Al diffusivity which was considered to be the rate-controlling step in the sintering of alumina.

2.5.4 Effect of MgO

Magnesium oxide is the additive most widely used in the sintering of alumina since it was discoverd about 25 years ago [2.117] that a small MgO addition prevents discontinuous grain growth and allows the material to be sintered to theoretical or nearly theoretical density. It has been pointed out repeatedly, for example by Brook [2.141], that the achievement of a fine-grained microstructure is an important aim in the sintering of ceramics for several reasons. Primarily, the grain size directly affects a series of physical and mechanical properties such as strength, hardness, creep, and electrical conductivity, and with respect to these properties a material of finer grain size is always superior to a material of coarser grain size. In the work of Cahoon and Christensen [2.117] it was clearly demonstrated how the onset of discontinuous grain growth deteriorated the mechanical strength of polycrystalline alumina. It led to a sudden drop to about 65% of the strength achieved immediately before the beginning of discontinuous grain growth. On the other hand, densification to full density can only proceed as long as the grain size is kept small. The production of fully dense materials, however, is another objective in the sintering of ceramics because the absence of pores also results in the improvement of certain properties such as strength and optical transparency.

It had been recognized early [2.56,2.117] that the beneficial effect of MgO on density is due to the fact that it eliminates discontinuous grain growth during the late stages of sintering. Discontinuous grain growth means that grain boundaries break away from the pores, thereby including the pores inside the new large grains. For pores trapped inside the grains the diffusion distance to the next grain boundary is large so that they are eliminated only very slowly during the further course of the sintering process. Therefore, once discontinuous grain growth has started high density will not be attained any more. The characteristics of discontinuous grain growth and the conditions for its occurrence in ceramics have been described in detail by Coble and Burke [2.120] and by Brook [2.141].

Even today, there does not exist full agreement, however, about the question by which mechanism discontinuous grain growth in alumina is avoided when MgO is used as a sintering additive. The decisive point is that the pore phase must remain attached to the grain boundaries, i.e., breakaway of the boundaries from the pores must be prevented during the final stage of sintering. This situation can be promoted, in principle, by two different means [2.141]: (1) Increasing the pore mobility will enable the pores to follow the grain-boundary movement more easily, thus reducing the probability of pore trapping. (2) Decreasing the grain-boundary mobility will decrease the absolute velocity of grain-boundary migration and reduce the probability of boundary breakaway.

Considering the effect of MgO in Al_2O_3 in detail, the possible mechanisms as discussed by Coble and Burke [2.120] and, more recently, by Johnson and Coble [2.138] are as follows: (1) second-phase pinning of grain boundaries to inhibit their breakaway and the entrapping of pores; (2) solute-impurity segregation to grain boundaries to inhibit discontinuous grain growth by a solute-drag effect; (3) pore-phase pinning of grain boundaries through a change of the shape of the pore phase due to a small change in the value of the grain-boundary energy to surface energy ratio; and (4) enhancement of the densification rate relative to the grain growth rate through a change in the defect structure due to the divalent Mg ion (solid-solution model).

Neither the second-phase pinning theory (1) nor the pore-phase change model (3) have ever been used to explain the effect of MgO in alumina because it has been argued [2.56,2.135] that a pinning mechanism that inhibits discontinuous grain growth by preventing the breakaway of boundaries also should prevent normal grain growth which, however, never has been observed. The solute-segregation mechanism (2), on the other hand, was employed to interpret the sintering kinetics of alumina in the final stage [2.142,2.143], but it has lost its importance since it was proved that Mg segregation does not occur in alumina [2.144 to 2.147] and that the segregation of other impurities does not affect the sintering kinetics either [2.138].

There is strong evidence, however, that discontinuous grain growth is eliminated in alumina through a solid-solution mechanism (4). Yet, Mg solubility in Al_2O_3 is very limited, as may be seen from the phase diagram. Figure 2.20 shows the system $MgAl_2O_4-Al_2O_3$ from the work of Viechnicky et al. [2.148]. More precise measurements of the Mg solubility in the range of usual sintering temperatures were carried out by Roy and Coble [2.81]. For 1830°C, for example, they reported a solubility of 1400 ppm, and for 1630°C the value is only 300 ppm. Figure 2.20 illustrates that the spinel phase, $MgAl_2O_4$, is formed when the MgO content exceeds the solubility limit. The spinel phase is normally found as precipitates on the Al_2O_3 grain boundaries or on triple points. Even with very small solute concentrations, however, a solid-solute mechanism may be expected to have a tremendous influence on the number of defects and the diffusion coefficient, as was outlined in Section 2.3.2.

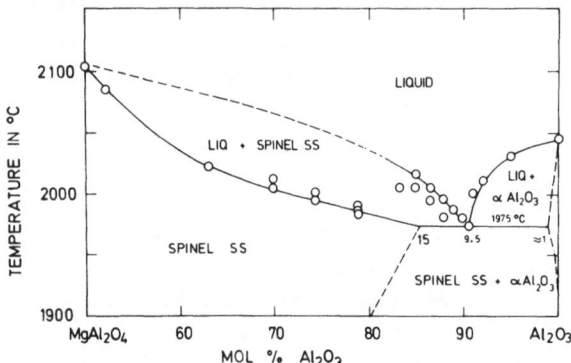

Fig. 2.20. The system $MgAl_2O_4$-Al_2O_3 [2.148]

Several experimental studies clearly demonstrate that alumina may be sintered to theoretical density using MgO additions even below the solubility limit. Figure 2.21 shows the density results from the work of Peelen [2.137], who sintered specimens containing between zero and 3000 ppm MgO. A maximum density was found at 300 ppm MgO, which is just the solubility limit reported in [2.81] at the sintering temperature of 1630°C. Peelen stated that this result is evidence that the densification rate in the absence of discontinuous grain growth is controlled by the solute concentration. In a similar study on Al_2O_3 doped with 200 ppm MgO, Harmer, Roberts, and Brook [2.51] observed a distinct increase in the sintering rate com-

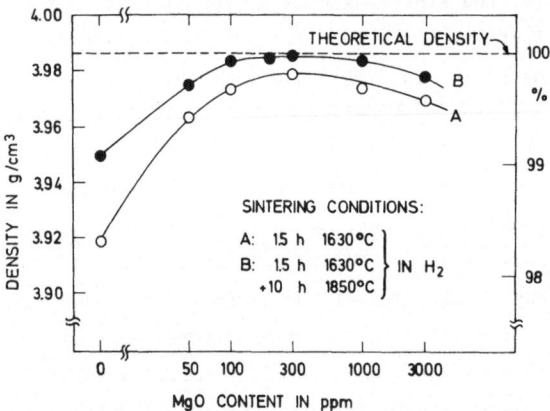

Fig. 2.21. Effect of MgO content on the density of sintered alumina. - After Peelen [2.137]

pared to pure Al_2O_3. Their interpretation was that the $Al_i^{...}$ concentration is increased by the presence of Mg'_{Al}, thereby increasing the Al diffusivity, which was considered the rate-controlling step in the sintering of alumina.

Impressive evidence that the second-phase pinning and solute-segregation mechanisms do not apply was given by Johnson and Coble [2.138]. Sintering of undoped Al_2O_3 to theoretical density was achieved by merely putting a mixture of spinel and alumina close together. They proved that during sintering no spinel precipitates could be present in the alumina specimen, and that neither did an impurity segregation exist which could have caused a solute-drag effect. They concluded, therefore, that the beneficial effect of Mg must be due to a solid-solution mechanism based on a change of the pore removal rate relative to the grain growth rate [2.138]. A quite similar experiment which led to the same result and a similar conclusion was also reported by Warman and Budworth [2.139].

Recently, a certain modification of the solid-solution mechanism has been proposed by Heuer [2.149] and by Harmer and Brooks [2.150]. As discussed above, the beneficial effect of the MgO additive is due to the increase of the densification rate (pore removal rate) with respect to the grain growth rate. In other words, the ratio of the pore mobility to the grain boundary mobility, M_p/M_b, must exceed a certain value. An increase in the pore mobility will favor the increase of the mobility ratio. In the case of pore movement due to surface diffusion, the pore mobility is given by the expression [2.141].

$$M_p = K \frac{D_s}{r^4} \tag{2.26}$$

where D_s is the surface diffusion coefficient, r is the pore radius, and K is a constant. It has been argued by Heuer [2.149] that the role of MgO in the sintering of alumina may consist in the fact that the presence of solute Mg increases the surface diffusion coefficient. The interpretation of Harmer and Brook [2.150], has been that at a given moment of the sintering process a MgO-doped alumina sample will have a smaller pore size than a sample made from pure alumina because the presence of the dopant increases the rate of pore removal. Therefore, the doped material will also be characterized by a greater pore mobility which is simply due to the size effect given by Eq. (2.26), and hence will show a decreased or retarded tendency for discontinuous grain growth.

2.5.5 Influence of Atmospheres

The atmosphere may show two different effects in the solid state sintering of ceramics, as has been discussed in detail by Coble [2.119]. On the one hand, changes
of atmosphere may change the densification rate because the stoichiometry and the
defect structure may be affected when the ambient oxygen pressure varies. According
to Coble [2.119], the dependence of sintering rate on atmosphere is due to a change
in the number of diffusing defects with changing oxygen pressure in the atmosphere. An increase in the number of defects of the slower diffusing species increases the lattice diffusion coefficient of that species and therefore the sintering rate. We saw examples of this behavior in the preceding section when we discussed the sintering kinetics of Al_2O_3:Mn [2.124] and Al_2O_3:Fe [2.42], where it was
found that the sintering rate is increased by decreasing p_{O2}. In both papers this
behavior was attributed to an increase in the number of the rate-controlling defects. A further indication that the defect structure of alumina is affected by
the atmosphere may be found in the paper of Warman and Budworth [2.139]. They reported that the final density and color of alumina specimens containing various
additives (MgO, NiO, CoO, ZnO, SnO_2, Cr_2O_3, and CaO) dependend on whether they
had been sintered in oxygen, vacuum, or hydrogen. In contrast to these results, however, no influence of the oxygen partial pressure on the densification rate of
MgO-doped alumina was found by Coble [2.119] when the material was sintered in O_2
or H_2.

On the other hand, the atmosphere may cause changes in limiting density when sintering proceeds until the final stage [2.119]. The final stage is characterized
by a discontinuous pore phase [2.134]. At the moment when the open pore channels
are pinched off the gas becomes trapped at the ambient pressure. When shrinkage
proceeds the gas is compressed, and therefore the driving force for sintering also has to overcome the increasing gas pressure inside the pores. Coble [2.119]
predicted that, for the case that no gas diffusion occurs, shrinkage will stop
when the internal gas pressure, P, is equal to that of the driving force, i.e.,
when $P = 2\gamma/r$ where γ is the surface tension and r the pore radius. Thus, a limiting density will be reached when the pores have shrunk to their stable size.
With increasing gas diffusity, shrinkage in the final stage is first controlled
by the kinetics of gas diffusion to the surface, and finally becomes independent
of the gas [2.119]. In an experimental study on final stage sintering of alumina containing 0.25 w/o MgO, Coble [2.119] has shown that the material could be
sintered to theoretical density if the ambient atmosphere was hydrogen, oxygen,
or vacuum, but that a residual porosity could not be eliminated when the ambient
atmosphere was helium, argon, or nitrogen. Roy and Coble [2.151] determined the

solubility of hydrogen in Al_2O_3 and demonstrated that the solubility at sintering temperature is large enough that all hydrogen contained in the pore phase at the beginning of final-stage sintering (5%) can be dissolved in the matrix. The solubility limits of Ar, He, and N_2 were found to be < 1% of the hydrogen solubility.

2.5.6 Grain Growth

Although small MgO additions eliminate discontinuous grain growth in alumina they do not inhibit normal grain growth. This conclusion has been drawn from the fact that extensive normal grain growth is observed during sintering. In the first quantitative study on grain growth in alumina during sintering, no effect at all of the dopant on the grain growth rate was observed until breakaway of grain boundaries occurred in the undoped material [2.56].

There are other investigations, however, which do find a reduction of normal grain growth due to the addition of MgO. Mocellin and Kingery [2.152] demonstrated that the grain growth constant K of the grain growth law (2.4)

$$GS^n - GS_o^n = K \cdot \exp(-Q/RT)t \qquad (2.27)$$

decreased with increasing MgO content above the solubility limit. Peelen [2.137] determined the grain size of sintered compacts as a function of the MgO content in the ppm range at constant sintering time and temperature. He found a steady increase in grain size for MgO concentrations from zero up to the solubility limit (300 ppm at his sintering temperature), but above this value the grain size decreased. The result was interpreted in terms of a solid-solute enhancement of the diffusion rate and a second-phase pinning effect. Monahan and Halloran [2.153] determined the apparent grain-boundary mobility in doped and undoped Al_2O_3 and found that it was decreased by a factor of 11 due to a 0.2 w/o MgO-doping.

Other additives are also known to reduce the grain growth of alumina. The reduction of grain growth by a factor of 3.5 due to the addition of 2.0 w/o metallic Ni powder was reported by Evans [2.154], and according to McHugh et al. [2.155] the dispersion of 5 v/o of finely divided Mo particles in the alumina matrix resulted in a very small grain size (2 μm) after sintering, and hence a very high strength. Wang and Kröger [2.156] reported that the presence of second-phase particles of $FeAl_2O_4$ in alumina samples doped with 3 w/o Fe slowed the grain growth.

Quantitative studies of grain growth in alumina were reported by several authors [2.56,2.130,2.135,2.152,2.154,2.156]. The grain growth parameters of Eq. (2.27), n and Q, obtained in these investigations are summarized in Table 2.12. The measurements of Coble [2.56], Bruch [2.135], and Mocellin and Kingery [2.152] were performed in the final stage of sintering and agree well as to the value of the grain growth exponent, n = 3. Figure 2.22 shows some results from these papers at various temperatures which demonstrate that even the agreement between the absolute values of the grain size in different materials is very close. The paper of Greskovich and Lay [2.130] is concerned with grain growth during the initial stage and reports a change of the growth kinetics with temperature. Experimental activation energies lie in the range of those for Al volume diffusion.

Table 2.12. Grain growth exponents, n, and activation energies, Q, of Eq. (2.27), for various doped aluminas

Authors	Ref. Year	Dopant	n		Q in kJ/mol
Coble	2.56 1961	0.25 w/o MgO	3		641 ± 63
Bruch	2.135 1962	0.1 w/o MgO	3		643.5
Evans	2.154 1966	2.0 w/o Ni	–		578 ± 29
Greskovich & Lay	2.130 1972	0.2 w/o MgO	2 3	at at	$1700^{\circ}C$ $1470^{\circ}C$
Mocellin & Kingery	2.152 1973	0.3 w/o MgO	3		–
Wang & Kröger	2.156 1980	3 w/o Fe	2		–

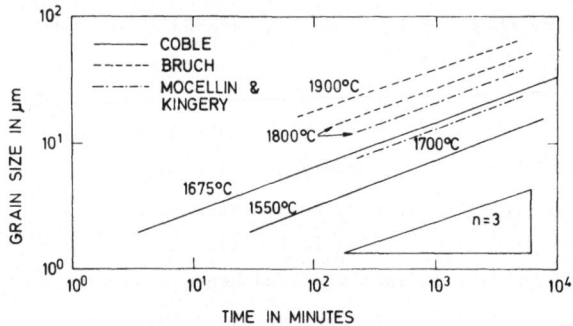

Fig. 2.22. Grain growth in MgO-doped alumina at various temperatures [2.56], 0.25 w/o MgO; [2.135], 0.1 w/o MgO; and [2.152], 0.3 w/o MgO. Individual data points were not plotted

As to the interpretation of these experimental findings, a series of possible rate-controlling mechanism which may operate have been reviewed by Mocellin and Kingery [2.152]. Accordingly, values of the exponent n of 2, 3, 4, or 5 may be expected for mechanisms based on intrinsic grain-boundary mobility, material transfer across a continuous boundary phase, or viscous drag of discontinuous second-phase inclusions moving with the boundary [2.152]. Mocellin and Kingery found experimentally that an increasing amount of MgO clearly decreased the grain-growth rate. Since an increase in the MgO concentration will lead to an increase in the number of spinel inclusions present on the grain boundaries and/or an increase in their average size the result was attributed to a pinning effect of the inclusions. On the other hand, pores lying on grain boundaries were also observed to decelerate boundary migration. It was concluded, therefore, that the most important growth-limiting factor is the dragging of intergranular inclusions, either particles or pores [2.152].

Another quantitative result on the slowing effect of second-phase particles was published by Wang and Kröger [2.156]. They demonstrated that the grain growth constant K of Eq. (2.27) in alumina containing $FeAl_2O_4$ particles on the grain boundaries decreased according to $K \propto C_{sp}^{-2/3}$ where C_{sp} is the number of second-phase particles per unit volume.

Gupta [2.157] pointed out the interesting observation that, for many materials including Al_2O_3, there exists a linear relationship between grain size and density during the intermediate stage of sintering, which means that the ratio of the rate of pore removal and grain growth is a constant. Thus, both processes appear to be caused by the same basic mechanism of matter transport, i.e., volume diffusion in Al_2O_3 [2.157]. This finding, however, contradicts the solid-solute mechanism of MgO in Al_2O_3, which rests on the assumption that the densification rate is enhanced by the presence of Mg relative to the grain growth rate.

2.6 Hot Pressing

Hot pressing or pressure sintering is a fabrication method which is widely used in ceramic processing in order to produce special ceramics or ceramics with special properties. It offers a series of advantages over conventional sintering techniques among which one may find such items as close control of the microstructure and increased final densities which can be achieved without the use of sintering additives, the possibility of densifying materials, hardly sinterable, and the possibility of producing multiphase materials such as, for example, metal-reinforced ceramics. Hot pressing has been used in numerous studies to produce pure or doped alumina samples for scientific investigations (see for example [2.30,2.41,2.76, 2.85,2.86]). It has also been employed successfully in the fabrication of molybdenum-reinforced alumina [2.158,2.159] and as a particularly rapid densification method in synergetic sintering of alumina where an extremely fast process rate is required [2.160].

A characteristic of hot pressing consists in producing compacts of very small grain size and essentially full density at a greatly reduced operating temperature compared with conventional sintering techniques. The reason is that the densification rate is increased by the applied pressure whereas other processes, such as grain growth and surface diffusion, occur at their normal rates. Hot pressing has been suggested, therefore, as a means to avoid discontinuous grain growth in alumina [2.141] because densification is so rapid compared to grain growth that the compaction is finished and the sample cooled down before discontinuous grain growth has a chance to start. Furthermore, hot pressing has also been proposed as a technique for investigating conventional pressureless sintering phenomena [2.150] because it favors selectively the rate of one (and of the most interesting) of otherwise simultaneously occurring processes. When we consider here hot pressing as a "property" and discuss its details, we do so for the same reason as we have already treated sintering as a material property: it involves a series of basic mechanisms of material behavior at elevated temperatures that it is worth treating in a separate section. Technological and practical aspects of hot pressing will be discussed in Section 4.

The present understanding of the mechanisms that operate during hot pressing has been outlined in an excellent review article by Vasilos and Spriggs [2.161]. In pressureless sintering, the only driving force is the surface tension, and the mechanisms of matter transport activated by this force are those illustrated in Fig. 2.15. Yet, when a pressure is applied during sintering, it acts as an additional driving force [2.161]. It has been demonstrated by Coble [2.162] that in all three stages of pressure sintering (initial, intermediate, and final stage) the driving force γ/r is augmented by a pressure term $\sigma_a/(1-P)$ where γ is the surface tension, r the particle radius, σ_a the applied stress, and P the porosity of the compact.

Simply speaking, the applied pressure (or stress) brings about two additional mechanisms of matter transport, i.e., plastic flow at the particle contacts and stress-induced diffusion. As to the first mechanism, it has been shown experimentally by Coble and Ellis [2.163] that for aluminum oxide, plastic flow accounts

66

for the enlargement of the particle contact area during the initial stage of hot pressing but that the contribution of plastic flow to the overall densification is small. At 1500°C, plastic flow was observed to cease when a certain value of the contact area was reached which corresponded to a constant final stress of 350 MN/m². Since this stress was greater by a factor of 3 to 5 than the flow stress of sapphire for basal glide found by Kronberg [2.164] for the same temperature, it was concluded by Coble and Ellis [2.163] that plastic flow was stopped due to strain hardening.

Similar results were found by Hamano and Kinoshita [2.165]. On measuring the densification rate during hot pressing of alumina compacts and evaluating the effective stress acting on the particle contact area as a function of the deformation rate, stress values were obtained which showed fairly good agreement with the lower yield stress values of sapphire on the basal glide system measured by Kronberg [2.164]. It was concluded, therefore, that the plastic flow at the particle contacts controlled the densification rate in the initial stage of hot pressing of alumina [2.165].

Stress-induced diffusion is the second transport mechanism activated by the applied pressure in hot pressing of powder compacts. But, whereas plastic flow evidently contributes only little to densification in the early stage of compaction, it is believed that densification during the intermediate and final stages of hot pressing occurs mainly by a stress-controlled diffusion mechanism. This interpretation has also been suggested for alumina by Coble and Ellis [2.163] and by Vasilos and Spriggs [2.166] in two early hot-pressing studies and has been supported further by Fulrath and coworkers [2.167,2.168], and recently by Harmer and Brook [2.150].

Stress-directed motion of point defects between grain boundaries causes macroscopic deformation which is known as diffusional or Nabarro-Herring creep [2.169, 2.170]. The deformation rate, $\dot{\varepsilon}$, is given by

$$\dot{\varepsilon} = \frac{14\Omega\sigma_a}{kTGS^2} D \qquad (2.28)$$

where Ω is the volume of an Al_2O_3 molecule in alumina, GS the grain size, and D the diffusion coefficient of the slower moving species, as for example, given by Eq. (2.16). Deformation rate and densification rate, $\dot{\rho}$, are related by the expression $\dot{\varepsilon} = \dot{\rho}/\rho$, where ρ is the density [2.168]. Thus, measuring the densification kinetics of compacts during hot-pressing is a simple means to check the validity of Eq. (2.28) and, hence, the validity of the assumption of densification by a stress-controlled diffusion mechanism.

Unfortunately, all quantitative studies on the densification kinetics during hot pressing have shown that diffusion coefficients calculated by means of Eq. (2.28) are orders of magnitude greater than diffusion data obtained from tracer diffusion or from normal diffusional creep studies, shown in Fig. 2.8. Nevertheless, there is much evidence that densification beyond the initial stages is a diffusion-controlled process. The linearity between $\dot\rho/\rho$ and σ was verified in [2.150,2.167,2.168]. Figure 2.23, which is taken from the work of Rossi et al. [2.168], shows the varia-

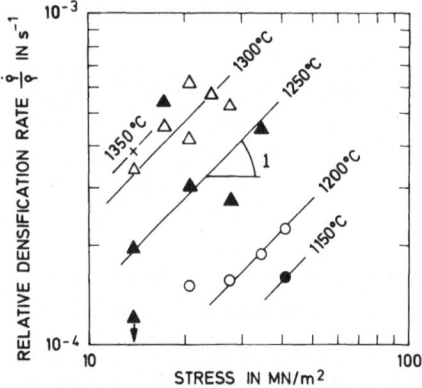

Fig. 2.23. Variation of relative densification rate, $\dot\rho/\rho$, during hot-pressing of alumina, with applied stress at various temperatures and porosity of 1%. - After Rossi et al. [2.168]

tion of $\dot\rho/\rho$ with the applied stress at various temperatures and at a constant porosity of 1%, and illustrates the linearity of deformation rate with stress. The grain size dependence of $\dot\varepsilon$ according to Eq. (2.28) was proved in [2.150], and the activation energy of the densification rate was shown to be very close to the activation energy of Al lattice diffusion in Al_2O_3 in [2.166 to 2.168]. On the basis of the latter result it was concluded by Fulrath and co-workers [2.167,2.168] that both intermediate and final stage densification is controlled by the stress-induced diffusion of the Al ion in alumina.

According to Vasilos and Spriggs [2.161] the processes involved in hot pressing may be summarized as follows:
(1) When pressure is applied compact density increases promptly by the same mechanisms as in cold compaction. Depending on the applied pressure, a density up to about 70% may be achieved.

(2) In a second step, sliding and rearrangement of powder particles occur under the action of the applied pressure, leading to a further increased density. This step may or may not be accompanied by particle fragmentation.

(3) In a third step, plastic flow takes place at the particle contacts. This stage seems to be limited by the occurrance of strain-hardening processes in alumina, thus leading to a limited density which was assessed by Coble and Ellis [2.163] to be about 84%. The steps 1 to 3 occur rapidly (within minutes) and may be conceived as the initial stage of densification.

(4) The intermediate and final stages of densification occur by a stress-controlled diffusion mechanism similarly to the Nabarro-Herring creep model. The densification rate is greater than in the case of pressureless sintering. The ratio of the rate of pressure-motivated sintering to the rate of surface-energy motivated sintering is given by $\sigma_a r/4\pi\gamma$ [2.163]. The time required for full densification may take from minutes to hours.

As to the very high values of the diffusion coefficient for Al lattice diffusion obtained from hot pressing kinetics using the Nabarro-Herring creep equation, Eq. (2.28), several explanations have been proposed. It has been argued by Rossi and Fulrath [2.167] that enhanced cation diffusion along grain boundaries may take place in the very fine-grained starting powder. Also, the stress-concentrating effect of porosity has been considered to change the diffusion geometry on a microscopic scale so that the actual diffusion path length may be much less than the grain size [2.167]. Finally, Rossi, Buch, and Fulrath [2.168] have also taken into consideration the possibility that a mechanism other than the Nabarro-Herring one, such as dislocation climb, may be operative in hot pressing.

Taking lattice and grain-boundary diffusion data of Al in Al_2O_3 at 1300°C from curve 12, Ref. [2.76] of Figs. 2.8 and 2.9, respectively, one obtains for a starting powder of grain size 0.5 μm: $D_{Al}^\ell \approx 10^{-14}$ cm^2/s and $\pi\delta_{Al}D_{Al}^b/GS \approx 10^{-12}$ cm^2/s. Thus, the **grain-boundary diffusion term** of Eq. (2.16) indeed dominates the diffusion process. Hence, assuming a densification process controlled by grain-boundary diffusion decreases the difference between observed and predicted behavior by a factor of 100.

2.7 Segregation

The composition that arises at grain boundaries of ceramics can have a strong influence on processing and properties. The fabrication behavior in hot pressing and sintering is often controlled by the processes that occur at grain boundaries, and properties like fracture strength and creep strength at elevated temperatures, as well as electrical properties, are known to be strongly dependent on the composition of grain boundaries. A review will be given, therefore, in the following paragraphs on what is known today about grain-boundary segregation in alumina. Grain-boundary segregation in ceramics, mainly in Al_2O_3 and MgO, was reviewed by Johnson [2.171] in 1977.

It has already been shown in the preceding section that the segregation of Mg as the most important sintering aid in alumina has attracted much attention, and that sintering models depend on whether Mg segregates or not. Yet, results on the segregation behavior of Mg in Al_2O_3 are contradictory. Segregation was reported to occur by Taylor, Coad and Brook [2.172], Taylor, Coad and Hughes [2.173], and Dufek et al. [2.174]. Their findings were questioned, however, by Johnson and Coble [2.138] and Johnson [2.146], mainly after a series of papers had appeared which reported that the search for any segregation of Mg was unsuccessful. Among the papers that did not find Mg segregation are, in chronological sequence, the works of Marcus and Fine [2.144], Johnson and Stein [2.145], Peelen [2.137], Johnson [2.146], and Clarke [2.147]. Johnson [2.146] has shown that a detection method of sufficient spatial resolution must be used in order not to misinterpret the presence of Mg containing precipitates (the spinel phase) at grain boundaries as a continuous solute Mg concentration. In view of this argument, the earlier findings of segregated Mg seem to be due to the limited resolution of the analysis technique.

In a very recent study, however, Franken and Gehring [2.175], on investigating the segregation of Mg in very weakly doped alumina, demonstrated that samples doped with MgO contents below and up to the solubility limit contained a constant Mg concentration of 0.4 at-% at the grain boundaries. Since spinel precipitation could be excluded, as the origin of the Mg signal obtained, the effect was interpreted as segregation. The corresponding enrichment factor was between 4 and 16. The authors concluded, therefore, that the sintering model in which discontinuous grain growth is inhibited by the solute drag effect of the segregated Mg still cannot be rejected [2.175].

Ca is an important impurity element in Al_2O_3 because it is known to affect considerably the mechanical strength. Funkenbusch and Smith [2.176] and Jupp et al. [2.177] demonstrated that a reduction of the Ca content in the grain boundary increased the fracture toughness of polycrystalline alumina, and an increase in fracture strength was reported by Sinharoy et al. [2.178] for decreasing Ca concentration at grain boundaries. The segregation of Ca was studied by many investigators, and they found unanimously that strong segregation occurs [2.137,

2.144,2.145,2.147,2.172,2.173,2.175 to 2.178]. The enrichment factor of Ca, i.e., the ratio of grain-boundary concentration, c_{gb}, to bulk concentration, c_{bulk}, was found to be as high as 3500 [2.144] (see also Table 2.13). The Ca concentration profile obtained in Ref.[2.145] by using an Auger Electron Spectroscopy technique and sputtering by inert ion bombardment is reproduced in Fig. 2.24. The figure shows a characteristic feature of the Ca segregation in alumina, i.e., a very high concentration (the enrichment factor in this example is 1800) within just a few atomic layers beneath the grain boundary.

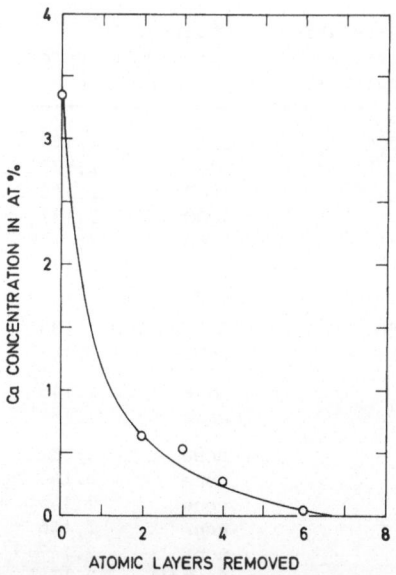

Fig. 2.24 Sputter profile of Ca segregation in sintered, MgO-doped alumina. - After Johnson and Stein [2.145]

In addition to Mg and Ca, the segregation behavior of a few other elements is also known from the literature. Aust et al. [2.179], using microhardness measurements, found a depletion of Cr in the grain boundary region. Tong and Williams [2.180] reported on small enrichments in the range from 2 to 9 of the elements Na, Si, K, Ca, and Fe, and Johnson et al. [2.181] detected F segregation at the grain boundaries of LiF-doped Al_2O_3. Ni was found by Johnson and Stein [2.145] and by Bender et al. [2.182] to behave quite similarly to Mg, i.e., no segregation was observed. Dufek et al. [2.174] reported that Fe and Cr did not segregate in alumina. Segregation of Y in Y_2O_3-doped alumina was found by Bender et al. [2.182] and by Nanni et al. [2.183]. Table 2.13 contains a summary of experimental results.

In the review paper of Johnson [2.171], two possible mechanisms for solute segregation in an ionic crystal were analyzed, i.e., the space charge model and the strain energy model. The study arrived at the conclusion that the space charge effect of a charged grain boundary in a ceramic is apparently not the principal driving force for segregation. Rather is was stated [2.171] that the experimentally observed segregation in oxides can be interpreted in terms of the same mechanism acting in metallic systems, i.e., the strain energy model.

Table 2.13. Misfits, segregation energies, and enrichment factors of several solutes in Al_2O_3 grain boundaries [2.171]

Solute element	Ionic radius in nm	Misfit ε	ΔG_{seg} in kJ/mol	Calculated enrichment factor *)	Observed enrichment factor and reference	
Ca^{2+}	0.099	+ 0.98	117	1300	3500	2.144
					2300	2.177
					1800	2.145
					1000	2.137
					200	2.173
					100	2.175
					60	2.172
Y^{3+}	0.093	+ 0.86	92	260	150 to 200	2.183
					240	2.182
Ni^{2+}	0.072	+ 0.44	23	4	none	2.145
					none	2.182
Mg^{2+}	0.065	+ 0.30	11	2	none	2.137
					none	2.144
					none	2.145
					none	2.146
					none	2.147
					4 to 16	2.175
Si^{4+}	0.041	- 0.18	4	1.3	2 to 12	2.180

*) at $1700^{\circ}C$

According to the treatment presented in [2.171], the main contribution to the change of free energy of segregation, ΔG_{seg}, stems from the strain energy, W, which is due to the size misfit, ε, between the solute impurity and the unoccupied cation site in the lattice. Johnson showed [2.171] that

$$\Delta G_{seg} \propto W \propto \varepsilon^2 = (r_1 - r_o)^2/r_o^2 , \tag{2.29}$$

where r_1 is the radius of the isolated solute and r_o is the radius of the unoccupied lattice site, and that, in thermodynamic equilibrium, the enrichment factor is given by a Langmuir-type relation

$$c_{gb}/c_{bulk} \simeq \exp(\Delta G_{seg}/RT) . \tag{2.30}$$

Furthermore, Johnson [2.171] demonstrated that Eq. (2.29) can be used to calculate ΔG_{seg} values for other solutes if ΔG_{seg} is known for one of them because the misfit parameter can be obtained from the ionic radius of the solute. Taking the experimental value for Ca segregation in alumina from [2.145], ΔG_{seg} = 117 kJ/mol, he calculated a series of other ΔG_{seg} values and the corresponding enrichments from Eq. (2.30). Table 2.13 contains, for some solute elements, ionic radii, ΔG_{seg} values, and calculated enrichments according to Johnson's study [2.171] as well as a comparison with experimental enrichments as given in [2.171] and completed by us. As can be seen from the table, the rather simple strain energy model based on Eqs. (2.29) and (2.30) gives good agreement between theoretical predictions and experimental observations of the segregation of various solutes in alumina.

3 Mechanical Properties

The utility of a ceramic material in an engineering application is critically
determined by its mechanical behavior. This applies particularly, of course, to
structural applications. However, the mechanical behavior is also of outstanding
importance in such uses where the material is employed mainly because of its
optical, electrical, or magnetic properties. In any case, the structural integrity of the components must be guaranteed, and the prevention of mechanical failure is almost always an important, if not critical, requirement.

In the subsequent sections of this chapter, the most important features of the
mechanical behavior of aluminum oxide will be discussed in detail. The following
properties will be dealt with: elastic properties, fracture strength and time-
dependent strength, behavior under mechanical and thermal shock, plastic flow
and creep behavior, and friction and wear. Finally, strengthening mechanisms
in alumina will be reviewed.

The knowledge of the mechanical behavior of ceramic materials has increased very
rapidly during the last 20 years. A series of summarizing publications on this
subject is available in the literature, for example, in the work of Kingery et
al. [3.1], Evans and Langdon [3.2], Rice [3.3], and Davidge [3.4]. To a great
extent, these books and articles are also concerned with the behavior of aluminum
oxide, and in them the reader may find valuable information on the mechanical
properties of this material.

Aluminum oxide is probably the most extensively studied of all structural ceramics. Many papers on the various aspects of its mechanical behavior have been
published. In a series of basic properties of mechanical behavior, aluminum oxide
turns out to be quite similar to other structural ceramics such as ZrO_2, Y_2O_3,
$MgAl_2O_4$, Si_3N_4, and SiC. Thus, Al_2O_3 may be regarded as a typical representative
of the class of modern structural ceramics.

74

Aluminum oxide stands out due to beneficial properties such as high hardness and, hence, high wear resistance and a low friction coefficient, high resistance to corrosion by practically all chemical reagents, a very high resistance to high-temperature corrosion in air, thermodynamic stability, i.e., the absence of phase transformations within the entire temperature range of solid state, and the fact that the material retains its strength even at very high temperatures ($\approx$1500-1700°C). Its unfavorable properties are also well known: its low strength as compared with the theoretical strength and with the strength of metallic alloys, the large statistical spread of its strength values, its great brittleness, i.e., the complete absence of any plastic deformation until about 1200°C, its low toughness and, hence, large susceptibility to thermal and mechanical shock loading, and the phenomenon of time-dependent strength. Due to these detrimental properties, the use of aluminum oxide in engineering applications, like that of many other structural ceramics, is still limited.

3.1 Elastic Properties

The determination of the elastic properties of a ceramic material is an important step in studying mechanical behavior. At a given stress, lattice strains and strain energies are controlled by the values of the elastic constants and, for a given strain, they determine the stresses that the material will resist. The anisotropy of the elastic constants accounts for the buildup of thermal and elastic stresses.

For a hexagonal material such as alumina, distinct anisotropic elastic behavior is expected. It can be shown that the complete description of the elastic properties of the corundum structure needs a total of six elastic constants of the general six-by-six elasticity matrix. These constants were determined experimentally for single-crystal alumina by Wachtman et al. [3.5]. Starting from these parameters, they also calculated the variation of Young's modulus, E, and of the shear modulus, G, with orientation. Extreme values were found for <0001> and <10$\overline{1}$1> directions. These are summarized in Table 3.1, which shows that the anisotropy is quite pronounced. The variation of single-crystal elastic constants with temperature in the range 77-850 K was measured by Wachtman, Tefft, and Lam [3.6] and by Tefft [3.7].

In a polycrystalline body, the orientation-dependent contributions of individual grains, which, of course, are anisotropic on a microscopic scale, balance each other so that an isotropic elastic behavior is obtained on a macroscopic scale. However, for the complete description of an isotropic body, only two elastic con-

Table 3.1. Elastic constants of alumina at room temperature and 0% porosity

Authors	Ref. Year	Method[1]		Young's modulus E GN/m^2	Shear modulus G GN/m^2	Bulk modulus K GN/m^2	Poisson's ratio ν
(A) Single-crystal alumina							
Wachtman et al.	3.5 1960	RV, PV	highest: lowest:	461.0^2 335.7^3	187.7^3 144.0^2		
(B) Polycrystalline alumina							
Coble & Kingery	3.8 1956	TB		386	163		0.27
Wachtman et al.	3.5 1960	SC	upper bound: lower bound: mean:	408.4 397.3 402.9	166.0 160.6 163.3		0.234
Crandall et al.	3.9 1961	RV		411			
Spriggs et al.	3.10 1964	TB		391			
Chung & Simmons	3.11 1968	PV		403.45^4	163.16	255.07	0.236

[1] RV resonance vibration; TB transverse bending; PV pulse velocity; SC calculated from single-crystal data;
[2]$<0001>$ direction; [3]$<10\bar{1}1>$ direction; [4]calculated from G and K.

stants are necessary, for example, Young's modulus and the shear modulus. There exist two different methods in the literature that permit one to calculate the two moduli of a polycrystalline body from the orientation-dependent values of the respective single-crystal moduli of its grains. These methods yield an upper and a lower bound, and these two bounds as calculated by Wachtman et al. [3.5] as well as their arithmetic average are also given in Table 3.1. The table also contains a series of experimental data on Young's modulus, shear modulus, bulk modulus, and Poisson's ratio of polycrystalline alumina at room temperature and 0% porosity. The compilation shows that the newest data by Chung and Simmons [3.11] agree very well with the mean value of the upper and lower bounds obtained from single-crystal data [3.5].

The elastic constants of a solid are known to be influenced by temperature and microstructure. In the following we shall focus attention on these two parameters. The variation of Young's modulus, shear modulus, and longitudinal modulus of po-

lycrystalline alumina with temperature, obtained in a series of experiments, is
shown in Fig. 3.1. The curves present some features which are common to refrac-

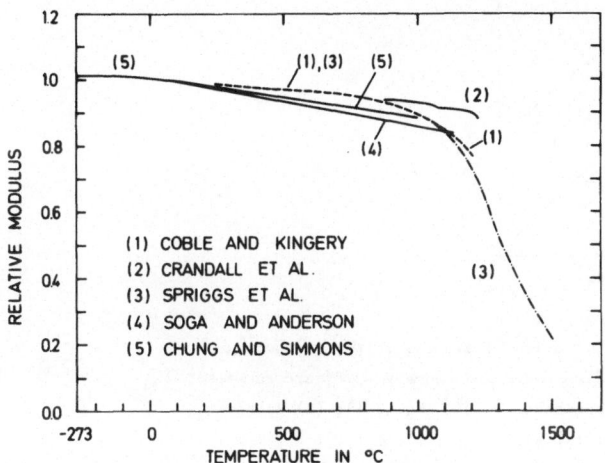

Fig. 3.1. Relative change of elastic moduli with temperature of polycrystalline
alumina. - Curves 1, 2, and 3: Young's modulus, after Coble and Kingery [3.8]
Crandall et al. [3.9], and Spriggs et al. [3.10]; curve 4: shear modulus, after
Soga and Anderson [3.12]; curve 5: longitudinal modulus, after Chung and Simmons
[3.11]

tory single-phase ceramics: a zero slope at absolute zero temperature [3.6], a
slight linear decrease with temperature in an intermediate temperature range up
to about 1000°C [3.4], and a strong and more than linear decrease at temperatures
above about 1000°C, which is attributed to the occurrence of inelastic strains and
the onset of grain-boundary sliding [3.1]. From the work of Soga and Anderson
[3.12], the mean logarithmic changes between room temperature and 1000°C are: for
Young's modulus $\Delta \ell nE/\Delta T = 1.4 \cdot 10^{-4} K^{-1}$, and for the shear modulus
$\Delta \ell nG/\Delta T = 1.5 \cdot 10^{-4} K^{-1}$.

As to the changes of elastic moduli with changes of the microstructure, it has
been stated by several authors [3.10,3.12,3.13] that no influence of the grain
size could be detected in single-phase alumina. Crandall et al. [3.9] determined
a 6% decrease of Young's modulus by the addition of 1% silica. However, a strong
variation of Young's modulus with porosity was observed by many investigators.
Data from eight references have been compiled by Knudsen [3.14] and are reproduced in Fig. 3.2 together with experimental data by Coble and Kingery [3.8].
Figure 3.2 shows that a 5% porosity reduces Young's modulus by about 20%.

77

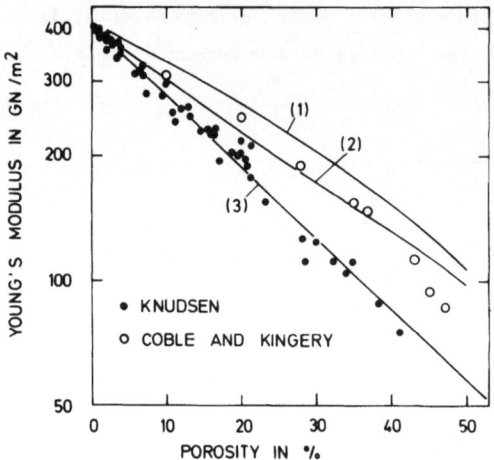

Fig. 3.2. Effect of porosity on Young's modulus of alumina at room temperature.
- Data points from Knudsen [3.15] and from Coble and Kingery [3.8] are compared
with three different predictions: (1) MacKenzie [3.16], Eq.(3.1); (2) Hashin
[3.17], Eq.(3.3); (3) Spriggs [3.18], Eq.(3.4)

The literature contains a series of approaches which describe theoretically the
effect of porosity on the elastic moduli. The various models have been reviewed
by Wachtman [3.15] and by Rice [3.3]. The individual solutions for the variation
of the modulus with porosity depend strongly on the assumptions made on the location, shape, and orientation of the pore phase. For an isotropic and homogeneous
body containing a large number of isolated spherical holes distributed at random,
the following relation has been given by MacKenzie [3.16]:

$$E/E_o = 1 - kP + (k - 1)P^2 \qquad (3.1)$$

with
$$k = 5(3K_o + 4G_o)/(9K_o + 8G_o), \qquad (3.2)$$

where E_o, K_o, and G_o are Young's modulus, bulk modulus, and shear modulus, respectively, of the perfect material without porosity, and P is the volume fraction
of pores. For the same assumptions, another solution has been derived by Hashin
[3.17]:

$$E/E_o = 1 - aP/[1 + (a-1)P], \qquad (3.3)$$

78

where $\underline{a}$ is a constant. Whereas Eqs. (3.1) and (3.3) are based on the stress concentration effect of the pore phase, Spriggs [3.18] has suggested an expression
which results from the principle of the remaining load-bearing cross section of
the matrix and the minimum distance between pores:

$$E/E_o = \exp(-bP),$$
(3.4)

where $\underline{b}$ is an empirical constant.

Knudsen [3.14] was the first to demonstrate that Sprigg's relation, Eq. (3.4),
could be fitted very well to experimental data of alumina between zero and 40%
porosity. His data fit gave $E_o = 410.2$ GN/m^2, very close to other values of Young's
modulus given in Table 3.1, and b = 3.95. It is plotted as curve 3 in Fig. 3.2.
Equations (3.1) and (3.3) are also shown for comparison using the same value for
E_o and k = 1.969, which follows from Eq. (3.2) using the data of Chung and Simmons
[3.11] for K_o and G_o, and $\underline{a}$ = 3.24. It can be seen from Fig. 3.2 that Mackenzie's
solution gives the poorest fit to the data, whereas the exponential expression of
Spriggs accurately reproduces the decrease of Young's modulus with porosity. Further evidence for the usefulness of Eq. (3.4) was reported by Soga and Anderson
[3.19] for alumina up to 5% porosity; the value found for b was 4.17. The b value
given by Rice [3.3] in his compilation of Young's modulus-porosity data of alumina
is 3.5 ± 0.1.

3.2 Fracture Strength

In this section, an overview of the strength and brittle fracture of Al_2O_3 will be
given. It is well known that the short-term strength of ceramics is controlled
mainly by two variables: fracture energy and flaw size [3.4]. Hence, after some general remarks on the strength of ceramics, fracture-energy data and flaw types observed in alumina will be reviewed, and a compilation of strength data will be given. Since strength is known to be highly dependent on microstructure, emphasis
will be put on the discussion of strength-controlling variables in terms of their
microstructural dependence, and models of the grain-size dependence of fracture
energy and of strength will be presented. The effects of environment and time-dependent strength phenomena are reserved for Section 3.3. Strengthening mechanisms
will be reviewed in Section 3.7.

In addition to the review literature on the mechanical behavior of ceramics cited above, there are also some introductory papers that particularly deal with the strength of ceramics, for example, those by Kelly [3.20], Davidge and Evans [3.21], Davidge [3.22], and Lawn and Wilshaw [3.23].

3.2.1 The Strength of Ceramics

The strength of ceramics can be understood in terms of the concept of brittleness of material's developed by Griffith [3.24]. He postulated that a brittle body contains small flaws which act as stress concentrators when an external stress is applied. Due to the absence of plastic deformation, stresses cannot relax by local plasticity, and the local stress in the vicinity of the most severe flaw may reach the theoretical strength of the material, thereby causing failure of the body, with the flaw acting as the failure origin. This concept is still accepted today to explain the strength of brittle ceramics. The Griffith equation of strength, as modified by Davidge and Evans [3.21], is given by

$$S = \frac{1}{Y} \left(\frac{2\gamma_I E}{C} \right)^{1/2},$$ (3.5)

where S is the strength, γ_I is the fracture energy per unit area necessary to initiate failure, E is Young's modulus, C is the flaw size, and Y is a dimensionless constant which depends on the flaw size-sample size ratio.

Fracture mechanics states that fracture occurs in a material when the stress intensity factor K_I at the tip of a crack or flaw reaches the critical material value K_{Ic}, the fracture toughness of the material, which is given by

$$K_{Ic} = Y S C^{1/2}.$$ (3.6)

Comparing Eqs. (3.5) and (3.6) shows that fracture toughness and fracture energy are related by

$$K_{Ic}^2 = 2\gamma_I E.$$ (3.7)

Thus, the two quantities may be considered as synonyms for the same basic idea, but ceramists usually prefer γ_I to K_{Ic}.

Equation (3.5) predicts that the strength of ceramics is mainly controlled by two variables, fracture energy and flaw size. Both parameters depend strongly in the microstructure, as will be seen in the following sections. But, whereas γ_I is a material parameter which is characteristic of a given type of ceramic material, the flaw size is a more complex quantity. It is common to distinguish between extrinsic and intrinsic flaw types. Extrinsic flaws are defects which are generated by a particular, perhaps defective, process of fabrication or surface finishing; their presence does not depend on the type of ceramic. Intrinsic flaws are thought to originate at pores, large grains, or second phase particles, or they may be generated due to the anisotropy of elasticity or thermal expansion. Their formation depends on the individual features of the actual microstructure and on the constants of the respective material. Thus the strength of a ceramic is not an unequivocal material parameter but a rather complex quantity that may be influenced by the details of the fabrication process and the microstructure, as well as by a series of material constants.

3.2.2 Fracture Energy

When a material fractures energy is required to create the new surfaces. For the case of ideal reversible fracture of a material in single-crystal form, it may be expected that the energy necessary to rupture the lattice bonds and generate two new surfaces is equal to twice the thermodynamic surface energy, γ_o. However, it has been found experimentally that in many cases single-crystal fracture energy values, γ_{sc}, are greater than the thermodynamic surface energy. This deviation has been attributed to the operation of additional energy-absorbing mechanisms, for example, plastic deformation near the crack tip or heat generation [3.4].

Single-crystal fracture energy values of aluminum oxide were determined by Wiederhorn [3.25]. He found preferential cleavage for $\{10\bar{1}0\}$ and $\{\bar{1}012\}$ planes in sapphire; the respective energy values are 7.3 and 6.0 J/m^2. These values are quite high compared with calculated ones or those extrapolated from high-temperature measurements of the thermodynamic surface energy (see Section 2.2), about 1.2 J/m^2. However, etch pit studies showed that plastic deformation could be excluded as a major energy-absorbing mechanism in sapphire [3.25]. In a later study by Wiederhorn et al. [3.26], no evidence of fracture-induced dislocation activity could be detected in sapphire up to 400°C by the TEM technique. From this discrepancy, some lack of understanding of the fundamental fracture mechanism in sapphire may be concluded.

The experimental data of γ_{sc} obtained by Wiederhorn [3.25] and a room temperature value of single-crystal K_{Ic} reported by Wiederhorn et al. [3.26] are shown in Table 3.2. The table also contains single-crystal fracture toughness data for various

Table 3.2. Fracture toughness K_{Ic} (in $MN/m^{3/2}$) and fracture energies γ (in J/m^2) of single-crystal and polycrystalline alumina at room temperature

(A) Single-crystal alumina

Authors Ref.	Wiederhorn 3.25	Wiederhorn et al. 3.26	Iwasa et al. 3.27
Crystal plane	γ_{sc}	K_{Ic}	K_{Ic}
$(1\bar{1}02)$	6.0		2.38 ± 0.14
$(11\bar{2}0)$		2.47^1	2.43 ± 0.26
$(1\bar{1}00)$	7.3		3.14 ± 0.30
(0001)	>40		4.54 ± 0.32

(B) Polycrystalline alumina

Ref.	3.29 to 3.31	Fig. 3.4	Fig. 3.5	Fig. 3.7	Fig. 3.10
	γ_{gb} ≈ 4	γ_{NB} 18 to 36	γ_{DCB} 15 to 47	γ_{WOF} 30 to 56	K_{Ic} 3.8 to 5.1
Grain size range, μm		5 to 50	3 to 50	10 to 50	

[1] Orientation of crystal plane not indicated

crystallographic planes obtained recently by Iwasa et al. [3.27]. The trend of their data is the same as the trend of Wiederhorn's fracture energy data, i.e., fracture is most easily obtained on $(1\bar{1}02)$ and $(11\bar{2}0)$ planes, but becomes increasingly difficult on the $(1\bar{1}00)$ and (0001) planes. Wiederhorn [3.25] was unsuccessful in producing cleavage along the basal plane. He concluded, therefore, that the surface fracture energy was greater than 40 J/m^2. Spontaneous cleavage cracking along the basal plane, however, was detected several years later by Anderson [3.28].

In polycrystalline ceramics the fracture mode can be transgranular or intergranular. For transgranular fracture, it may be expected that the fracture process is controlled by γ_{sc}. However, when the material is separated along a grain boundary, γ_{sc} should be replaced by the grain-boundary fracture energy γ_{gb}. Since the atomic bonds are weaker for atoms located at or near the grain boundary, it may be expected that $\gamma_{gb} < \gamma_{sc}$. Nothing is known, however, about the experimental value of γ_{gb} in polycrystalline alumina. It has been assumed repeatedly [3.29-3.31] that γ_{gb} is approximately half of γ_{sc}, i.e., $\gamma_{gb} \approx 4$ J/m^2. This assumption may be justified by the results of dihedral angle measurements made by Kingery [3.32], who found a value of 0.48 for the ratio of the thermodynamic grain-boundary energy to the thermodynamic surface energy (see Section 2.2).

Fracture of a polycrystalline body is a complicated process compared with the cleavage of a single crystal or the separation of a bi-crystal. In most cermics, experimental values of the fracture energy of polycrystals, γ_{pc}, exceed γ_{sc} values by a factor of 3-10. The reasons for this behavior, as discussed, for example, by Rice [3.3] and Davidge [3.4], are as follows:

(1) The true fracture surface of a polycrystal is greater than the projected surface used for calculating of the fracture energy per unit area. The effect of roughness may reach a factor of 2 to, at most, 4.

(2) The tortuosity of the fracture surface causes many microscopic deviation of the local crack front from the path of the macrocrack. As a consequence, local crack pinning and crack bowing occurs, thereby consuming additional energy.

(3) Increased local stresses along the tortuous crack front may cause increased occurrence of microplasticity.

(4) Crack propagation may be accompanied by other forms of non-conservative energy consumption, such as generation of heat or emission of sound and light.

(5) Subsidiary cracking (i.e. crack branching) may occur in the stress field ahead of the main crack, i.e., the propagation of the main crack may be accompanied by the formation of a "microcracked process zone".

It is quite clear that the contribution of most, if not all, of these processes depends on details of the microstructure such as location, size and shape of pores, grain size, and the presence and location of second phases, as well as on the temperature. These effects will now be reviewed in detail.

Figure 3.3 shows fracture energy data from the work of Evans and Tappin [3.33], Simpson [3.34], Pabst [3.35], and Claussen et al. [3.36] as a function of the volume fraction porosity P. The fracture energy decreases with increasing porosity, which is to be expected, since in most cases of slightly porous ceramics the fracture mode is intergranular. Therefore, an increase in porosity reduces the cross-sectional area to be fractured. Furthermore, even spherical pores at boundaries will not be able to trap cracks, because they contain sharp cusps which develop at pore-grain boundary intersections [3.2].

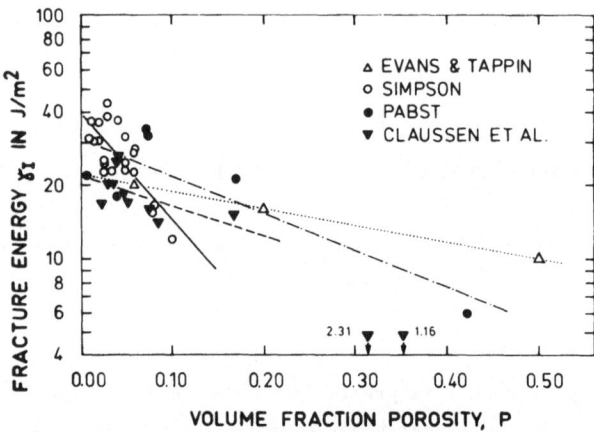

Fig. 3.3. Effect of porosity on fracture energy. - Data from Evans and Tappin [3.33], Simpson [3.34], Pabst [3.35], and Claussen et al. [3.36]

It has been suggested by Rice [3.3] that fracture energy data should show an exponential dependence on porosity,

$$\gamma_I = \gamma_o \exp(-bP), \tag{3.8}$$

similar to the porosity dependence of Young's modulus, Eq.(3.4), and strength. The b values of the three relations should be identical. Rice's suggestion follows from the principle that the crack will seek the path with the minimum cross-sectional area, i.e., the maximum fraction of porosity, not an average fraction of porosity [3.3]. Figure 3.3 indeed shows that experimental data sets fit the exponential relation well, but that there are large differences between the slopes. Values for b reach from 1.5 (dotted line) to 10 (continuous line). This fact may be attributed to different types of porosity. The b values of Pabst [3.35] and Claussen et al. [3.36] are in between, being 3.5 and 2.8, respectively. Thus, their values are quite close to those for the porosity dependence of Young's modulus given in Section 3.1, i.e., 3.95 [3.14] and 3.5 ± 0.1 [3.3].

The fracture energy also depends on grain size, but contradictory trends have been observed. It has been found that γ may decrease or increase with increasing grain size, depending on the testing method and on the type of specimen used to determine K_{Ic} or γ. Specimen types employed to measure fracture mechanics parameters in ceramics are the notched-beam (NB), double-cantilever beam (DCB), and the double-torsion (DT) specimens [3.37]. The fracture energy is determined either by a K_{Ic} test and subsequent use of Eq. (3.7) or by direct measurement of the work of fracture (WOF test) which is needed to break the sample into two pieces [3.38].

In the following figures, fracture energy data are compiled which were obtained by NB, DCB and DT, and WOF test, respectively. Only materials with an Al_2O_3 content > 95% were considered. When original data were reported as K_{Ic} values they were transformed into fracture energies by means of Eq. (3.7). All data were corrected for 0% porosity using Eq. (3.8) and b = 3.5. Figure 3.4 shows γ_{NB} data from 19 references. In the grain-size range below 10 μm, the fracture energy seems to increase with decreasing grain size, but the trend is not very pronounced. At intermediate grain sizes the data lie within a band between about 18 and 36 J/m^2, showing a spread by a factor of 2. Individual data series, for example, the data points of Claussen, Mussler, and Swain [3.31], exhibit a distinct increase when

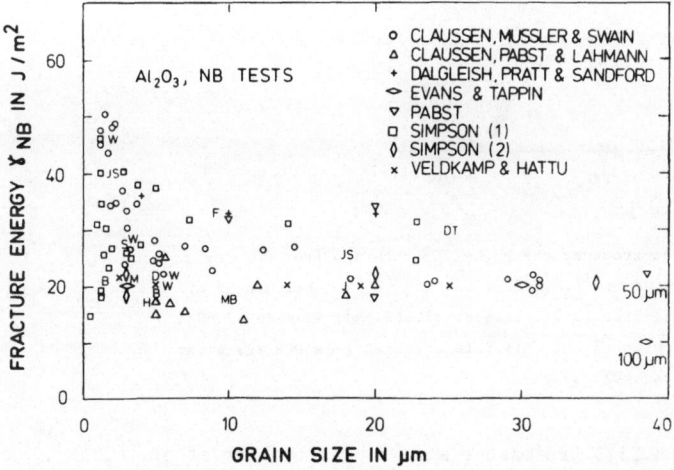

Fig. 3.4. Grain-size dependence of fracture energy, γ_{NB}, obtained from NB tests.-
Data from Claussen, Mussler and Swain [3.31]; Claussen, Pabst, and Lahmann [3.36];
Dalgleish, Pratt, and Sandford [3.39]; Evans and Tappin [3.33]; Pabst [3.35]; Simpson (1) [3.34]; Simpson (2) [3.40]; and Veldkamp and Hattu [3.41]. - Characters:
B - Bansal and Duckworth [3.42]; D - Dalgleish et al. [3.43]; DT - Davidge and Tappin [3.44]; F - Funkenbusch and Smith [3.45]; H - Hübner [3.46]; J - Hübner and Jillek [3.47]; JS - Jupp, Stein and Smith [3.48]; M - Mai [3.49]; MB - Munz, Bubsey and Shannon [3.50]; S - Simpson [3.51]; W - de With and Hattu [3.52]

the grain size is reduced to a few microns. But, since nearly all NB measurements are performed on samples containing sawed notches of finite width, the risk exists that the experimentally measured value of K_{Ic} will be larger than the actual K_{Ic} [3.53], and this tendency increases as the grain size is decreased, as has been shown by Simpson et al. [3.54]. Therefore, the increase in fracture energy should be taken with some reservation unless it is proved that it is not an effect of notch geometry.

Figure 3.5 shows γ_{DCB} data from eight references as well as some results from DT and short-rod tests. It is obvious from Fig. 3.5 that γ_{DCB} data, on the average, are somewhat greater than γ_{NB} data, and that individual sets of data points, for

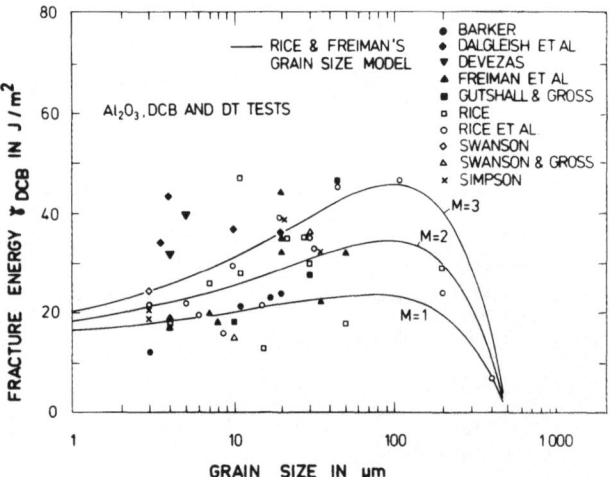

Fig. 3.5. Grain-size dependence of fracture energy, γ_{DCB}, obtained from DCB tests. - Data from Barker [3.55]; Dalgleish et al. [3.39]; Devezas [3.56]; Freiman et al. [3.57]; Gutshall and Gross [3.58]; Rice [3.3]; Rice et al. [3.59]; Swanson [3.60]; Swanson and Gross [3.61]; and Simpson [3.40]. All data obtained from DCB tests except Barker (Short Rod) and Devezas (DT)

example those from Barker [3.55], Freiman et al. [3.57], or Rice et al. [3.59], exhibit an increasing trend as the grain size increases in the range up to 100 μm. The only study which employs grain sizes above 100 μm is the very recent paper by Rice et al. [3.59], which demonstrates that γ_{DCB} finally decreases at very large grain sizes. Thus, as a function of grain size, γ_{DCB} of alumina seems to go through a maximum, thereby exhibiting a behavior quite similar to other noncubic ceramics such as TiO_2 and Nb_2O_5 where a maximum of the grain-size dependence of DCB fracture energy was also observed by the same authors [3.59]. This behavior would agree very well with recently presented predictions of models of the grain-size dependence of fracture energy, which will be discussed below [3.30,3.62,3.63]. The curves shown in Fig. 3.5 are such predicitons for alumina from the model of Rice and Freiman [3.30].

The other experimental method for determining the fracture energy of ceramics employs the direct measurement of the energy expended during cracking. This work-of-fracture (WOF) method goes back to a proposal made by Nakayama [3.38] and was first applied to brittle materials including alumina by Tattersall and Tappin [3.64] and then by Davidge and Tappin [3.44]. The method consists in breaking precracked or notched bend specimens in a slow and controlled way so that it can be assured that all the work expended by the testing machine is converted into fracture energy. The mechanical work is determined from the area under the load-deflection record. Figure 3.6 is an example of a completely controlled fracture experiment on an alumina sample [3.47].

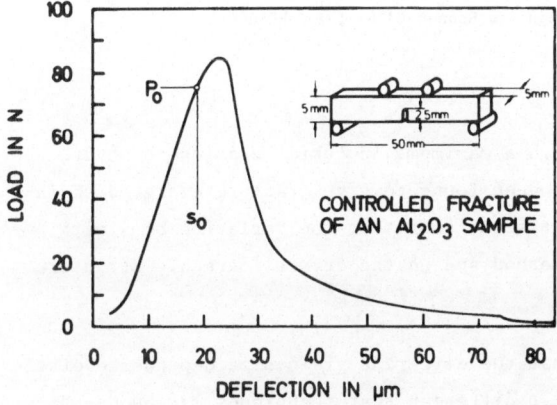

Fig. 3.6. Controlled fracture experiment of an alumina sample. – From Hübner and Jillek [3.46]. The values P_o and s_o define the end of the purely elastic region

Results on γ_{WOF} from eight references are compiled in Fig. 3.7. In the grain-size range investigated, the data show a strongly increasing trend, reaching values of 50 J/m^2 and more, clearly higher than both γ_{NB} and γ_{DCB} data at a comparable grain size. The grain-size dependence of γ_{WOF} energies coincides with that obtained from DCB and DT test in the same grain-size range.

This fact has already been noted by other authors. Simpson [3.40], Rice [3.3], Rice et al. [3.59], and Claussen et al. [3.31] have pointed out that fracture energy data of alumina and other noncubic ceramics show a decreasing or constant trend with increasing grain size when determined in NB tests, but that DCB, DT, and WOF

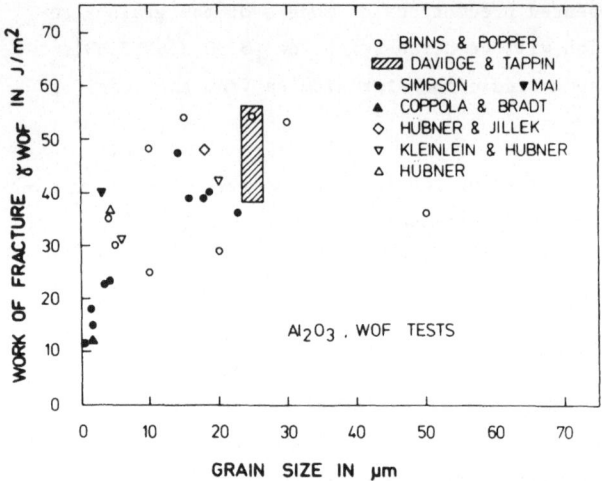

Fig. 3.7. Grain-size dependence of work of fracture, γ_{WOF}. - Data from Binns and Popper [3.65]; Davidge and Tappin [3.44]; Simpson [3.34]; Coppola and Bradt [3.66]; Mai [3.49]; Hübner and Jillek [3.47]; Kleinlein and Hübner [3.67]; and Hübner [3.46]

test data increase or even pass through a maximum. In cubic ceramics, however, the DCB, DT, and WOF test techniques have been found to give fracture energy data independent of grain size [3.3,3.59]. It seems thus that the variation of γ with grain size depends both on the test method and on the crystallographic structure of the material.

It has been pointed out [3.3,3.31] that the different grain-size dependence of the fracture energy of noncubic ceramics in different test techniques may be due to the microscopic details of crack propagation. In an NB test, fracture usually originates from a single more serious flaw which is small compared to the specimen dimensions, and results in catastrophic failure of the test piece. Therefore, the NB test measures the fracture energy at the moment of crack initiation regardless of all later interactions between crack front and microstructure. On the other hand, the other test techniques, such as DCB, DT, and WOF tests, are thought to promote microcracking effects in ceramics susceptible to such cracking. There is sufficient time for a microcracked zone to be developed so that interactions between the main crack and subsidiary cracks may occur as the main crack propagates through the material. These latter techniques, therefore, measure the resistance to stable crack propagation rather than to catastrophic failure [3.3,3.31]. To facilitate comparison, Table 3.2 summarizes the data ranges of the fracture energies γ_{NB}, γ_{DCB}, and γ_{WOF} found in polycrystalline alumina together with the grain-size range, as taken from the respective figures. Also given is a common value for γ_{gb}.

Microcracking has been recognized to be a mechanism which contributes to increasing the fracture energy and thus the toughness of ceramic materials. Microcracking is assumed to occur within a small cylindrical zone of radius R_{mc} ahead of the crack tip, the so-called microcracked zone or process zone. According to Rice, the propensity of a ceramic material to microcracking may be due to high termal expansion anisotropy (TEA) [3.68] or to substantial porosity [3.3]. Rice et al. [3.62] and Rice and Freiman [3.30] have presented a model in which microcracking accounts for the grain-size dependence of the fracture energy in noncubic ceramics. This model is based on the assumption that in these ceramics the thermal expansion anisotropy causes stresses across the grain boundaries which increase with increasing grain size and promote grain-boundary microcracking in the vicinity of the main crack. Such TEA stresses may be quite high. For aluminum oxide, values of 140-210 MN/m^2 were assessed by Rice [3.68]. Using an X-ray fluorescence technique, Grabner [3.69] determined the residual stress of several sintered aluminas having a grain size of < 7 μm and a flexural strength of 400 MN/m^2 to vary between 170 and 190 MN/m^2. The model of Rice and coworkers [3.30,3.62] predicts that, as a function of grain size G, the fracture energy first increases because the number of microcracks increases, thereby absorbing an increasing amount of energy. As the grain size approaches G_s, the grain size for spontaneous cracking due to TEA stresses, crack linking between the microcracks and the main crack gains increasing importance, thus facilitating the propagation of the main crack front and reducing γ, which reaches zero at $G = G_s$. The quantitative relationship derived by Rice and coworkers [3.30,3.62] is given by

$$\gamma = \gamma_{pc} \left(1 - \frac{G}{G_s} \right) + 12 M \gamma_{gb} \left(\frac{G}{G_s} \right)^{1/2} \left[1 - \left(\frac{G}{G_s} \right)^{1/2} \right] \tag{3.9}$$

where γ_{pc} is the fracture energy of the polycrystal in the absence of microcracking (i.e., at very small grain size), γ_{gb} is the grain-boundary fracture energy, and M is a numerical factor which virtually indicates the energy expended per microcrack in multiples of γ_{gb} (typically between 1 and 3). G_s is given by

$$G_s = 12 \, E \, \gamma_{gb} / \sigma_i^2 \tag{3.10}$$

with $\sigma_i = E \overline{\Delta \alpha} \Delta T$ the internal stress, where $\overline{\Delta \alpha}$ is half the difference between the maximum and minimum single-crystal thermal expansion coefficient and ΔT is the difference between the temperature at the onset of plastic relaxation and room temperature.

Equation (3.9) is plotted in Fig. 3.5 for three different M values (M = 1, 2, and 3) and using the following values of the various constants: γ_{pc} = 15 J/m^2, γ_{gb} = 3.75 J/m^2, E = 410 GN/m^2, $\overline{\Delta\alpha}$ = 3.7 · 10^{-7} K^{-1}, and ΔT = 1200°C. This yields G_s = 500 µm. As may be seen from Fig. 3.5, M = 2-3 gives a reasonable fit, although the spread of the data points is remarkable. For M = 3, the maximum of γ lies at G = 100 µm and is about 2.2 times the value at very small grain size, i.e., the contribution of microcracking to the total fracture energy is about 55%.

A somewhat different microcracking model has been proposed very recently by Gesing and Bradt [3.63]. In this model, flaws are thought to be located at three-grain junctions which, in the stress field of the main crack, extend along grain boundaries to microcracks having the size of a single grain facet. The process which leads to the main crack propagation is assumed to be a linking of these microcracks ahead of the main crack front. For aluminum oxide the model predicts a maximum contribution of microcracking to the total fracture energy of 41% at a grain facet size of 15 µm and a triple-point flaw size of 5 µm, the microcracked zone size at these conditions being 10 times the grain facet size.

Experimental evidence of microcracking has been found for many ceramics. It was detected for the first time in Salem limestone by Hoagland et al. [3.70]. Optical evidence of microcracks in alumina was given by Noone and Mehan [3.71], Hübner and Jillek [3.47], Wu et al. [3.72], Evans [3.73], and Rice et al [3.62]. As was pointed out by Wu et al. [3.72], microcracking in noncubic ceramics was found to be much more pronounced than in cubic ceramics, and wandering and branching of bracks occurred on a multi-grain scale. Figure 3.8 shows crack branching and re-joining on the side face of an alumina bend specimen [3.47]. The distance between the upper and the lower branching point equals about 50 times the grain size. The occurrance of microcracking in alumina was also inferred from the results of ac-oustic emission studies. The observation of acoustic emission events at loads far below the fracture load was attributed to microcracking by Simpson et al. [3.54], Noone and Mehan [3.71], and Evans et al. [3.74]. Claussen et al. [3.31] calcula-ted the microcracked zone size in alumina from fracture energy data obtained by various test techniques. They showed that the ratio of the microcracked zone size to grain size for NB tests, which is about 2, is virtually independent of grain size, but increases with grain size for DCB and WOF tests. At a grain size of 10 µm, the ratio from DCB and WOF tests is about the same as the ratio from NB tests, but at a grain size of 100 µm the former exceeds the latter by a factor

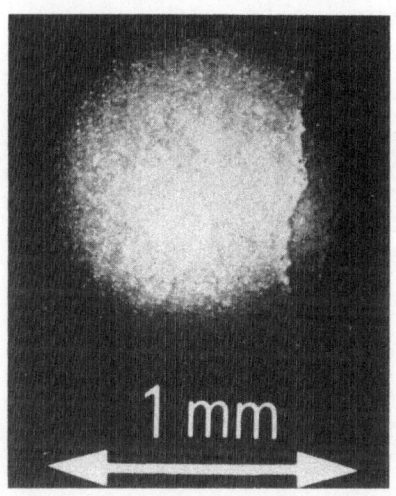

Fig. 3.8. Crack branching and rejoining on the side face of an alumina bend specimen. - From Hübner and Jillek [3.47]

of 10, thereby indicating that microcracking in the DCB and WOF tests strongly
increases with grain size [3.31].

A direct consequence of the occurrence of microcracking in a ceramic material is the
R curve behavior. The R curve of a material is the variation of the resistance to
crack propagation, R, measured in J/m^2, with crack length a. The R curve is obtained
experimentally from a WOF test, assuming that a balance is maintained between
the crack propagation resistance R and the energy release rate $\mathcal{G}$ at any moment
of the test, which gives $R = \mathcal{G}$. Since $\mathcal{G} = (P^2/2B)(dC/da)$, where P is the instanta-
neous load and B and C are the breadth and compliance of the specimen, respective-
ly, R is obtained point-by-point by evaluating the load-deflection record.

The R curve of an ideally brittle material is a step function that jumps from $R = 0$ to a characteristic value $R = R_o$ at $a = a_o$, where a_o is the initial crack length
of the precracked sample. An increasing R curve means that a material fractures on-
ly at an increasing stress intensity factor. This type of failure is characteristic
for tough metals, but it has also been met in some ceramic materials [3.70,3.75].
An increasing R curve in brittle materials has been explained by the increase in
the size of the microcracked zone as the main crack advances [3.70]. Hoagland and
Embury [3.75] recently have presented a model for microcracking which permits one
to calculate microcrack densities as a function of the applied stress intensity
factor and to assess the microcracked zone size. Allowing for the interaction between microcracks and the main crack, the model also predicts the R curve, which

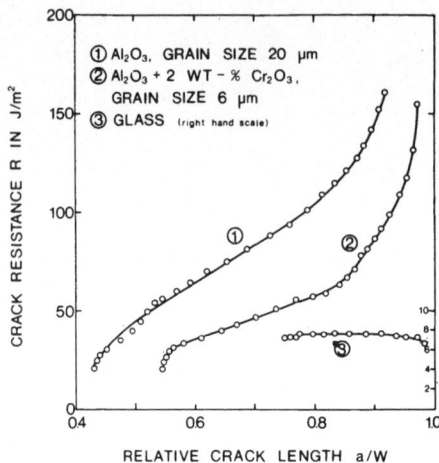

CRACK RESISTANCE R IN J/m²

① Al₂O₃, GRAIN SIZE 20 μm
② Al₂O₃ + 2 WT - % Cr₂O₃,
   GRAIN SIZE 6 μm
③ GLASS (right hand scale)

RELATIVE CRACK LENGTH a/W

Fig. 3.9. R curves of two aluminas compared with soda-lime glass. - From Kleinlein and Hübner [3.67]

has been found to increase with the length of the main crack. R curves of alumina were determined by Hübner and coworkers [3.47,3.67]. Pronounced R-curve behavior was found for grain sizes of 6 and 20 μm. Figure 3.9 shows R curves of two aluminas and a soda-lime glass from [3.67]. Whereas the R curve of the glass is virtually independent of a, those of alumina show a steady increase, the coarse-grained material having the higher R value and the steeper slope. This R-curve behavior is a further evidence of increased microcracking of alumina as the grain size is increased.

The size of the microcracked zone, R_{mc}, in a brittle material can be assessed by the expression

$$R_{mc} = \alpha (K_{Ic}/\sigma_c)^2,$$
(3.11)

as has been shown by several authors [3.76 to 3.78]. In Eq. (3.11), σ_c is a characteristic stress at which microcracking starts, and α is a dimensionless number which was reported to be equal to 0.831 [3.76], to $4/\pi$ [3.77], or to 1/3 [3.78].

Table 3.3 contains a summary of available data of the microcracked zone size and the ratio of the microcracked zone size to grain size in aluminum oxide. The values from several experimental studies [3.31,3.42,3.47,3.54,3.77,3.79] may be compared with the results from analytical calculations using Eq. (3.11) [3.76,3.78] and from the models of Hoagland and Embury [3.75] and of Gesing and Bradt [3.63]. From Table 3.3 it may be seen that experimental and theoretical results of the microcracked zone size agree within an order of magnitude. The microcracked zone size in alumina typically has an extension of several hundred microns and comprises some 10-30 grains.

Table 3.3. Measured and calculated microcracked zone size in polycrystalline alumina

Authors	Ref. Year	Microcracked zone size R_{mc} in μm	Grain size G in μm	R_{mc}/G	Method
(A) Experimental data					
Simpson et al.	3.54 1975	100 to 600 ≈ 50	20 3	5 to 30 15	Penetration depth of dye penetrant
Hübner & Jillek	3.47 1977	350 to 1000	18	20 to 50	Inferred from K_{Ic} increase of pre-cracked specimens
Bansal & Duckworth	3.42 1978	41	1–2	≈ 30	As above
Buresch	3.77 1978	114 to 275	2 to 9	≈ 40	Notch-width dependence of K_{Ic}
Pabst et al.	3.79 1978	45	11	4	Inferred from an energy dissipation model
Claussen et al.	3.31 1981		1.5 to 30 3 to 120	2 1 to 150	Calcul. from γ_{NB} data Calcul. from γ_{DCB} data of Rice et al. [3.59]
(B) Analytical calculations					
Kesler et al.	3.76 1972	≈ 130[4]			$R_{mc} = 0.831(K_{Ic}/\sigma_c)^2$
Kreher & Pompe	3.78 1981	≈ 50[4]			$R_{mc} = 0.333(K_{Ic}/\sigma_c)^2$
(C) Model calculations					
Hoagland & Embury	3.75 1980			9 to 11[1]	Two-dimensional computer simulation
Gesing & Bradt	3.63 1983	min. 16 max. 230	10 to 60[2]	3 to 10[3]	Two-dimensional analytical model

[1] in multiples of microcrack length
[2] grain facet size
[3] depending on assumptions of flaw size distribution
[4] using K_{Ic} = 4.5 MN/m$^{3/2}$ and σ_c = 350 MN/m^2

From the foregoing discussion it follows that the dependence of the fracture energy on grain size can have practical consequences when Al_2O_3 is used as a structural material. For the purpose of high strength, where the fracture initiation threshold plays the predominant role, materials of small grain size should be developed because γ_{NB} attains high values, thereby increasing the strength. For applications, however, where a large resistance to crack propagation is desired, i.e., when the material is exposed to mechanical or thermal shocks, a large grain size up to about 100 μm should be envisaged in order to take advantage of the maximum of γ_{DCB} and γ_{WOF}.

The temperature dependence of K_{Ic} of polycrystalline and single-crystal alumina is shown in Fig. 3.10. As long as the temperature dependence of Young's modulus is not exactly known, the representation of toughness data should be preferred in terms of

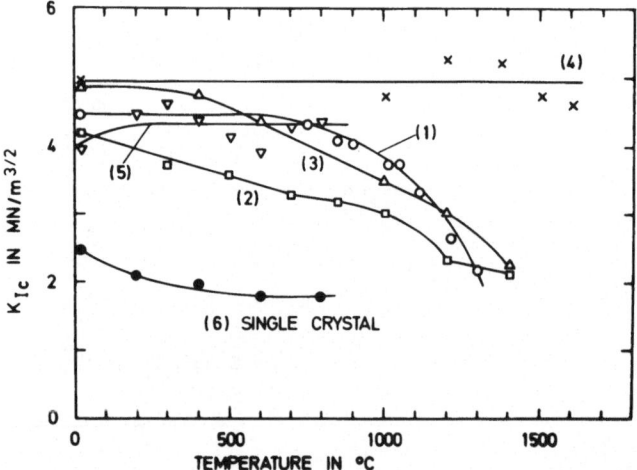

Fig. 3.10. Temperature dependence of fracture toughness. - Curves (1) to (5): polycrystalline alumina. (1) Claussen et al. [3.36], grain size 5 to 20 μm; (2) Davidge and Tappin [3.80], 11 μm; (3) Evans, Linzer and Russell [3.74], 30 μm; (4) Kobayashi et al. [3.81], grain size not reported; (5) Mai [3.49], 3.5 μm. - Curve (6) single-crystal alumina, Wiederhorn, Hockey and Roberts [3.26]

K_{Ic} rather than γ_I. Much, but not all, of the K_{Ic} decrease with increasing temperature shown in Fig. 3.10 seems to be due to the decrease of Young's modulus. Curves 1-3 agree that the decrease becomes more pronounced at higher temperatures, typically above $1000^\circ C$. This number coincides with the value of the transition temperature of strength between regions A and B in alumina, which has been indicated by Davidge and Evans to be about $1000^\circ C$ [3.21]. Typical room temperature fracture toughness values of polycrystalline alumina lie between 3.8 and 5.1 $MN/m^{3/2}$. Table 3.2 contains these values for comparison.

Little is known about the effect of impurities on the fracture energy. Two experimental investigations [3.45,3.48] which studied the influence of Ca impurities on K_{Ic} agree that Ca is strongly enriched at the grain boundaries (see also Section 2.7), and that an increasing Ca content deteriorates the fracture toughness of materials that fail predominantly by intergranular fracture. A 20% increase in K_{Ic} was found by Funkenbusch and Smith [3.45] when the total Ca concentration was reduced from 0.20 w/o to nominally zero. Jupp et al. [3.48] also reported a 20% increase in K_{Ic} when the Ca content on the fracture surface was reduced from 1.6 to < 0.1 at %. Whereas in these two cases the decrease of K_{Ic} seems to be caused by the weakening of the grain-boundary adhesive strength due to impurity segregation, no effect of varying Ca additions on K_{Ic} was found by de With and Hattu [3.52] in an alumina whose fracture mode was predominantly transgranular.

3.2.3 Types of Flaws

According to Eq. (3.5), the second important variable that controls the strength of ceramics is the flaw size C. In the following paragraphs a survey will be given of strength-controlling flaws found in alumina. The classification of flaw types generally adopted for ceramics will be used here as well, i.e., depending on the failure mechanism, flaws are considered as pre-existing or as stress-induced. Pre-existing flaws are usually divided into extrinsic flaws, which are caused by external conditions such as machining and sintering, and intrinsic flaws, the origin of which is due to material properties and microstructural features. In any case, the most severe of these flaws, whatever the type, will initiate the failure of a ceramic component. A summary of the following description of flaw types observed in alumina is given in Table 3.4.

There is a series of pre-existing, extrinsic flaws which may control the strength of alumina. The machining process can cause various defects in the surface and the near subsurface region. Fracture starting from partially broken edges was identified to account for 20% of failed samples by Rice [3.82]. Stepped flaws were found at the fracture origins in alumina by Kirchner et al. [3.83]. Stepped flaws are defects with a steplike appearance. They are formed by the linking up of flaws in two or more different planes parallel to the fracture surface, and are thought to be caused by surface finishing problems [3.83]. In the range of small grain sizes where the flaw size is greater than the grain size, C > G, the flaw size is believed to be independent of the grain size and determined by the machining conditions only [3.3, 3.29,3.57,3.82,3.84]. Therefore, when the parameters of the surface treatment are kept constant, a constant flaw size is produced and the strength becomes independent of the microstructure. This fact will be discussed in more detail in Section 3.2.5. In the range of small grain sizes, it was also observed that strengths of alumina samples having tensile stresses normal to the machining grooves are substantially lower than those with tensile stresses parallel with the machining striations [3.82].

Residual pores due to incomplete sintering may be another type of extrinsic flaws in polycrystalline alumina. Fracture origins at large surface pores were detected by Rice [3.82]. Large internal pores were also found by the same author [3.85] to originate fracture. He observed that, in many cases, fracture-initiating pores had a size between 20 and 200 μm and were located between 100 and 200 μm beneath the surface. Rice also pointed out [3.85] that pores must have a sufficient size to act as fracture origins because they must compete with other strength-controlling defects, i.e., surface flaws induced by machining. Thus, a uniformly distributed porosity consisting of many small pores is much less dangerous than a small number of large pores. Since fracture follows the weakest path through the material, strength can be correlated much more easily with local extremes of the porosity than with its average value [3.85].

In the following examples of flaw types, stress-assisted mechanisms are assumed to operate during loading to form the fracture-originating flaw. But, since flaw formation is supposed to start from a pre-existing defect, these cases will be discussed here together with other failure causes which also start from pre-existing flaws. Pore groups beneath the surface can act as fracture origins as was observed by Evans and Tappin [3.33], Rice [3.68], and Meredith and Pratt [3.86]. These authors agree that crack linking during loading between the individual pores of the group is the mechanism which produces a fracture-originating large flaw. Another mechanism of flaw generation caused by the presence of pores may be the cracking of a grain boundary adjacent to a pore during the stressing period. Flaw sizes in alumina equal to the sum of the pore diameter plus one grain size were reported by Davidge and Tappin [3.80], Evans and Tappin [3.33], Rice [3.82], and Kirchner et al. [3.83]. In fractographic examinations of the area in the vicinity of the fracture origin, Kirchner and co-workers [3.83,3.87,3.88] observed that the original, pre-existing defect, which could be a pore or a machining flaw, was surrounded by a flat circular cleavage region consisting of many grains which had undergone transgranular fracture. Taking the size of the cleavage region as the flaw size that originated fracture, Kirchner et. al. [3.83] obtained close coincidence between

96

measured and calculated strengths. Therefore, they interpreted their findings as stress-induced subcritical cleavage crack growth prior to failure.

In the range of increased grain sizes, where the flaw size approximately equals the grain size, $C \approx G$, or is less than the grain size, $C < G$, the strength is independent of a variation of the machining process, because the size of the largest grains present in the microstructure determines the critical flaw size [3.29,3.57, 3.82]. This grain-size range is the region where intrinsic flaws control strength, It has been emphasized repeatedly [3.29,3.68,3.84,3.85,3.89] that in the large-grained region fracture initiation is associated with one of the largest grains that occur in the stressed region, but not with the mean grain size of the material. However, for fracture to be originated by larger grains (and not by other intrinsic or extrinsic inhomogeneities), a significant grain size distribution must exist [3.85].

In particular, large grains were identified as strength-controlling flaws in alumina by Rice [3.68] and by Tressler and coworkers [3.84]. In a later work, Rice [3.85] reported on a fractographic study that indicated large grains or groups of large grains acting as preferred sources of failure in dense hot-pressed alumina. Kirchner et al. [3.83] also observed fracture originating at large crystals lying both on the surface and beneath the surface. The combination of a large crystal plus a large pore was indicated as the fracture origin by Davidge and Tappin [3.80] and Rice [3.3,3.85]. A further source of fracture origin may be impurity particles, e.g., small alumina particles chipped from milling bodies and containers, as has been indicated by Rice [3.85]. Other second-phase particles, however, obviously do not play an important role in high alumina ceramics or pure aluminum oxide.

There are also some intrinsic stress-induced flaws others than those caused by linking up of pre-existing small extrinsic flaws that have been found to control strength in aluminum oxide. They may be divided into two general groups: those generated by dislocation interaction and those generated by mechanical twinning. Stofel and Conrad [3.90] interpreted their results on the deformation-dependent tensile strength of sapphire in the temperature range 1100–1500°C by the dislocation-assisted formation of Griffith cracks. Bayer and Cooper [3.91,3.92] reported that the failure mechanism of sapphire whiskers at room temperature depended on the surface treatment. Unpolished samples fractured from pre-existing flaws through the Griffith mechanism, whereas for chemically polished samples a mechanism based on the pile-up or interaction of dislocations was suggested as the cause of fracture. In an experimental study on the strain-rate dependence of the strength of sapphire whiskers at room temperature performed by Pollock and Hurley [3.93], failure was attributed to the Orowan mechanism of crack extension in brittle materials, i.e.,

to dislocation-assisted crack growth caused by dislocation pile-ups as stress concentrators. This interpretation was contradicted, however, by Evans et al. [3.94],
who demonstrated that the observed behavior could be explained by subcritical crack
growth. Actually, the lack of any dislocation activity in the vicinity of the
crack tip in sapphire up to 400°C could be proved by Wiederhorn et al. [3.26].

Fracture of aluminum oxide can also be caused by mechanical twinning. The formation of deformation twins in sapphire compression specimens at temperatures between
1100 and 1500°C was reported by Stofel and Contrad [3.90]. Twins on both basal and
rhombohedral planes were found, and preferential cracking along the twin band-crys-
tal interfaces was observed. Deformation twinning in Al_2O_3 single crystals fractured in bending was also found by Heuer [3.95]. Twins formed on rhombohedral planes at temperatures as low as -196°C and on basal planes at 1000°C or more. From
microscopic observations of twin-subgrain and twin-twin interactions Heuer concluded that twins can form at low temperatures in Al_2O_3 and can initiate fracture
[3.95]. Also in polycrystalline aluminum oxide loaded in both compression and tension mechanical twins were detected to act as failure-initiating defects. Lankford
[3.96] reported on twin bands formed in alumina specimens loaded in compression to
near failure. Microcracks originated at the intersections of these bands with grain
boundaries, and failure was caused by the coalescence of such microcracks. In an
experimental study on the temperature dependence of the bend strength of coarse-
grained as-sintered alumina specimens (which did not contain machining-induced
flaws), Lankford [3.97] observed pronounced twin formation at tensile stresses far
below the fracture stress. SEM examinations of fracture surfaces and tensile surfaces revealed that many large grains contained microcracks which originated at
twin-grain boundary and twin-surface intersections. Therefore, the temperature-dependent strength was interpreted in terms of a twin-nucleated microcrack formation
process [3.97].

Resuming the discussion of flaw types that cause failure in polycrystalline alumina (see Table 3.4), it may be stated that in fine-grained materials extrinsic
flaws like machining-induced defects and large pores are the major source of failure. To the extent that these flaws can be avoided (e.g., by an optimization of
the surface treatment or by a proper adjustment of the sintering conditions), and
with creasing grain size, intrinsic flaws gain importance. Microstructural inhomogeneities such as large grains, clusters of large grains, and combinations of large
grains and pores can act as strength-controlling flaws. Large grains can fracture spontaneously by cleavage or along the grain boundary, presumably due to high
TEA stresses. The presence of large grains can also facilitate the occurrence of
twin-nucleated microcracking. The objective in developing a high-strength alumina,
therefore, should be the fabrication of a pure material of high density and a ho-

mogeneous microstructure characterized by a uniformly distributed residual porosity and a narrow grain size range and finished by an optimized surface treatment.

Table 3.4. Flaw types found in aluminum oxide (numbers are references)

(A) Pre-existing flaws

Extrinsic flaws		Intrinsic flaws
Machining-induced flaws	Fabrication-induced flaws	
edge flaws 3.82	large pores 3.82	large surface grains 3.68,
stepped flaws 3.83	pore groups 3.33,	3.83 to 3.85
regular machining-	3.68,3.86	large sub-surface grains 3.83
induced flaws 3.3,3.29,	pore plus one grain 3.33,	groups of large grains 3.85
3.57,3.82,3.84	3.80,3.82,3.83	large grain plus one pore
	pore plus cleavage	3.3,3.80,3.85
	region 3.83,3.87,3.88	impurity particles 3.85

(B) Stress-induced flaws (all intrinsic)

dislocation-nucleated flaws in single-crystal aluminum oxide 3.90 to 3.93

twin-nucleated flaws in single-crystal aluminum oxide 3.90,3.95
twin-nucleated flaws in polycrystalline aluminum oxide 3.96,3.97

3.2.4 Strength Data

In the following section, a compilation of fracture strength data of polycrystalline alumina is given. Both tensile and compressive strength will be reviewed, and
the effects of porosity, grain size, and temperature will be considered.

The dependence of the strength of ceramic materials on porosity has been reviewed
extensively by Rice [3.3]. He has also discussed existing strength-porosity models
and has emphasized the fact that many, if not all, of the relations which have been
developed to describe elasticity-porosity interactions can also be used for
strength, because of the similarity between strength and elastic properties. Therefore, the relationship most widely used to describe the porosity dependence of
strength is the exponential relation

$$S = S_o \exp(-bP), \qquad (3.12)$$

where S is the strength at porosity P, S_o the strength at zero porosity, and b an empirical constant. Equation (3.12), which is based on the principle of the remaining load-bearing cross section of the matrix and the minimum distance between pores, was suggested for the first time by Duckworth [3.98]. It has the same form as Eqs. (3.4) and (3.8), which describe the porosity dependence of Young's modulus and the fracture energy, respectively. Following Rice [3.3], the b value depends strongly on the type of porosity, i.e., on pore location, size, and shape. For a material of a given type of porosity, the values of Eqs. (3.4) and (3.8) should be equal, so that the same b value also results for the porosity dependence of strength [3.3] from the Griffith equation, Eq. (3.5).

Figure 3.11 is a semilogarithmic plot of room temperature strength-porosity data obtained in bending of polycrystalline alumina from five different sources [3.8, 3.33,3.65,3.99,3.100]. The figure shows that porosity has a strong influence on

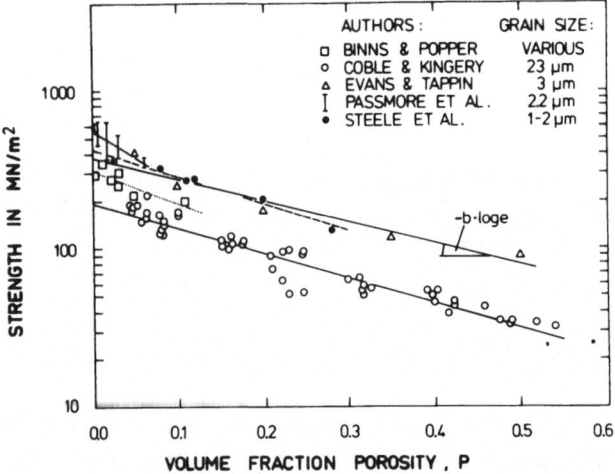

Fig. 3.11. Effect of porosity on the strength of polycrystalline alumina.- Data from Binns and Popper [3.65], Coble and Kingery [3.8], Evans and Tappin [3.33], Passmore et al. [3.99], and Steele et al. [3.100]

strength. A 5% increase in porosity causes a decrease of strength of about 20%. Figure 3.11 also demonstrates that individual data sets obey very well the exponential relation of Eq. (3.12) up to high values of porosity. The b values of the various data sets lie between 3.1 and 4.6 and are listed in Table 3.5 for comparison. Table 3.5 also corroborates the hypothesis of Rice [3.3], which states that the three b values of the exponential relations for the porosity dependence of Young's modulus, fracture energy, and strength should be the same. Most of the b values found in polycrystalline alumina are close to or slightly smaller than 4.

100

Table 3.5. Comparison of porosity dependence of Young's modulus, fracture energy, and strength

Young's modulus $E = E_o \exp(-bP)$			Fracture energy $\gamma = \gamma_o \exp(-bP)$			Strength $S = S_o \exp(-bP)$		
Authors	Ref.	b	Authors	Ref.	b	Authors	Ref.	b
Rice	3.3	3.5	Evans & Tappin	3.33	1.5	Coble & Kingery	3.8	3.6
Knudsen	3.14	4.0	Simpson	3.34	10	Evans & Tappin	3.33	3.1
Soga & Anderson	3.19	4.2	Pabst	3.35	3.5	Binns & Popper	3.65	4.6
			Claussen et al.	3.36	2.8	Steele et al.	3.100	4.0

Strength-grain size data of alumina at room temperature taken from seven different references [3.9, 3.10, 3.33, 3.65, 3.84, 3.99, 3.101] are shown in Fig. 3.12. Before plotting, the data were corrected for 0% porosity by means of Eq. (3.12), using the following b values. Data of Evans and Tappin [3.33], Binns and Popper [3.65], and Passmore et al. [3.99] were corrected by taking the b values obtained from a

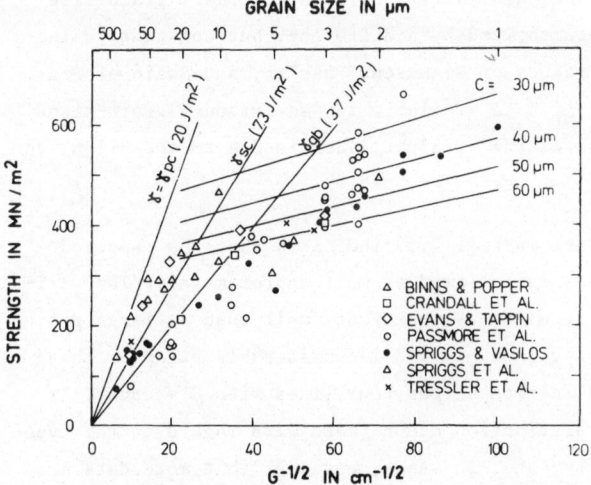

Fig. 3.12. Grain-size dependence of strength at room temperature. - Data from Binns and Popper [3.65], Crandall et al. [3.9], Evans and Tappin [3.33], Passmore et al. [3.99], Spriggs and Vasilos [3.101], Spriggs et al. [3.10], and Tressler et al. [3.84]

101

linear regression of their strength-porosity data plotted in Fig. 3.11 (see also the b values in Table 3.5). For the correction of the data from Crandall et al. [3.9] and Spriggs et al. [3.10], b = 4 was used. No correction was necessary for the data of Refs. [3.84] and [3.101] because Spriggs and Vasilos gave already corrected values and Tressler et al. reported that their material was a dense alumina.

In Fig. 3.12, strength-grain size data are represented as a function of the inverse square root of the grain size because it is often assumed that, in the range of large grain sizes, the flaw size equals the grain size. Thus, a proportionality between strength and $G^{-1/2}$ is expected at large grain sizes, as indicated in Fig. 3.12 by the three straight lines that pass through the origin. However, there is some disagreement in the literature on which type of fracture energy controls strength when failure originates at large grains. Therefore, in order to see which type of fracture energy predicts the right trend of the data, three different slopes were used tentatively in Fig. 3.12, which correspond to three different types of fracture energy, i.e., the fracture energy for grain-boundary fracture, γ_{gb}, the single-crystal fracture energy, γ_{sc}, and the polycrystalline fracture energy, γ_{pc}. The numerical values of the fracture energies were taken from Table 3.2. On the other hand, according to the discussion of flaw types in the foregoing section, the flaw size at small grain size is expected to be extrinsic (independent of microstructure) and a function of the machining conditions only. In order to prove this fractographic observation, a set of four nearly horizontal straight lines was calculated by means of Eq. (3.5) and included in Fig. 3.12 in the small grain-size region. Each of these lines is characterized by a different, but constant, value of the flaw size. For γ_T, the γ_{ND} values of Clauosen, Mussler, and Swain of Fig. 3.4 were used, which are the only γ_{NB} data available for an extended grain-size range [3.31]. The small inclination of the straight lines is due to the slight variation of γ_{NB} with grain size.

The position of the experimental data in Fig. 3.12 indicates that the expected trend depicted by the straight lines is followed at most approximately. The variation of strength with grain size seems to be weaker at small than at large grain sizes but in the small grain-size region it is still considerably stronger than the variation predicted by the inclination of the four lines with C = const. At large grain sizes, no unequivocal distinction can be made with regard to the type of fracture energy that controls strength. It seems, however, that more data are localized in the vicinity of the γ_{gb} and γ_{sc} curves than in the vicinity of the γ_{pc} curve. But, it should also be kept in mind that a strict comparison of strength data from different sources may not be possible in some cases, because the data may differ due to experimental conditions and material composition.

Strength-temperature data of polycrystalline alumina from five sources [3.9,3.10, 3.49,3.97,3.102] are shown in Fig. 3.13. Except for the data of Meredith et al. [3.102], all measurements were obtained on high-purity alumina and exhibit a similar variation with temperature which is characterized by a region of nearly constant strength over a broad temperature range followed by a pronounced decrease that starts at about 1000°C.

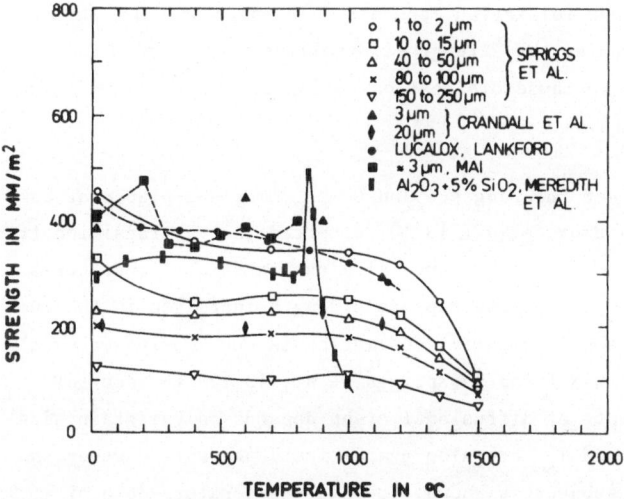

Fig. 3.13. Temperature dependence of the strength of polycrystalline alumina. - Data from Spriggs et al. [3.10], Crandall et al. [3.9]. Lankford [3.97], Mai [3.49], and Meredith et al. [3.102]

The temperature dependence of the strength of ceramics is usually explained in terms of a model originally proposed by Davidge and Evans [3.21] and further elaborated by Davidge [3.4,3.22,3.103]. In this model it is assumed that there exist three neighboring temperature ranges, A, B, and C, in which different strength-controlling mechanisms predominate. In region A the fracture is completely brittle and the strength is controlled by pre-existing flaws and varies little with temperature. In region B the fracture is still brittle but the strength is smaller than the Griffith strength. Pre-existing flaws are extended and new flaws are nucleated by limited plastic deformation processes which may be due to dislocation motion or to a grain-boundary sliding mechanism. The temperature dependence of the strength is the same as that of the flow stress, i.e., strong. Region C is characterized by the occurrence of appreciable macroscopic plastic deformation, but is rarely observed in polycrystalline ceramics [3.4,3.21,3.22,3.103].

The temperature variation of the strength of pure alumina can easily be understood
in terms of this model. Looking at the data points of Spriggs et al. [3.10] and
Crandall et al. [3.9] in Fig. 3.13, the transition temperature between ranges A
and B, T_{AB}, may be localized between 1000 and 1200°C. This value coincides well
with $T_{AB} \approx 1000^{\circ}$C as proposed by Davidge and Evans [3.21] for polycrystalline alumina, and agrees with observations on the beginning of limited plasticity at 1100°C
[3.104]. Thus it may be argued that the fracture strength of alumina above T_{AB} is
controlled by a grain-boundary sliding mechanism [3.4], perhaps combined with
some diffusional creep as a stress relaxation mechanism. In any case, dislocation-
assisted fracture can be excluded as the strength-controlling mechanism of range
B because the stresses required to cause dislocation motion are much greater than
the observed fracture stresses [3.4].

The transition temperature between the ranges B and C, T_{BC}, must be placed at a
somewhat higher temperature. Passmore et al. [3.10] reported that a transition from
brittleness to ductility occurred in fine-grained aluminum oxide at temperatures
from 1300 to 1400°C and that the transverse rupture strength increased in the ductile region. The onset of macroscopic plastic deformation in the alumina material
of smallest grain size (1-2 µm) was found by Spriggs et al. [3.10] to occur at
1500°C. Since the deformation rate of diffusional creep depends on the grain size
it should be expected that T_{AB} and T_{BC} are also grain-size dependent. However, no
detailed investigation of this subject is known. The high-temperature data of
Spriggs et al. [3.10] in Fig. 3.13 confirm this prediction at least qualitatively,
indicating that the smaller the grain size the greater the propensity of the material to lose its strength, as the temperature is increased.

Former models of the temperature dependence of strength of alumina, for example,
those by Congleton and Petch [3.106] and by Petch [3.107], which interpreted the
decrease of strength with increasing temperature already in the range between room
temperature and 1000°C by dislocation-nucleated formation of microcracks, are no
longer accepted today because no dislocations have been observed. The sharp peak of
the strength-temperature variation of the data of Meredith et al. [3.102] at 850°C
was interpreted as the behavior of 5% of a glassy phase which, at the temperature
of the peak, becomes viscous enough to relax stress concentrations at flaw tips by
increasing the effective crack tip radius.

The fracture behavior of a material can be represented in a "fracture mechanism
map", as has been shown by Gandhi and Ashby [3.108]. A fracture mechanism map is a
diagram of fracture stress versus temperature in which a series of neighboring
fields that define the regime of the predominating failure mechanism are indicated.

The limits that confine the field of a fracture mechanism also depend on the grain size and the strain rate. Seven different mechanisms of fracture have been distinguished by Gandhi and Ashby [3.108]. Figure 3.14 is the fracture mechanism map for polycrystalline alumina (grain size 10 μm) taken from [3.108]. The map shows that,

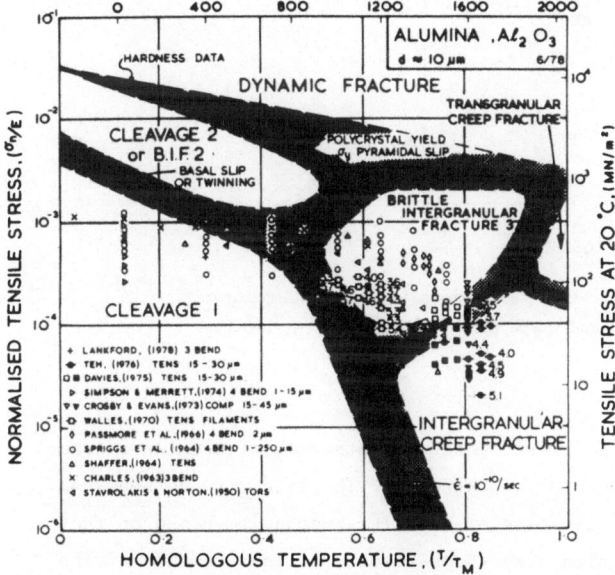

Fig. 3.14. Fracture mechanism map for alumina, Ref. [3.108]. Reprinted with permission from Acta Met. Vol.27, Gandhi, C. and Ashby, M.F.: Fracture-mechanism maps for materials which cleave: F.C.C., B.C.C. and H.C.P. metals and ceramics, Copyright 1979, Pergamon Press Ltd.

depending on the temperature, experimental strength data are located in three different regimes named "cleavage 1", "brittle intergranular fracture 3" and "intergranular creep fracture". These regimes are identical with the temperature regions A, B, and C defined earlier by Davidge and Evans [3.21]. According to Gandhi and Ashby [3.108], the regime of cleavage 1 is characterized by the complete absence of microplasticity, and failure occurs from pre-existing flaws. In the regime of brittle intergranular fracture 3, plasticity is still limited but sufficient to blunt and extend pre-existing cracks and to nucleate new cracks, and in the regime of intergranular creep fracture failure is due to the nucleation and linking of voids on grain boundaries caused by substantial diffusion-induced plasticity [3.108]. The two limiting temperatures lie between 900 and 1000°C and between 1400 and 1600°C, respectively, and agree well with T_{AB} and T_{BC} given above. From this discussion it may be concluded that the basic mechanisms that control the temperature-dependent strength of alumina are well understood.

Compressive strength data of alumina from the investigations of Evans [3.109], Dawihl and Dörre [3.110], Rice [3.111], and Lankford [3.112] are compared in Fig. 3.15. The figure shows values obtained at temperatures between room temperature

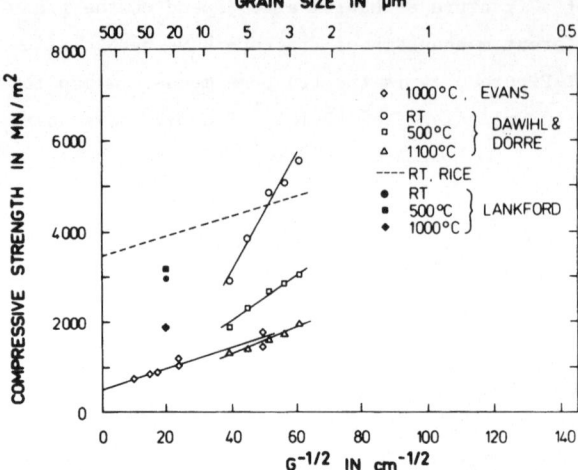

Fig. 3.15. Compressive strength of polycrystalline alumina at various temperatures.
- Data taken from Evans [3.109], Dawihl and Dörre [3.110], Rice [3.111], and Lankford [3.112]

and $1100^{\circ}C$ with varying grain size. The compressive strength of alumina at room temperature is about ten times the tensile strength. The linearity between compressive strength and $G^{-1/2}$ is indicative of a failure mechanism based on the equality between flaw size and grain size. Some results of measurements of the compressive strength of alumina as a function of temperature at various strain rates [3.109,3.110,3.112] are shown in Fig. 3.16. Both the temperature and strain-

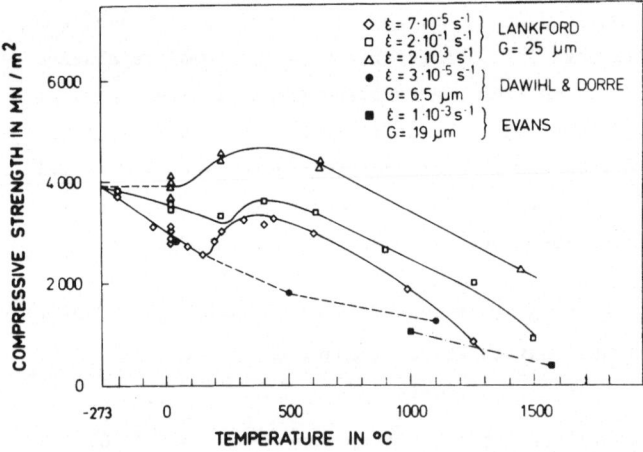

Fig. 3.16. Temperature and strain-rate dependence of compressive strength. - After Lankford [3.112], Dawihl and Dörre [3.110], and Evans [3.109]

106

rate dependence of the compressive strength of alumina have been investigated intensively by Lankford [3.96,3.112 to 3.114]. In these papers he has presented a model of compressive failure in alumina which is able to explain the strength minimum and the strain-rate dependence of strength shown in Fig. 3.16. According to Lankford, compressive failure in alumina is a thermally activated process based on localized plasticity in the form of twinning. Twin bands are formed at stresses well below the fracture stress which can interact with grain boundaries or the specimen surface, thereby causing the initiation of microcracks. A further increase in stress causes an increasing number of microcracks, which finally coalesce at failure [3.96,3.112 to 3.114]. Since the typical length of a twin-nucleated microcrack is that of a grain dimension, the model also explains the variation of compressive strength with the inverse square root of grain size.

3.2.5 Strength-Grain Size Relationships

In the foregoing sections, the main variables that control the strength of a ceramic, i.e., fracture energy and flaw size, have been discussed for alumina, and it has become evident that the strength depends in a complicated way on the details of the microstructure. Figure 3.12 demonstrates that the room-temperature strength of alumina is scattered between 100 and 600 MN/m^2 as the grain size varies between 2 and 100 µm. Many attempts have been made to develop strength models for ceramics and to establish a relation between flaw size and grain size. Some of them are reviewed in the following.

It was recognized early that the strength-grain size variation of ceramics, when represented in a strength vs $G^{-1/2}$ plot, is characterized by a two-branch curve, each branch being linear in $G^{-1/2}$, as indicated in Fig. 3.12. Analogously to the strength-grain size behavior of metals, the branch through the origin has been called the Orowan branch and the weakly inclined branch at small grain sizes the Petch branch [3.115].

In a first attempt to establish a strength-grain size model for ceramics, Carniglia [3.115,3.116] interpreted the Orowan branch he had found to be valid for many coarse-grained ceramics as due to pure Griffith behavior, i.e., fracture was assumed to be caused by pre-existing flaws of a size approximately equal to the grain size. In the Petch branch, where the fracture stress is smaller than the Griffith stress, fracture was thought to be initiated by flaws that are extended to a critical size prior to fracture by a plasticity-induced concentration of local stresses which may be caused by dislocation pile-ups at grain boundaries or by accommodation problems at triple points due to grain-boundary sliding [3.115,3.116].

In a comprehensive review of the grain-size dependence of strength of about 30 different brittle materials, Rice [3.68] proposed that Orowan-type behavior should only predominate in the hardest brittle materials, whereas the strength of "softer" ceramics (among them also alumina) should follow a pure Petch-type behavior, without branching at large grain size, of the form $S = M + HG^{-1/2}$, where M is the intercept and H the slope of the Petch branch. According to Rice [3.68], a nonzero intercept (M > 0) is due to microplastic strength control. Microplasticity in individual grains can build up stresses that cause the formation of fracture-initiating flaws situated within single grains or between grains. Hence, the propagation of these flaws during catastrophic failure should be controlled by the single-crystal fracture energy, γ_{sc}, or by the grain-boundary fracture energy, γ_{gb} [3.68].

So far, the interpretation of the Petch branch proposed by Rice is quite similar to Carniglia's explanation described above. However, Rice was the first to emphasize the role of deformation twinning as a microplastic mechanism in relatively brittle materials like alumina and to discuss the effects of machining-induced stresses and other internal stresses, like TEA and EA (elastic anisotropy) stresses, on the occurrance of the Petch branch [3.68].

Today, the concept of a two-branch curve for the grain-size dependence of the tensile strength of ceramics is still maintained, and for historic reasons, the two branches are still being called the Orowan and the Petch branches. The transition between the two regions in alumina has been localized repeatedly to lie in an intermediate grain-size range at about 40-60 μm [3.33,3.84,3.116]. However, contrary to former explanations, the interpretation of both branches has changed. Since no dislocation activity could be detected in alumina at least up to 400°C [3.26], there is full agreement today (see, e.g., [3.3,3.29,3.57,3.62,3.89]) that the Petch branch is also a completely brittle region, caused by the presence of extrinsic, machining-induced flaws the size of which is independent of, and large compared with, the grain size, C > G. Since a machining flaw is surrounded by many grains, the fracture energy that controls strength is thought to be that of the polycrystalline body, γ_{pc}. The small increase in strength with decreasing grain size is attributed to the change of γ_{pc} with grain size.

The Orowan branch, on the other hand, is thought to be the region where strength is controlled by intrinsic flaws. But there is substantial disagreement on the details of failure initiation in this grain-size range. One of two alternative micromechanisms currently being discussed states that small flaws of sizes smaller than the grain size can easily be propagated to a size equal to the grain size [3.21,3.84, 3.87 to 3.89,3.116,3.117]. This propagation will require a fracture energy equal to γ_{sc} or γ_{gb}, depending on whether the flaw is contained within a grain or at a boundary, but small compared with the fracture energy of the polycrystal. After arriving at the boundary, the flaw will encounter a region of increased fracture energy, i.e., γ_{pc}, so that it is temporarily arrested until loading is increased to a level where the energy release rate at the onset of catastrophic fracture is high enough to overcome the increased fracture energy. The variables that control strength, therefore, are a flaw size equal to the largest grain size, $C \approx G$, and the fracture energy of the polycrystalline material, γ_{pc}. Hence, the first micromechanism suggests that the strength of a given polycrystalline ceramic is characterized by a single value of γ_{pc} throughout the whole grain-size range.

The second micromechanism of flaw propagation put forward by authors like Rice [3.3, 3.29, 3.82, 3.85, 3.118], Rice et al. [3.62], and Freiman et al. [3.57] is based on the idea that, in the region of large grain sizes, even flaws a size smaller than the grain size, $C < G$, can act as failure-initiating flaws. In their strength model, small flaws need not be arrested at the first grain boundary, but may overshoot and lead immediately to catastrophic failure. The strength-controlling variables along the Orowan branch, therefore, are a flaw size that can ammount to some fraction of the grain size up to the full value of the grain size, and the fracture energy for single-crystal or grain boundary fracture, γ_{sc} or γ_{gb}. This model predicts a transition from γ_{sc} (or γ_{gb}) to γ_{pc} strength control as the grain size decreases. The use of γ_{sc} in the Griffith equation, Eq. (3.5), instead of γ_{pc} at large grain sizes, has also been suggested by Wiederhorn [3.119], McKinney and Herbert [3.120], Sedlacek et al. [3.121], and Rhodes and Cannon [3.122]. Moreover, it has already been pointed out in the foregoing section that the strength data of alumina at large grain sizes (Fig. 3.12) are grouped more closely along the Orowan lines represented by γ_{sc} and γ_{gb} than by γ_{pc}. This observation gives some support to the idea of γ_{sc} strength control in the coarse-grain-size region.

Very recently, a series of model calculations has been published (Singh et al. [3.123], Evans [3.124], Kirchner and Ragosta [3.125], Virkar et al. [3.126]) concerning the problem of the extension of small flaws in large grains prior to catastrophic failure. The objective of these theoretical studies was to obtain an assessment of whether such flaws are arrested at the first grain boundary, where they meet with an area of increased fracture energy, or are not arrested as a result of overshooting from an area where crack propagation is controlled by γ_{sc}. Although the criteria used to assess overshooting or arresting of flaws are very different (i.e., based on an energy balance, a stress balance, or subcritical crack growth), all four models arrive at two common conclusions [3.123 to 3.126]: (1) For a ratio of $\gamma_{pc} : \gamma_{sc}$ which is not too close to unity, crack arrest does occur at the first grain boundary. (2) Only for very small flaw sizes, i.e., a flaw-size to grain-size ratio of less than about 0.15-0.30, and for high loading rates, overshooting and, hence, γ_{sc} strength control is to be expected.

3.3 Time-Dependent Strength and Subcritical Crack Growth

In the following section, time-dependent strength in alumina will be reviewed. It is a well known fact that most ceramic materials undergo a loss in strength when subjected to tensile loading in a corrosive enviroment [3.2,3.4]. This strength degradation is already observed in the temperature range of purely brittle fracture and has nothing to do with creep-induced,time-dependent rupture. Rather, it is attributed to the slow growth of subcritical flaws to a size large enough to cause catastrophic failure. The time-dependent strength degradation, which is also known as static and dynamic fatigue, imposes serious restrictions on the use of ceramics as structural materials unless precautions that account for the strength loss under service conditions are taken [3.2]. After a brief presentation of the model of time-dependent strength of ceramics, available data on subcritical crack growth and strength degradation of alumina will be summarized. Finally, crack growth mechanism in alumina will be discussed.

3.3.1 The Model of Time-Dependent Strength

Most ceramic materials show the phenomenon of time-dependent strength when subjected to mechanical loading and exposed to a corrosive environment, such as water. The time-dependent strength may be observed as a dynamic fatigue effect, i.e., a stress-rate sensitivity of strength (strength decreases when the stressing rate is decreased), or as a static fatigue effect which means that the lifetime of the ceramic under a constant stress is limited and decreases with increasing stress level. In aluminum oxide, time-dependent strength was observed for the first time in the works of Pearson [3.127] and Williams [3.128] in 1956. The strength degradation of ceramics is explained in terms of the slow, stress-dependent growth of pre-existing flaws at values of the stress intensity factor smaller than the critical one. This subcritical crack growth may continue until a critical flaw size is reached which is determined by Eq. (3.6). Obviously, according to this equation, experimental conditions that promote subcritical crack growth (increasing C) will decrease the strength S, because K_{Ic} is a material constant.

For the variation of the crack growth rate with the stress intensity factor, a three-stage curve has been found experimentally for many ceramics, shown schematically in Fig. 3.17. Crack growth in region I is thought to be controlled by the reaction rate of the chemical attack of water near the crack tip [3.129,3.130], whereas in region II the diffusion of the corrosive species to the crack tip should be rate-controlling [3.130]. The reasons for the steep rise of the crack velocity

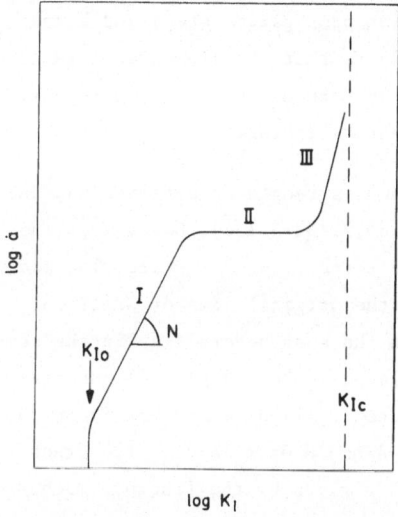

Fig. 3.17. Variation of crack velocity, $\dot{a}$, with stress intensity factor, K_I, for a ceramic material in a corrosive environment. - After Evans [3.130]

in region III as the critical stress intensity factor is approached are still unknown. In general, a ceramic component loaded in a corrosive environment will spend most of its lifetime in region I, passing through regions II and III rapidly at the end of the total loading history. Thus the details of region I are of special importance.

For the variation of the crack velocity $\dot{a}$ with the stress intensity factor K_I in region I, a power law of the form

$$\dot{a} = AK_I^N$$

(3.13)

has been proposed by Evans [3.130], where A and N are material parameters. Equation (3.13), although of minor physical meaning, has proved very useful in the description of experimental data and in calculating lifetime predictions.

However, since brittle fracture is thought to be a thermally activated process, an Arrhenius type equation for the crack propagation rate should be more appropriate, as has been first suggested by Zhurkov [3.131] and later by Wiederhorn et al. [3.26]. The relationship proposed in [3.26] has the form

$$\dot{a} = \dot{a}_o \exp[-(Q - bK_I)/RT]$$ (3.14)

where Q is the activation energy of the rate-limiting propagation step, and b and $\dot{a}_o$ are constants. The physical meaning of Eq. (3.14) is that the elementary fracture event is characterized by a temperature-assisted surmounting of an energy barrier Q, which is lowered by the applied stress intensity factor.

During the last ten years, the model of time-dependent strength of ceramics has been elaborated by several authors, mainly by Evans [3.130,3.132-3.134], Davidge [3.135], Wiederhorn [3.136], Ritter [3.137,3.138], and their respective co-workers. The model combines three basic assumptions: (1) According to the original idea of Griffith, failure occurs when a critical stress is reached at the most severe flaw in the sample. This stress and the critical flaw size are related by Eq. (3.6). (2) Subcritical crack growth occurs prior to failure. The growth rate is given by Eq. (3.13) or (3.1 and thus determined by the two material parameters A and N or $\dot{a}_o$ and b. (3) Crack tips are loaded by a stress intensity factor which is given by the fracture mechanics relationship

$$K_I = Y\sigma a^{1/2}$$ (3.15)

where σ is the applied stress and a is the actual crack length. All three quantities of the right-hand side of Eq. (3.15) may vary with time. The stress intensity factor causes crack growth, thereby increasing the crack length which further increases K_I. This leads to an unstable process of ever accelerating crack growth resulting finally in catastrophic failure.

Substituting for K_I in Eq. (3.13) yields a differential equation for a, which can be integrated until failure (a = C) if the "loading history" $\sigma(t)$ is known. For the two simple cases of static and dynamic loading, detailed analysis are available in the literature. Under a static load the time to failure t_f is given by [3.133,3.135, 3.139,3.140]

$$t_f = BS^{N-2}\sigma^{-N}$$ (3.16)

where S is the inert strength from Eq. (3.6) and B is an abbreviation for $B = 2/(AY^2(N-2)K_{Ic}^{N-2})$. For dynamic loading it has been shown [3.132-3.135,3.139-3.141] that the fracture stress σ_f and the stress rate $\dot{\sigma}$ are correlated by

$$\sigma_f^{N+1} = B(N+1)S^{N-2}\dot{\sigma}.$$ (3.17)

112

Since N in general is large (N>20), Eq. (3.16) predicts a very strong dependence of the lifetime on the stress, and Eq. (3.17) predicts a weak effect of the stressing rate on the strength.

To predict the time-dependent mechanical behavior of a ceramic, the two parameters of subcritical crack growth, A and N, must be known for a given temperature and environment. There are two different ways of obtaining them. The first one employs the direct measurement of subcritical crack velocities on samples containing a macrocrack using optical observation or a compliance method, as for example reported in [3.130]. Such crack growth data will be reviewed in Section 3.3.2. The second way uses static or dynamic loading of unnotched samples. Measuring data pairs of t_f and σ or of σ_f and $\dot{\sigma}$, respectively, and using Eqs. (3.16) and (3.17), gives values of A and N. Such fatigue data will be presented in Section 3.3.3.

3.3.2 Subcritical Crack Growth Data

The first quantitative crack velocity measurements in aluminum oxide were performed by Wiederhorn [3.142] in 1968 on sapphire. Using double cantilever beam specimens under a constant mechanical load and a travelling microscope, he obtained ($\dot{a}$,K_I) curves of the type shown in Fig. 3.17 for a range of relative humidities in air and at room temperature. A pronounced dependence of the crack velocity of regions I and II on the water vapor concentration was found, and it was argued that subcritical crack propagation of sapphire is due to a similar moisture-assisted stress corrosion process as in glass [3.142].

Figure 3.18 is a compilation of crack velocity data obtained at room temperature and in various environments for polycrystalline alumina [3.56,3.74,3.130,3.143]. All measurements, including those shown in Figs. 3.19 and 3.20, were carried out by tracing the propagation of macrocracks in double-torsion specimens, using direct optical observation or compliance measurements. From Fig. 3.18, some common features of subcritical crack propagation in alumina can be recognized, although the plotted data show considerable spread: (1) The variation of the crack growth rate with K_I is very strong. In region I, Eq. (3.13) is obeyed, the value of N lying between 30 and 50. (2) The environment has a distinct effect on the crack propagation. Passing through the sequence dry - humid air - water, an increase in the crack velocity by three orders of magnitude was observed by Devezas [3.56]. (3) According to Evans [3.130], slow crack growth occurs even at values of $K_I \lesssim 0.5\ K_{Ic}$. If there is a stress corrosion limit for K_I in alumina it should still be smaller.

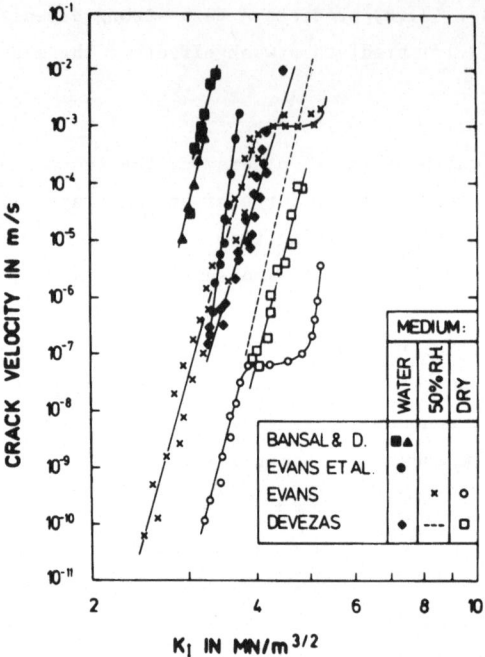

Fig. 3.18. Crack velocity vs. K_I data of polycrystalline alumina in various environments. - From Bansal and Duckworth [3.143], Evans et al. [3.74], Evans [3.130], and Devezas [3.56]

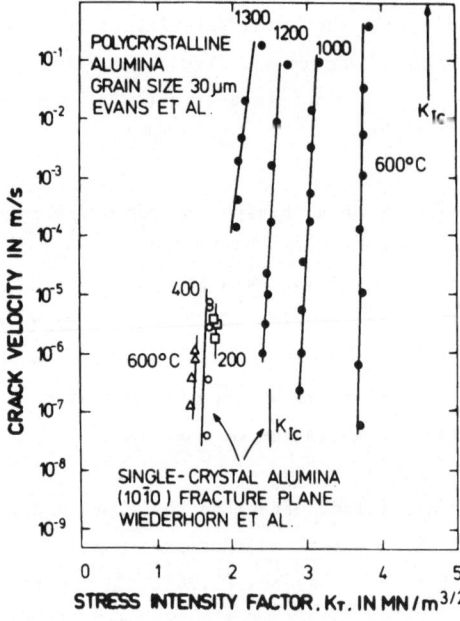

Fig. 3.19. Temperature dependence of $\dot{a},K_I$-data of single-crystal and polycrystalline alumina. - Data from Wiederhorn et al. [3.26] and Evans et al. [3.74]

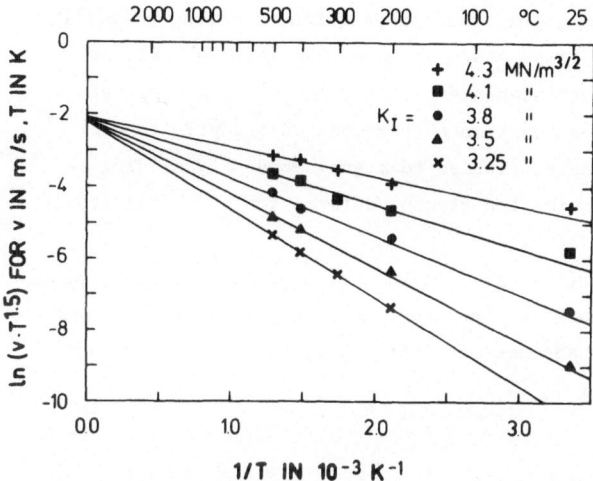

Fig. 3.20. Variation of crack velocity with temperature of polycrystalline alumina at intermediate temperatures in an Arrhenius plot. - From Devezas [3.56]

The temperature variation of slow crack growth in both single-crystal and polycrys-talline alumina is shown in Fig. 3.19. Single-crystal crack velocity data were ob-tained in vacuo [3.26], and polycrystals were tested in air [3.74]. The stress in-tensity factor necessary to achieve a certain crack velocity decreases as the tem-perature is increased. Furthermore, the decreasing slope of the (ȧ, K_I) curves with increasing temperature is indicative of a K_I dependence of the effective ac-tivation energy, as indicated by Eq. (3.14). By fitting the crack propagation data of single-crystal alumina to this equation, Wiederhorn et al. [3.26] obtained Q = 890 kJ/mol. This figure may appear somewhat high for alumina, because it exceeds the average value of the activation energy of self-diffusion by about 50% (see Table 2.6 A).

In a recent investigation, Devezas [3.56] studied the subcritical crack growth of alumina up to 1000°C in detail. The material contained 1.3% of silica which was present as a SiO_2 rich glassy phase at the grain boundaries. Crack velocity mea-surements were performed on DT specimens in air. By determining activation energies as a function of temperature, Devezas was able to discern three temperature regimes of crack propagation. In the low temperature range, between room temperature and 500°C, crack growth data were found to obey Eq. (3.14) with Q = 66 kJ/mol. When the crack velocity was normalized to $T^{-3/2}$ and represented in an Arrhenius plot, a fan-shaped set of straight lines was obtained, Fig. 3.20, with each line corre-sponding to a different K_I. From the $T^{-3/2}$ dependence of the crack velocity and the value of Q, it was concluded that crack growth was controlled by the adsorption rate of water molecules at the crack front, the crack propagation occurring primarily through the glassy phase [3.56]. At temperatures between 600 and 800°C the activation energy was about the same, but independent of K_I. In the high temperature region, between 850 and 1000°C, the activation energy changed abruptly to 480-520 kJ/mol, and crack propagation was attributed to a grain-boundary sliding mechanism, the rate-controlling step being the viscous deformation of the glassy phase

So far, it has been proven that aluminum oxide, like many other ceramics, is prone to stress corrosion cracking in a moist environment. Moreover, slow crack growth has also been found to occur in the absence of water vapor, but at much smaller velocities.

3.3.3 Fatigue Data

Apart from their usefulness to assess the time-dependent strength behavior of a ceramic, static and dynamic fatigue data may also be used for determining the crack growth parameters through Eqs. (3.16) and (3.17). In particular, dynamic tests can be performed very easily since they only need simple bend specimens and a testing machine with a constant cross-head speed. Moreover, when crack propagation parameters are to be determined for the purpose of failure predictions, static and dynamic tests are thought to be a more consistent method than direct observation, because the parameters are measured for the true strength-controlling flaws in their natural surrounding, rather than for artificial macrocracks [3.133].

Dynamic strength data of alumina may be found in several sources. One of the first papers on the subject was the work of Davidge et al. [3.135] who performed strength measurements at various stress rates to prove the validity of Eq. (3.17). Further work was published by Evans [3.132], Bansal and Duckworth [3.143], Kotchick and Tressler (on single crystals) [3.144], McLaren and Davidge [3.145], Rockar and Pletka [3.146], Jakus et al. [3.147], and Xavier and Hübner [3.148]. All authors agree that the strength of alumina is degraded when the material is tested in a moist environment, and that the degradation depends on the stress rate, a lower rate leading to an enhanced degradation. An example is shown in Fig. 3.21 where the degraded strength of alumina at room temperature and in water is compared with the inert strength in a Weibull plot [3.148]. Decreasing $\dot{\sigma}$ causes a parallel shift of the Weibull distributions to smaller values, as would be predicted by Eq. (3.17) when Weibull statistics are included [3.135]. Testing in water at the highest stress rate of Fig. 3.21 results in a strength drop to 83% of the inert strength, and at the lowest stress rate the strength drops to 71%.

116

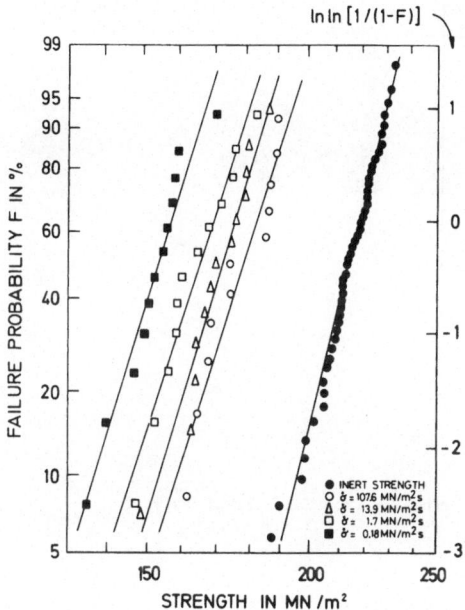

Fig. 3.21. Strength degradation of alumina due to dynamic fatigue at room temperature in water. - From Xavier and Hübner [3.148]

Figure 3.22 shows the stress rate dependence of the median fracture strength of a 99.5% alumina at room temperature and 1000°C after Jakus et al. [3.147]. The straight lines in this log-log plot mean that Eq. (3.17) is obeyed. The parameters N and A can be determined from the slope and the intercept, respectively. At high temperature, the strength is lower, but the stress rate dependence is much more pronounced than at room temperature. This indicates a change in the mechanism that controls slow crack growth. The authors suggested a change from moisture-assisted crack growth at room temperature to a mechanism that involves plastic deformation at high temperature, probably grain-boundary sliding due to the softening of very small amounts of a glassy phase at the grain boundaries [3.147].

Static fatigue data of polycrystalline alumina at room temperature and in air of varying content of humidity are compared in Fig. 3.23. The data were taken from the papers of Pearson [3.127], Williams [3.128], Dawihl and Klingler [3.149], Kirchner and Walker [3.150], Krohn and Hasselman [3.151], and Guiu [3.152]. The experimental results differ with respect to the absolute strength of the respective materials, but there is a common trend in all studies: the time to fracture is limited, and increases considerably as the applied stress is decreased. Not all of the data sets of Fig. 3.23 can be approximated by a straight line, i.e., the validity of Eq. (3.16) is not established in all cases. It is also clear from the fig-

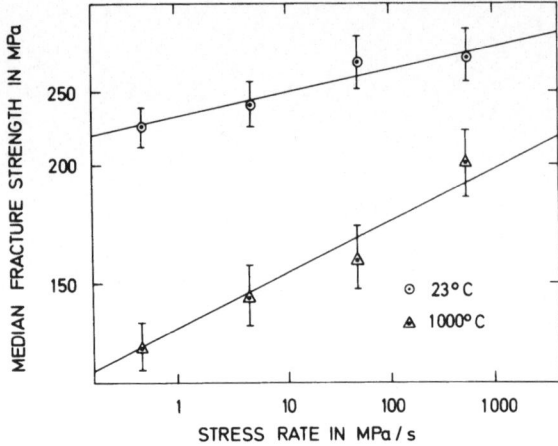

Fig. 3.22. Effect of stress rate on the strength of alumina. - From Jakus et al. [3.147]

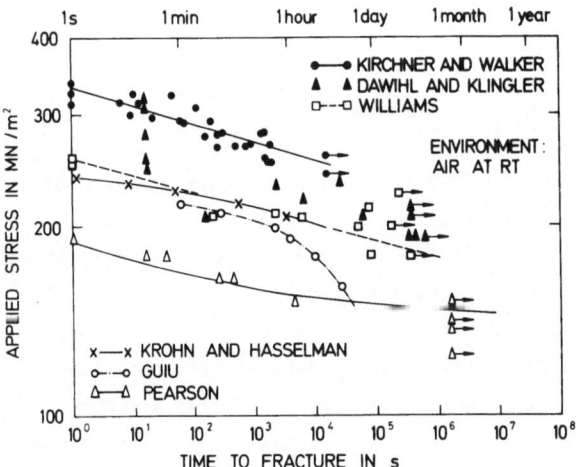

Fig. 3.23. Static fatigue data of alumina from various sources [3.127,3.128,3.149-3.152]

ure that there is a need for experimental data for lifetimes longer than several weeks, in order to enhance the accuracy of lifetime predictions.

A summary of subcritical crack growth data of single-crystal and polycrystalline alumina at room temperature is given in Table 3.6. The experimental method and some material propteries are indicated. The last column contains values of the crack growth exponent N. If no N value was reported in the original paper, it was assessed by us, for example in the somewhat older publications [3.127] and [3.128]. The spread of N extends from 24 to 52, but about 60% of the values listed in Table 3.6 fall within the range between N=38 and N=46. Furthermore, it seems that there is no systematic difference between crack growth exponents determined directly and those derivated from static and dynamic tests.

Table 3.6. Summary of subcritical crack growth data of alumina at room temperature

Authors	Ref.	Year	Method	Material Composition %	Grain Size μm	K_{Ic} MN/m$^{3/2}$	N
(A) Single-crystal alumina							
Wiederhorn	3.142	1968	direct, DCB	100	–	–	32
Kotchick & Tressler	3.144	1975	dynam. tests	100	–	–	28
(B) Polycrystalline alumina							
Pearson	3.127	1956	static tests	98.9	20	–	52
Williams	3.128	1956	static tests	99	50	–	41
Kirchner & Walker	3.150	1971	static tests	96+2.5 SiO$_2$	5	–	33
Rockar & Pletka	3.146	1978	dynam. tests	pure	3.5-19	–	26-39
Ritter & Wulff	3.140	1978	static tests dynam. tests	Al$_2$O$_3$+v.p.	n.i.	–	38.3 46.2
Ritter & Humenik	3.153	1979	static tests dynam. tests	90+10 v.p.	4-5	–	37.70 37.53
Jakus et al.	3.147	1980	dynam. tests	99.5	20	–	29.17
Xavier & Hübner	3.148	1982	static tests dynam. tests	99.5	15	3.47	41.3 42.5
Evans	3.130	1972	direct, DT	95	3	5.3	31
Freiman et al.	3.154	1973	direct, DCB	95	n.i.	–	24
Bansal & Duckworth	3.143	1978	direct, DT	95	n.i.	3.84	42
Ferber & Brown	3.155	1980	direct, DT	pure	n.i.	4.65	26.2- 42.4
Devezas	3.56	1981	direct, DT	98+1 SiO$_2$ 99.7	4 5	4.86 5.60	41 49

Legend: DT double torsion specimen; DCB double cantilever beam specimen; v.p. vitreous phase; n.i. not indicated.

The next two Figs. are examples of lifetime predictions for alumina by means of the model of time-dependent strength. Figure 3.24 shows such predictions of the Weibull distribution of failure times in static loading at three different stress levels [3.153]. Knowing the Weibull distribution of the inert strength and the crack propagation parameters of the material, Eq. (3.16) was used to predict the lifetimes. The comparison of the predictions with failure times obtained experimentally at the same stress levels shows fairly good agreement. From such studies it can be concluded that the flaw population that controls the inert strength of the ceramic is

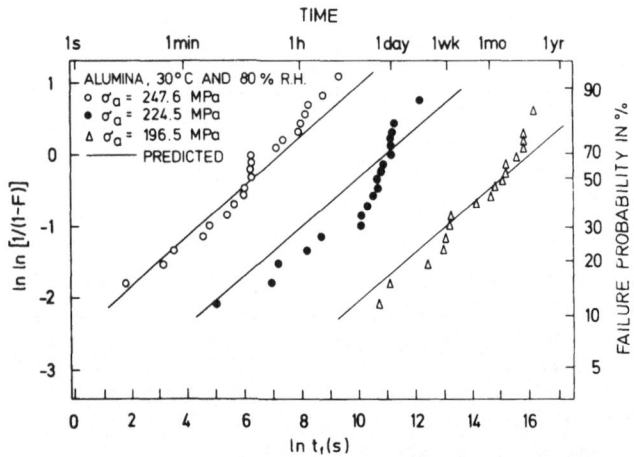

Fig. 3.24. Lifetime predictions from Eq. (3.16) and comparison with experimental data. - From Ritter and Humenik [3.153]

the same that controls the lifetime in static fatigue [3.153]. The second example of lifetime predictions concerns the fracture time of proof-test survivors. Since a proof test eliminates all samples which contain flaws of a size equal or greater than a certain size C_p, the surviving samples should sustain the applied stress in a static test at least a minimum lifetime t_{min}, which can be calculated from Eq. (3.16) by substituting S by the proof stress σ_p. If the strength-controlling flaw in a proof-test survivor is smaller than C_p, then t_{min} should be even exceeded. This prediction of the model may be verified by comparing the theoretical distribution of the time-to-failure after proof testing with fracture times of proof-test survivors, as shown in Fig. 3.25 [3.156]. This investigation demonstrated that in no case did the fracture time fall below the predicted minimum lifetime, and that the real lifetime of proof-test survivors may exceed the minimum lifetime by orders of magnitude. Thus it seems that the combination of the model of time-dependent strength with appropriate testing methods may provide a safe design of ceramic components that are susceptible to subcritical crack growth.

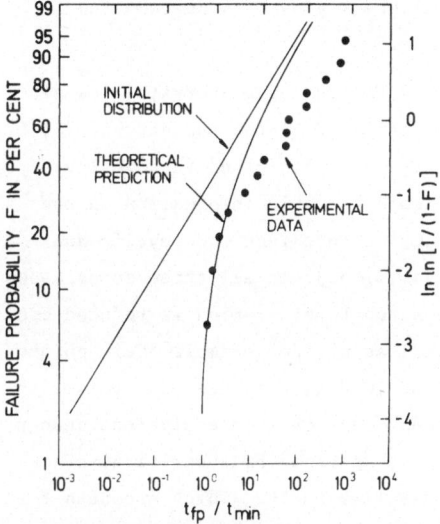

Fig. 3.25. Experimental time-to-failure data and theoretical prediction for proof-test survivors. - From Xavier and Hübner [3.156]

A series of investigations of the subcritical crack growth of alumina has also been performed in specific physiological environments [3.146,3.155,3.157,3.158, 3.158a]. This is because the use of alumina in human prosthetic devices is becoming increasing importance, as will be discussed in detail in Section 5.3. Alumina is an excellent implant material due to its high compressive strength, high wear resistance, and outstanding biocampatibility. However, to assure safe operation in the human body, its long-term strength behavior in body fluid must also be known. Crack propagation studies in various physiological media have been carried out and compared with the behavior in distilled water: in Ringer solution [3.146,3.158], in bovine serum [3.146,3.157], and in tissue fluid [3.155]. Slow crack growth was studied using direct observation in DT specimens [3.155,3.158] and dynamic [3.146] and static loading [3.157]. No influence of the biological environment as compared with water was found in [3.146]. In other studies [3.155, 3.157,3.158], however, it was observed that the stress-corrosion rate in the physiological environment was distinctly greater than in distilled water. The effect of preload and environment on the failure load of femoral heads made of pure aluminum oxide was studied in [3.158a]. Preloads of 10, 20, and 30 kN were applied in air, Ringer's solution, and sheep tissue during one year. No variation of the fracture load with preloading conditions was found, which was invariably close to 80 kN, independent of the environment. It was concluded, therefore, that these preload levels did not cause sufficient subcritical crack growth to degrade the strength of the ball heads.

At the end of this section, studies on the cyclic fatigue of polycrystalline alumina are reviewed. Only a few papers have been published on the subject [3.128, 3.151,3.152,3.159-3.161]. A S-N type fatigue curve was determined in some investigations [3.128,3.152,3.159,3.161], but still more valuable information on the material behavior under cyclic loading conditions was obtained when cyclic and static fatigue results were compared [3.128,3.151,3.152]. In all three comparisons the authors arrived at the conclusion that the strength of alumina is reduced considerably more by cyclic loading than by static loading. For example, Guiu pointed out [3.152] that specimens stressed at a constant load took a much longer time to fracture than specimens cycled up to the same peak load over an equivalent number of cycles. The cyclic fatigue effect could not be explained by subcritical crack growth only, i.e., by the g factor of Evans and Fuller [3.162] which accounts for the acccumulation of slow crack growth damage through an integration over the cyclic "loading history". Krohn and Hasselman found [3.151] that the strength is not only determined by the duration of loading, but also by the number of cycles, as for metals. They argued that static and cyclic fatigue mechanisms must be operative simultaneously to explain the observed stress and frequency dependence of cyclic loading, the cyclic fatigue mechanisms predominating at the higher stresses and frequencies. As to the nature of a cyclic fatigue mechanism in a brittle material like alumina, three suggestions have been made by Krohn and Hasselman [3.151]:

(1) The generation of heat due to nonlinear deformation at the tips of microcracks may increase the temperature, thereby assisting the breakage of lattice bonds by thermal activation. (2) The adsorption of water molecules and reaction products behind the crack tip during the opening cycle may prevent the crack from closing, thus increasing the rate of crack growth due to a "wedge action". (3) The generation of vacancies and other point defects by dislocation motion at the crack tip may put the material around the crack tip in a higher energy state, thereby lowering the activation energy for the stress-corrosion reaction. This latter mechanism was thought to be the principal reason for the cyclic fatigue effect in alumina [3.151].

3.3.4 Mechanisms of Slow Crack Growth in Alumina

After s short overview of mechanisms which have been suggested to predominate during slow crack growth in brittle materials, those mechanisms presumably acting in alumina will be discussed. High-temperature failure mechanisms will be treated in Section 3.6.

Crack growth mechanisms in ceramics have been reviewed by Evans and Langdon [3.2] and, more recently, by Wiederhorn et al. [3.163]. Three groups of mechanisms are being discerned in the literature, continuum models, diffusional models, and atomistic models of fracture. In the continuum models, chemical reaction rate theory is applied to the processes occurring around the crack tip, i.e., intrinsic bond rupture, adsorption of chemical reactants, and creation of new surface. By making some fracture-specific modifications, expressions are derived for the crack growth rate in terms of the thermodynamic variables of the system (reaction enthalpy and entropy, activation volume) to describe the effects of temperature, stress intensity factor, and water vapor concentration [3.163-3.165]. The change from intrinsic fracture due to bond rupture to chemically assisted fracture can easily be done by allowing for a reduction of the activation energy barrier by the corrosive species.

Diffusional models of crack growth attempt to explain slow cracking by the stress-directed diffusion of vacancies in the stress field ahead the crack tip [3.166], or by an evaporation/condensation process [3.2,3.166]. Crack growth exponents N between 3 and 8.5 have been calculated for several diffusion mechanisms such as volume diffusion and surface diffusion [3.166], but they do not agree with experimental findings at elevated temperatures [3.147].

Atomistic models of fracture are able to consider the discrete nature of the elementary fracture event, which consists of the rupture of a stretched lattice bond. The potential energy barrier between two lattice sites is the physical basis of the crack resistance of the material. By superimposing the work done by an external applied force to the potential curve, an expression for the crack extension force of fracture mechanics can be derived, and the phenomenon of "lattice trapping" of the crack can easily be accounted for. Thermally activated bond rupture as a universal atomistic fracture process was suggested for the first time by Zhurkov [3.131]. The first model calculation of lattice trapping in a one-dimensional atomic lattice model was presented by Thomson et al. [3.167]. Lawn [3.168] included the idea of activated crack sites ("kinks") into the lattice trapping model, and further work on atomistic fracture models was done by Lawn and Wilshaw [3.23], Fuller and Thomson [3.169], and Fuller et al. [3.170]. Even quantum me-

chanics tunneling was studied as a possible fracture mechanism on an atomistic
scale [3.169,3.171]. However, all these atomistic models are not able to calculate
the velocity of subcritical crack growth for a given K_I, temperature, and environment from basic material properties, or to assess the value of the activation energy of crack propagation from the height of the potential energy of the lattice.
At best, expressions of the type of Eq. (3.14) can be derived that predict the
variation of the crack velocity with temperature and K_I qualitatively.

Very few comments have been made in the literature on the interpretation of subcritical crack growth of alumina in the brittle temperature regime. The slow crack
growth of sapphire in a moist atmosphere was attributed by Wiederhorn [3.142] to
a stress corrosion process of fracture, similar to that in glass. Evans et al.
[3.94] interpreted the strain-rate sensitivity of sapphire filaments as caused by
thermally activated bond rupture at the crack tip. For the thermally activated
crack growth of sapphire in vacuum between 200 and 600°C, a mechanism based on
stress-induced vacancy diffusion or thermal fluctuations of atomic bonds was suggested by Wiederhorn et al. [3.26]. From his b value of Eq. (3.14), $b=0.423m^{5/2}/mol$,
the activation volume V^* may be assessed using the relationship $b=2V^*/(\pi\rho)^{1/2}$ from
[3.163], where ρ is the radius of the crack tip. Taking $\rho=0.5$ nm as in [3.163] and
the lattice constant $a_o=0.476$ nm from Table 2.1, the ratio V^*/a_o^3 is about 0.13
which seems to be a reasonable number for the activation volume. Finally, in the
only investigation known to us which studied the temperature and K_I variation of
the activation energy of slow crack growth in polycrystalline alumina, Devezas
[3.56] suggested that in the brittle temperature regime (up to 500°C) crack propagation mainly occurred through a very thin grain-boundary layer of an amorphous
phase, and that the rate-limiting step was the adsorption of water molecules at
the crack tip.

In summarizing this chapter on time-dependent strength, aluminum oxide turns out
to be a ceramic which is susceptible to slow crack growth, both in moist environments and in vacuo. However, very little is known about the mechanisms that control
the propagation of flaws and cracks, and nearly nothing about microstructural features that might influence the subcritical crack propagation, such as grain size,
pores, and second phases. Thus the assessment of safe service conditions in an
environment which promotes slow crack growth is still difficult.

3.4 Thermal and Mechanical Shock Resistance

One of the disadvantages of modern structural ceramics is their susceptibility to thermal and mechanical shocks. For many structural applications, the strength at room temperature or even at elevated temperatures may be sufficiently high. On heating and cooling, however, or under thermal gradients, thermal stresses can arise that may cause shock damage or catastrophic failure of the ceramic structural component. The two following sections will deal with studies of thermal shock behavior and other transient loading situations such as impact fatigue and indentation properties. In a later section, recovery from damage introduced by transient thermal and mechanical stresses will be reviewed, also known as crack healing effects.

3.4.1 Thermal Shock Properties

Before reviewing thermal fracture data of alumina, some basic ideas and models on thermal shock behavior of ceramics will be touched. The reason for the high susceptibility of ceramics to thermal shock is that they have very limited toughness. A first attempt at characterizing the shock behavior of ceramics quantitatively was made by Clarke et al. [3.172] who proposed a toughness parameter T given by

$$T = \log(6\gamma_{eff}/U)^3 \tag{3.18}$$

where γ_{eff} is the effective fracture energy and U is the elastic energy stored at the moment of fracture. They showed that different ceramics have different T values, and that the greater the toughness parameter, the greater is the shock resistance of the ceramic.

Davidge and Tappin [3.173] studied the thermal fracture of ceramics using two different approaches which are still valid today. In the fracture stress approach, fracture occurs when the thermal stress in the sample reaches the fracture stress of the material, S_t. In a very fast quench, this is the case when the temperature difference reaches a critical value ΔT_c which is given by the "thermal stress resistance parameter" R,

$$R = S_t(1-\nu)/(E\alpha) \tag{3.19}$$

where ν is Poisson's ratio, E is Young's modulus, and α is the coefficient of linear thermal expansion. Thus, the fracture stress approach states that fracture is initiated when a sufficiently high thermal stress is built up by temperature differences. In contrast to that, in the crack propagation approach it is assumed that thermal stress damage occurs due to the propagation of pre-existing cracks, when the elastic energy stored at the moment of fracture exceeds the effective energy of crack propagation. The corresponding thermal damage resistance parameter is

$$R'''' = \gamma_{eff}/U = \gamma_{eff}E/\left[S_t^2(1-\nu)\right] \tag{3.20}$$

since the elastic energy is given by $U = S_t^2(1-\nu)/E$. A comparison of Eqs. (3.19) and (3.20) shows that the two resistance parameters are affected by the material properties in an opposite manner, i.e., it is not possible to provide both high R and R'''' by selecting a suitable material. If fracture initiation is to be avoided, high R values must be envisaged which are associated with high values of strength and low values of Young's modulus and the coefficient of thermal expansion. On the other hand, once cracking has been initiated, or if the avoidance of any cracks is not a major aim, a material of high R'''' will show low thermal shock damage, and a high R'''' value corresponds to a high value of Young's modulus and a low value of strength.

The crack propagation approach was elaborated further by Hasselman [3.174]. He assumed that the ceramic body contains a number of uniformly sized and distributed microcracks which propagate simultaneously when the change of the stored elastic energy with increasing crack length exceeds the fracture energy expended per unit crack length. The model yields expressions for the critical temperature difference ΔT_c, at which damage first occurs, and for the crack length and the degraded strength as a function of the temperature difference of the quench, ΔT. The predicted variation of the strength after a sustained thermal shock, i.e., the retained strength S_a, with ΔT is that shown in the three lower plots of Fig. 3.26: at small temperature differences, up to ΔT_c, no damage is expected, and the retained strength equals the initial strength S_t. At ΔT_c, a sudden strength drop should occur to a level which depends on S_t, a high initial strength resulting in a more pronounced fall-off. By further increasing ΔT, Hasselman's model predicts a region of constant strength, where no additional damage should be observed, followed by a final region of gradual and ever increasing strength diminution with increasing severity of thermal shock.

The reader interested in further details on thermal shock models is referred to the original papers [3.173,3.174] and to textbooks on strength of ceramics [3.1, 3.4]. More thermal crack resistance parameters which have been defined for other thermal environments and material requirements may be found in [3.175] . An alternative model of thermal stress fracture based on fracture mechanics principles has been presented by Evans [3.176].

Results of thermal shock studies on polycrystalline alumina from [3.49,3.173,3.177-3.182] are summarized in Table 3.7. There is agreement among most experimental investigations on the critical quenching temperature difference, ΔT_c, for quenching in water at room temperature, which was found to lie between about 160 and 190 K. However, results on the extent of thermal shock damage are ambiguous, as indicated by the broad range of F_a values. According to Hasselman [3.174], the fractional retained strength F_a should vary with R'''' and S_t as

$$F_a = S_a/S_t \propto (R'''')^{3/4} \propto (\gamma_{eff}E)^{3/4}S_t^{-3/2} \qquad (3.21)$$

Table 3.7. Thermal shock properties of polycrystalline alumina

Authors	Ref.	Material characteristics					Thermal shock prop.		
	Year	Type	Compo- sition %	Poro- sity %	Grain size µm	Strength S_t MPa	Sample diam. mm	ΔT_c [4] K	F_a
Davidge & Tappin	3.173 1967					249	≈5x5	180	0.52
Ainsworth & Moore	3.177 1969		99.5	≈0	24	213	12.7	175	0.25
Hasselman	3.178 1970	AD-94[1]				307 348 320	9.5 4.8 2.0	250 300 300	0.22 0.21 0.30
Gupta	3.179 1972		99.8 99.8 99.8 99.9	2.2 0.6 0.4 0.3	10 34 40 85	329 198 187 156	≈3x3	170 175 175 150	0.27 0.51 0.56 -[3]
Krohn et al.	3.180 1973	AL-300[2]				269	4.8	212	0.48
Bertsch et al.	3.181 1974	AL 995[2]				290	6.4	185	0.14
Mai	3.49 1976			0.2		463	5-6	170	0.38
Smith et al.	3.182 1976		99.9	1.2 1.7 2.4 3.9 5.7	1.9 2.0 1.8 2.0 2.2	226 207 178 118 103	12	155 165 165 190 185	0.26 0.31 0.43 0.78 0.84

[1] Coors Porcelain Co., U.S.A.
[2] Western Gold and Platinum Co., U.S.A.
[3] no abrupt strength drop found
[4] for quenching in water at room temperature

As shown by the table, considerable different F_a values are obtained by different workers for only weakly differing initial strength values. This is undoubtedly due to variations in toughness among different materials.

A typical result of thermal fracture data of polycrystalline alumina from the work of Gupta [3.179] is shown in Fig. 3.26. After quenching from the temperature indicated in the figure, the retained strength was measured for materials of four different grain sizes. The data obtained at the three smaller grain sizes reproduce exactly the curve predicted by Hasselman [3.174]. At the largest grain size studied, however, only a gradual strength loss was found without exhibiting the strength plateau. Other experimental studies of alumina also confirmed the Hasselman model, as for example [3.173,3.177,3.180,3.182].

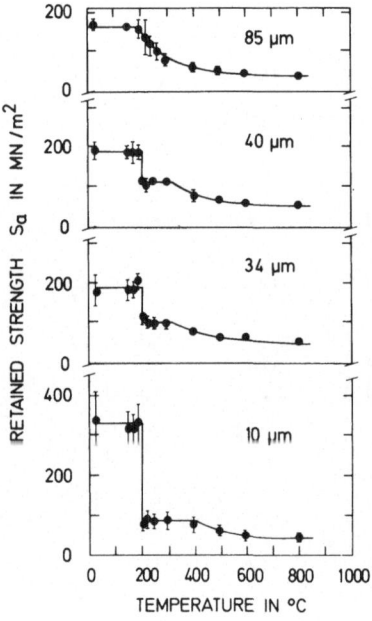

Fig. 3.26. Retained strength S_a of alumina specimens of various grain sizes as a function of quench temperature. – From Gupta [3.179]

Another prediction of the Hasselman model, the increasing degree of damage with increasing initial strength according to Eq. (3.21), could also be verified experimentally in polycrystalline alumina. Several grades of alumina materials which differed only in strength were produced in [3.179] by varying the grain size at nearly constant porosity, and in [3.182] by varying the porosity at nearly constant

128

grain size. As may be seen from Table 3.7, decreasing initial strength causes steadily increasing F_a values in both investigations. This is also demonstrated in Fig. 3.27 which further shows that the slope of -3/2 of Eq. (3.21) is reproduced very well by the experimental data.

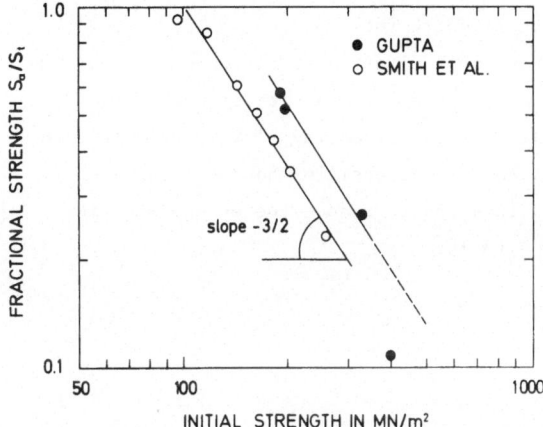

Fig. 3.27. Fractional retained strength S_a/S_t as a function of initial strength S_t. — Data from Gupta [3.179] and Smith et al. [3.182]

On comparing fractional retained strength data with Eq. (3.21), the question has been discussed in the literature which type of fracture energy controls the extension of cracks at the moment of fracture and, hence, the extent of thermal shock damage. The different meaning of γ_I or γ_{NB} on the one hand and of γ_{DCB} or γ_{WOF} on the other, was already outlined in Section 3.2.2. Several authors, as for example Clarke et al. [3.172], Davidge and Tappin [3.173], Hasselman [3.178], and Larson et al. [3.183], have proposed that γ_{WOF} should be used rather than γ_I, because considerably better quantitative agreement between theory and experiment can thereby be achieved. It seems therefore that the strength degradation of alumina subjected to severe thermal shock is controlled by the propagation rather than initiation of thermal cracks.

From the available data on strength and thermal shock behavior of alumina, some conclusions for the selection of material characteristics for different thermal environments can be drawn. Two opposite kinds of operating requirements are conceivable, in terms of the two approaches of the thermal shock resistance of ceramics described above. When thermal stress failure is to be avoided, the fracture stress approach applies, and the thermal stress resistance parameter of Eq. (3.19) controls the behavior under thermal shock. This means that a material of high strength should be selected. High strength is associated with small grain size (Fig. 3.12) and low porosity (Fig. 3.11). However, once failure occurs, the strength loss is large, as shown by [3.179] and [3.182]. In applications where repeated thermal cycling prevails, the crack propagation approach is more adequate, with R'''' of Eq. (3.20) controlling the strength loss. Thermal fracture should be tolerated, and a "soft" material of high strength retention capability should be chosen rather than a strong material. Favorable material characteristics would be a large grain size and some porosity. Large grain size results in both large γ_{WOF} (Fig. 3.7) and low strength (Fig. 3.12), thereby increasing R''''. Furthermore, the presence of pores and other microstructural heterogeneities has been found to improve the resistance to catastrophic failure and to reduce the extent of crack propagation in several ceramic compounds [3.1].

3.4.2 Mechanical Shock Properties

Knowledge of the behavior of alumina under mechanical shock loading is very poor. Only a few studies have been published on the subject. Sarkar and Glinn [3.184] and Huffine and Berger [3.185] reported on investigations on the impact fatigue of polycrystalline alumina. In both works, an S-N type fatigue curve was found for the decrease of strength with the number of impacts which caused failure. In [3.184], the impact energy sustained by the sample decreased asymptotically to about 0.3 of the initial value after 10^3 impacts, whereas in [3.185] a gradual diminution of the impact stress to 0.4 was found after about 10^4 impact cycles. However, no detailed explanation of the fatigue effect was given.

Instrumented impacted tests on alumina were reported by Bertolotti [3.186]. He pointed out that because of the high stiffness and brittleness of the material, the energy excess of the moving impact pendulum at the moment of fracture was so high (about 100 times the fracture energy) that no conclusions as to the fracture behavior under fast crack propagation conditions could be drawn.

Studies on the effects of solid particle impact and liquid droplet impact have been carried out mainly on glass. Nothing is yet known about these effects in alumina. This also holds for indentation fracture experiments, the principle area of which is also glass. The exception is the work of Lankford and Davidson [3.187], who studied the onset of crack propagation in alumina under a sharp indentor of varying mechanical load using acoustic emission. First acoustic activities were observed at a load of about 25 g.

3.4.3 Crack Healing

Thermal and mechanical shock loading causes numerous surface cracks in ceramics.
The idea that such cracks could heal at temperatures close to the sintering temperature was tested already in the early sixties, firstly on nuclear ceramic fuel
pellets. A number of investigators have also studied crack healing effects in polycrystalline alumina.

The direct determination of healing rates of cracks in a pure polycrystalline alumina at 1400°C was reported by Evans [3.188]. Figure 3.28 shows healing rates and
growth rates in precracked double-torsion specimens as a function of the applied
stress intensity factor. It was pointed out in [3.188] that the healing rate di-
minished very rapidly at quite small stress-intensity factors, i.e., at very small
crack opening displacement.

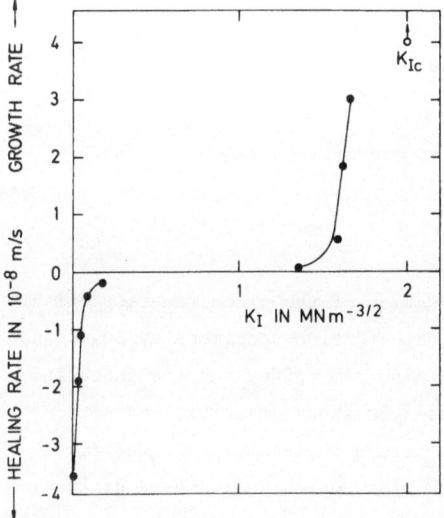

Fig. 3.28. Healing rate and growth rate of cracks in aluminum oxide at 1400°C. –
From Evans [3.188]

Strength recovery after a sustained thermal shock was used by Lange and Radford
[3.189], Dawihl and Altmeyer [3.190], and Gupta [3.191] to trace the effect of
crack healing. Quite short annealing times were found to be sufficient to achieve
a large amount of strength recovery. Samples degraded to $F_a=0.36$ recovered 0.86 of
the initial strength during a thermal treatment of 7 hours at 1700°C [3.189], and
annealing 8 hours at 1450°C resulted in a complete recovery in Al_2O_3 + 3% SiO_2 in

131

[3.190]. Recovery curves for various annealing temperatures after [3.191] are shown in Fig. 3.29. Maximum recovery was achieved from F_a=0.3 to F_a=0.9 by annealing 0.5 hours at 1700°C. The strength diminution after the recovery maximum found at 1700°C in [3.191] and [3.189] was attributed to simultaneously occurring grain growth.

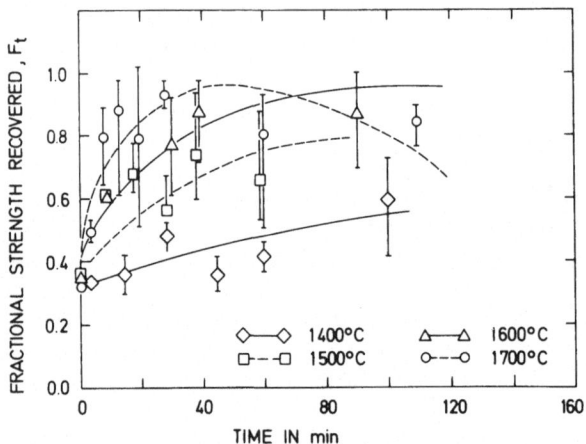

Fig. 3.29. Strength recovery of alumina samples annealed at various temperatures. – From Gupta [3.191]

Crack healing was also studied by optical observation of the crack tip retraction of macrocracks 20 mm in length in precracked double-torsion specimens by Evans and Charles [3.192]. The material was a pure, dense aluminum oxide, and the annealing temperatures were 1400, 1600, and 1800°C. Figure 3.30 shows crack tip retraction data vs. annealing time for these temperatures. Straight lines gave a good fit to the experimental data in a double-logarithmic plot, which indicates a diminution of the healing rate with time, as may be expected from reaction rate theory.

Strength increase of machined samples after annealing was used by Lino and Hübner [3.193] to establish the kinetics of crack healing in a pure aluminum oxide. As discussed in Section 3.2.3, machining is known to cause surface defects in alumina. The heat treatment of ground samples performed at temperatures between 1200 and 1600°C and times between 1 and 16 hours resulted in a shift of the Weibull strength distribution to higher values. The maximum strength increase, from 276 MPa to 400 MPa, occurred after annealing 1 hour at 1500°C. The crack tip retraction was calculated from the measured strength by means of Eq. (3.6). Experimental data at 1400°C are shown in Fig. 3.30. However, compared with the data of Evans and Charles

132

[3.192], the healing effect of machining cracks at the same temperature is much less pronounced, the difference amounting to a factor of 20-40.

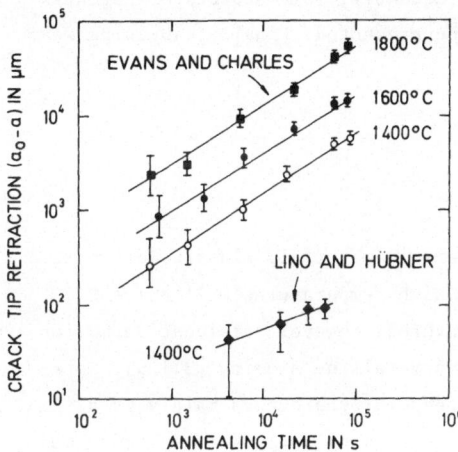

Fig. 3.30. Crack tip retraction during heat treatment. - From Evans and Charles [3.192] and Lino and Hübner [3.193]

Several mechanisms have been suggested that might control crack healing in alumina. Lange and Radford [3.189] argued that crack healing probably occurs by the growth of grains from one surface of the crack to the opposite surface, which fills and eliminates the crack. From the value of the activation energy of the crack tip retraction, Q=272 kJ/mol, surface diffusion was inferred to be the rate-controlling transport mechanism by Evans and Charles [3.192]. Based on a much higher value of the activation energy found in [3.193], Q=565 kJ/mole, crack healing was thought to be controlled by volume diffusion. In any case, the remark made by Lange and Radford [3.189], namely that crack healing by heat treatment is a phenomenon which is believed to occur in all materials that sinter, is worth considering. However, few efforts have been made until now to make use of this phenomenon technologically in aluminum oxide.

3.5 Plastic Deformation

Plastic deformation phenomena have been studied intensively in ceramic oxides, and especially in aluminum oxide. A number a review articles on plastic deformation in single-crystal and polycrystalline aluminum oxide is available in the literature. One of the first is the excellent review by Conrad [3.194]. Other review and introductory papers on the subject which the reader is referred to have been published by Heuer [3.195], Snow and Heuer [3.196], Chin [3.197], and, more recently, Mitchell [3.198]. In the following two sections we shall discuss the phenomena of slip and twinning in Al_2O_3 in some detail. Subsequently, some aspects of hardness as a technologically important parameter will be presented. Finally, abrasive wear behavior will be discussed.

3.5.1 Slip

Aluminum oxide is known to be a very brittle ceramic material. Plastic deformation due to dislocation activity is limited to very high temperatures well above 0.5 of the homologous temperature. At these temperatures, however, dislocation motion can be an important mode of deformation in polycrystalline alumina [3.199]. Moreover, even if macroscopic deformation at high temperatures occurs mainly by a gliding mechanism such as grain-boundary sliding or viscous glide of an amorphous grain-boundary phase, some localized dislocation-induced plasticity may be necessary to provide deformation continuity at triple points and grain-boundary ledges. Limited dislocation motion may also occur at lower temperatures, thereby strongly affecting the macroscopic material behavior. Slip on a favorably oriented slip system may be the cause of the formation of cracks and thus controls the strength at room temperature [3.91,3.92]. Under hydrostatic constraint during abrasion, surface deformation by slip may control the wear behavior at room temperature [3.200,3.201]. Thus a detailed study of deformation properties is necessary when various mechanical phenomena are to be understood.

3.5.1.1 Geometry of Slip

Slip in aluminum oxide occurs on essentially three different slip systems, i.e., the basal and prismatic slip system and several pyramidal slip systems. A summary of observed slip planes and slip directions is given in Table 3.8. The table also contains the mineralogical symbol of the slip plane after [3.196], the length of the Burgers vector, and the minimum temperature of operation under an uniaxial tensile stress, without any hydrostatic constraint. As expected from the brittle mechanical behavior, the temperature to activate slip on the various slip systems is quite high (0.50-0.63 T/T_m). As in other metallic and ceramic materials of hexagonal structure, slip occurs most easily on the basal slip system, that is, the temperature of operation is the lowest and the tensile flow stress the smallest of the three possible slip systems (see also Fig. 3.33). To activate slip on the se-

Table 3.8. Slip systems in aluminum oxide

Authors	Ref.	Year	Crystallographic plane and direction	Miner. symbol slip plane	Length of Burgers vector nm	Minimum temperature of operation °C
Basal slip system						
Wachtman & Maxwell	3.202	1954	$(0001)1/3<11\bar{2}0>$	c	0.475	900
Wachtman & Maxwell	3.203	1957				
Kronberg	3.204	1957				
Chang	3.205	1960				
Kronberg	3.206	1962				
Conrad et al.	3.207	1965				
Prismatic slip system						
Scheuplein & Gibbs	3.208	1960	$\{11\bar{2}0\}<10\bar{1}0>$	a	0.822	1150
Bayer & Cooper	3.91	1967				
Gooch & Groves	3.209	1972				
Gooch & Groves	3.210	1973				
Bilde-Sørensen et al.	3.211	1976				
Kotchick & Tressler	3.212	1980				
Gulden	3.213	1967	$\{11\bar{2}0\}1/3<1\bar{1}01>$	a	0.512	
Pyramidal and rhombohedral slip systems						
Bayer & Cooper	3.214	1967	$\{\bar{1}012\}1/3<10\bar{1}1>$	r	0.512	1200
Heuer et al.	3.215	1971				
Cadoz & Pellisier	3.216	1976				
Hockey	3.217	1971	$\{10\bar{1}1\}-$	s		
Gooch & Groves	3.218	1973	$\{10\bar{1}1\}1/3<1\bar{1}01>$	s	0.512	1600
Michael & Tressler	3.219	1974				1800
Tressler & Barber	3.220	1974				1760
Snow & Heuer	3.196	1973	$\{10\bar{1}1\}1/3<11\bar{2}0>$	s	0.475	1150
Hockey	3.221	1975	$\{11\bar{2}3\}<\bar{1}100>$	n	0.822	

cond and third (prismatic and pyramidal) slip system, the temperature has to be
raised considerably. This fact, together with the large difference between the
stresses to activate the various slip systems, gives rise to a pronounced deformation anisotropy, as will be discussed below. Table 3.8 also shows that the length
of the Burgers vector varies greatly, and that in several cases slip does not occur
along the shortest Burgers vector, as might be expected from energy conservation

principles. Rather, less energetically favorable slip directions are observed, thereby avoiding electrostatic interactions of charged half-plane edges with the lattice.

Both etch pits and transmission electron microscopy (TEM) have been employed to give experimental evidence for dislocations in alumina. Etch pits have often been used to determine slip directions and dislocation densities, as for example in [3.208,3.222-3.224]. However, TEM is a more suitable method for an exact determination of the crystallographic slip direction. Therefore, TEM was used to give evidence for the $1/3<11\bar{2}0>$ Burgers vector of the basal slip system [3.199,3.224, 3.225], as well as the $<01\bar{1}0>$ slip direction of the prismatic slip system [3.209, 3.211] and the $1/3<10\bar{1}1>$ slip direction of the pyramidal slip system [3.199,3.213, 3.216,3.224,3.225].

TEM is also a useful means to determine dislocation reactions and dissociations. The dissociation of the $<11\bar{2}0>$ basal dislocation into four partials was postulated by Kronberg [3.204] for energetic reasons. However, no such dislocation in the glide plane could be detected by Caslavsky et al. [3.226]. On the other hand, the dissociation of $<11\bar{2}0>$ dipoles due to self-climb out of the glide plane was observed by Mitchell et al. [3.227], according to the reaction $1/3[11\bar{2}0] \rightarrow 1/3[10\bar{1}0]+1/3[01\bar{1}0]$. In a TEM study of the prismatic slip system, Bilde-Sørensen et al. [3.211] found that $<01\bar{1}0>$ dislocations dissociated into three partial dislocations. The partials were found to have a Burgers vector of $1/3<01\bar{1}0>$ and to be separated by two identical faults. The distance between the partials was determined to vary between 7.5 and 13.5 nm.

Dislocation reactions, in particular the formation of sessile dislocations, play an important role in work hardening. Caslavsky et al. [3.226] and May and Shah [3.228] observed the reaction $[2\bar{1}\bar{1}0]+[\bar{1}2\bar{1}0]+[\bar{1}\bar{1}20]=0$ between basal dislocations. This reaction means the formation of a dislocation node and, hence, the self-pinning of dislocations in the basal plane. Dislocation reactions between $<10\bar{1}1>$ pyramidal dislocations have also been reported. According to Gooch and Groves [3.210] and Cadoz and Pellisier [3.216], the reaction $1/3[10\bar{1}\bar{1}]+1/3[\bar{1}\bar{1}01] \rightarrow 1/3[2\bar{1}\bar{1}0]$ could be established.

3.5.1.2 Yielding and Flow

The deformation behavior of single-crystal aluminum oxide at elevated temperatures is characterized by a marked upper and lower yield point which depends strongly on the strain rate and the temperature. Figure 3.31 shows stress-strain curves of

136

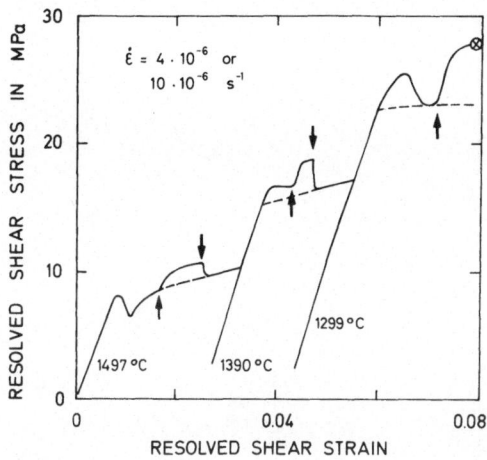

Fig. 3.31. Effect of temperature and strain-rate changes on the tensile flow
stress of sapphire. - From Conrad et al. [3.207]

Al_2O_3 single crystals deformed by basal slip in tension from the work of Conrad et
al. [3.207]. The arrows indicate changes of the strain rate between the two values
given in the figure. The pronounced yield drop of the basal slip system was reported by other authors as well [3.203,3.206,3.229]. It was also found for the prismatic slip system [3.212] and the pyramidal slip system [3.219,3.220]. In accordance
with this observation, an initial increase in the deformation rate (a kind of "work
softening") was found in creep experiments [3.203,3.218]. The phenomenon has been
explained repeatedly as being caused by dislocation multiplication rather than
unpinning of dislocations.

The strong dependence of the upper yield stress of the basal slip system on temperature and strain rate is demonstrated in Fig. 3.32. Figure 3.33 is a compilation
of several deformation studies of single-crystal alumina in tension. It compares
tensile flow stresses to activate the various slip systems as a function of temperature, except for the data of Castaign et al. [3.230] which were obtained in
compression and under hydrostatic constraint. Not considering these data points,
Fig. 3.33 illustrates some important findings. (1) The tensile flow stress of all
slip systems is strongly temperature-dependent. (2) The basal slip system can be
activated most easily, followed by the prismatic and pyramidal slip system. The
stresses to activate the three systems differ vastly. For example, at $1500^{\circ}C$, the
two stresses to activate the prismatic and the pyramidal slip system exceed the
stress to activate the basal slip system by a factor of 8 and 16, respectively.
(3) The minimum temperature of operation for the various slip systems increases
in the sequence basal-prismatic-pyramidal from $900^{\circ}C$ to $1150^{\circ}C$ and $1200^{\circ}C$.

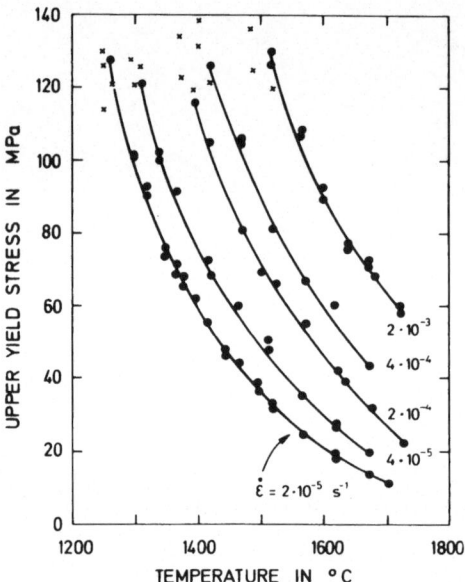

Fig. 3.32. Dependence of the upper yield stress of sapphire on temperature and strain rate. – From Kronberg [3.206]

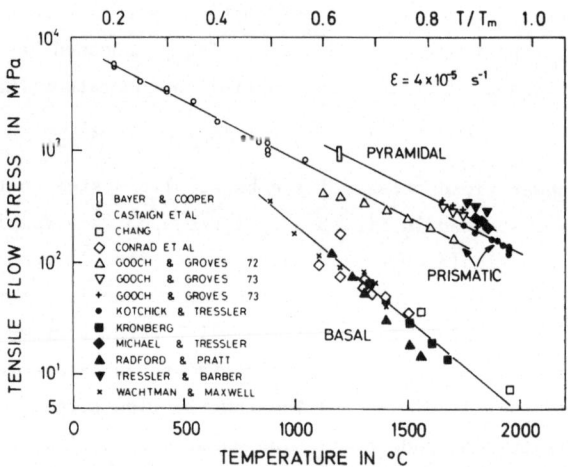

Fig. 3.33. Tensile flow stress to activate various slip systems in single-crystal aluminum oxide. – Data points from Bayer and Cooper [3.214], Castaing et al. [3.230], Chang [3.205], Conrad et al. [3.207], Gooch and Groves [3.209,3.210,3.218], Kotchick and Tressler [3.212], Kronberg [3.206], Michael and Tressler [3.219], Radford and Pratt [3.231], Tressler and Barber [3.220], and Wachtman and Maxwell [3.202]

From these observations it has been concluded that aluminum oxide is a material characterized by a very strong deformation anisotropy [3.196,3.210,3.218], and this property of the single crystal has an important consequence for the deformation behavior of polycrystals. To achieve macroscopic deformation of a polycrystalline body by slip, the Taylor-von Mises criterion has to be fulfilled, which states that five independent slip systems must operate. It has been shown by Snow and Heuer [3.196] that, in aluminum oxide, the simultaneous operation of basal and prismatic slip systems alone does not satisfy this criterion, but that pyramidal slip also needs to be activated to achieve homogeneous deformation. Since the stresses to activate prismatic and pyramidal slip are comparatively high, basal slip will always be initiated first. This will lead to dislocation pile-ups at obstacles in favorably oriented grains and, hence, to deformation-induced local stresses that may result in the formation of cracks. Thus, failure by plasticity-induced cracks is more likely to occur at high temperatures than homogeneous deformation by slip [3.196]. At even higher temperatures or smaller strain rates, macroscopic deformation is more likely to be accomplished by diffusional creep or by sliding processes than by homogeneous slip [3.196,3.210].

Dynamic and steady state flow properties of single crystals at elevated temperatures have been studied by many authors. In particular, the deformation data have been described by the general flow equation.

$$\dot{\varepsilon} = A\sigma^{n}\exp(-Q/RT) \qquad\qquad (3.22)$$

where $\dot{\varepsilon}$ is the deformation rate, A is a constant, σ is the stress, n is the stress exponent, Q is the apparent activation energy of plastic deformation, and R and T have their usual meaning. In Table 3.9, the results of a number of investigations are compiled. Separated into the three slip systems, the table compares values of the activation energy and the stress exponent found and quotes the suggested rate-controlling mechanism. The type of experiment, i.e., constant load or constant strain-rate tests, is also indicated. Both activation energies and stress exponents cover a very wide range of data, so that no unequivocal conclusion as to the deformation mechanism can be drawn. In many cases activation energies were found to depend on stress and to vary with temperature.

A number of mechanisms has been suggested that might control high-temperature deformation of single-crystal alumina. Weertman climb suggested in [3.205] is a creep model in which deformation occurs by dislocation glide, rate-controlled by a climb process [3.234]. It predicts a stress exponent of n=4.5. The Peierls stress model which has been suggested repeatedly [3.207,3.210,3.219,3.232,3.233], assumes

Table 3.9. Deformation parameters and rate-controlling mechanisms in single-crystal aluminum oxide at elevated temperatures

Authors	Ref.	Year	Type of experiment	Activ. energy Q kJ/mol	Stress exponent n	Rate-controlling mechanism
Basal slip system						
Chang	3.205	1960	static[1]	750	4.5	Weertman climb
Kronberg	3.206	1962	dynamic[2]	400		
Conrad et al.	3.207	1965	dynamic	570	4-7	Peierls stress
Bertolotti & Scott	3.232	1971	static	390-540[3]		Peierls stress
Prismatic slip system						
Gooch & Groves	3.210	1973	static	630-920[3]	19±4	
Kotchick & Tressler	3.212	1980	dynamic	210-590[3]	4-20	point defect interaction
Pyramidal slip system						
Heuer et al.	3.215	1971	static	375-490	3	Nabarro climb
Gooch & Groves	3.210	1973	static	590	4.2-4.8	Peierls stress
Gooch & Groves	3.218	1973	static	840-1300[3]	6-7	identification not possible
Michael & Tressler	3.219	1974	dynamic		6-7	Peierls stress
Tressler & Barber	3.220	1974	dynamic	330[3]	8-12	
Tressler & Michael	3.233	1975	dynamic	480-710[3]	6-7	Peierls stress
Firestone & Heuer	3.224	1976	static	330	3	Nabarro climb

Legend: [1]creep tests at constant load, tensile tests at constant strain rate,
[3]dependent on stress

a creep mechanism by dislocation glide, but rate-controlled by the thermally activated overcoming of a Peierls barrier to slip on a difficult slip system. Heuer
and coworkers [3.215,3.224] suggested the Nabarro climb model [3.235] which is a
modified model of diffusional creep where dislocations act as sources and sinks of
vacancies. It predicts n = 3 for lattice and n = 5 for dislocation core diffusion.
Gooch and Groves [3.218] and Kotchick and Tressler [3.212] pointed out that their

results did not conform to either of the available creep models, the latter giving support to an explanation based on the creation of a supersaturation of vacancies generated by the motion of jogs on gliding screw dislocations. However, agreement exists that, if a diffusional model applies, the activation energy of deformation should be that of the self-diffusion of the slower moving ion, which, according to Fig. 2.8, is expected to be the oxygen ion with $Q \approx 640 \pm 100$ kJ/mol.

Closing the discussion of plastic flow behavior of single-crystal alumina it should be noted that, in spite of the numerous papers published on the subject, the experimental deformation parameters are not well known and that much uncertainty still exists as to the mechanism that controls the deformation behavior. This may be due to difficult experimental conditions such as high testing temperatures and the problem of deforming a material of very limited plasticity in tension. Moreover, in the case where a diffusional mechanism controls deformation, very small impurity contents may disturb the development of steady-state conditions by masking the intrinsic defect structure, as has been shown in Section 2.3.3.

3.5.1.3 Hardening and Recovery Phenomena

To improve the mechanical performance of a material at high temperatures, one would like to know in detail which mechanisms dominate the dynamics of deformation. Contrary to metals, few efforts have been made to study work hardening and recovery in ceramics. For sapphire, however, a work-hardening model has been developed which will be described briefly in the following. Subsequently, some remarks on what is known about solution hardening and precipitation hardening in aluminum oxide will be made.

The work-hardening model was presented by Pletka, Heuer and Mitchell [3.236] after detailed investigations of the density and structure of dislocations carried out by the same authors [3.237,3.238]. The study of dislocation structures in single-crystal aluminum oxide which had been deformed by basal slip revealed a number of features which could be closely correlated to the form of the stress-strain curve. This curve was found to exhibit three characteristic regions: a region A immediately beyond the upper and lower yield point, which is characterized by a constant high work-hardening rate, followed by a region B of a gradually decreasing work-hardening rate, and ending in a region C characterized by a plateau flow stress and a zero work-hardening rate [3.237]. The dislocation structures associated with these three regions were as follows. In region A, long dislocation dipoles of edge character prevailed which were thought to be formed by "edge-trapping" of dislocations on parallel basal planes, the screw components being annihilated by cross-slip. The dislocation structure in region B also consisted of edge dipoles, however, greater in density, and, in addition to the previous structure, of a number of small dislocation loops which were thought to be formed through the breakup of dipoles by climb due to the reduction in the dislocation line energy. The dislocation structure in region C was characterized by an increase in density of both long dipoles and small loops, but no new features appeared [3.237,3.238].

Based on these experimental findings, the model of work-hardening and recovery in sapphire [3.236] assumes that work hardening is caused by the increasing number of dipoles in region A, because they serve as obstacles to glide dislocations. Entering region B, processes are increasingly activated which tend to counteract the accumulation of dipoles and to decrease the obstacle density, thereby causing recovery. These processes consist in the breakup of dipoles into loops and the annihilation of these loops by diffusion of point defects. Finally, a steady state of zero work hardening is reached in region C, where work hardening and recovery balance each other, when the rate of annihilation of dipoles by breakup and climb becomes equal to the rate of accumulation by edge-trapping. The model permits one to calculate the form of the stress-strain curve, the work-hardening rate in region A, and the plateau stress in region C which turns out to be proportional to $(\dot{\varepsilon}/D)^{1/3}$, where D is the diffusion coefficient of the slower diffusing ion. The model thus predicts a third-power stress dependence of the stationary deformation rate $\dot{\varepsilon}$. However, looking at Table 3.9, a stress exponent of n=3 has not been found experimentally until now for basal glide. On the other hand, the diffusion coefficients calculated from plateau stress data were found to be in good agreement with experiment, as well as the value of the work-hardening rate. Therefore, the authors concluded that their model gives an acceptable description of work hardening and recovery in sapphire [3.236].

Very little is known about solution hardening and precipitation hardening in sapphire, just as little as in other oxides, so that it has been pointed out previously that knowledge about solution hardening in oxides is still in its infancy [3.239] Only two alloying elements have been investigated in alumina, i.e., Cr and Ti. Both are soluble in Al_2O_3 and substitute Al^{3+} on its regular lattice sites. Chromium is always present as an isovalent ion, but titanium atoms may be incorparated as iso-valent Ti^{3+} or as aliovalent Ti^{4+}, depending on the ambient oxygen partial pressure. Since the solubility of Ti in Al_2O_3 is limited, it may also form precipitates.

A solution hardening effect of Cr^{3+} was reported briefly by Wachtman and Maxwell [3.203], Chang [3.205], and Radford and Pratt [3.231]. A more detailed study on solution hardening of sapphire was carried out by Pletka et al. [3.238] who compared the hardening effect of Cr^{3+}, Ti^{3+}, and Ti^{4+}. All three ions were found to increase the critical resolved shear stress of sapphire at 1400 and $1500^{o}C$. The hardening effect of the two isovalent ions increased with the ionic radius and was thus attributed to elastic interactions between glide dislocations and the symmetric stress field of the solute ions. In comparing the effect of valence of the solute ion on hardening, an opposite result was found, the smaller Ti^{4+} being a more potent hardener than Ti^{3+}. This was attributed to the fact that a charge-compensating defect must be created to maintain charge neutrality, when Ti^{4+} is added to Al_2O_3 (see Section 2.4). The Ti^{4+} ion and its associated defect cause an asymmetric stress field which is believed to interact with glide dislocations even more strongly [3.238].

The precipitation hardening behavior of Ti-doped sapphire was studied by Pletka et al. [3.238] and by Busovne et al. [3.240]. However, only a weak effect was detected in both investigations. In [3.238] it was observed that the strength increase due to precipitation hardening after an appropriate aging treatment did not exceed the increase due to solution hardening alone, and in [3.240] it was found that the doping level had little effect on room temperature hardness.

3.5.2 Twinning

Twinning has been observed frequently in single-crystal and polycrystalline alumina. It has been recognized that twinning is an important mode of plastic deformation at elevated temperatures [3.232,3.241], and that it plays an essential role in crack initiation at room-temperature. Twin-nucleated fracture has been observed in bending [3.95], in compression [3.96], and in tension [3.97].

Twinning can occur on basal and on rhombohedral planes. The crystallographic description of the basal twinning system, i.e. $(0001)<1\bar{1}00>$, has been given by Kronberg [3.204], while the crystallographic indices of the rhombohedral twinning system have been established by Heuer [3.95] to be $\{10\bar{1}1\}<10\bar{1}2>$ (based on the morphological unit cell notation). Rhombohedral twinning was observed in single-crystal Al_2O_3 under compressive loading at elevated temperatures by Stofel and Conrad [3.90], Conrad et al. [3.241], Becher and Palmour [3.242], Bertolotti and Scott [3.232], Achutaramayya and Scott [3.243], and Scott and Orr [3.244], and in bending at temperatures as low as $-196^{o}C$ by Heuer [3.95]. Rhombohedral twinning was also found to occur in alumina polycrystals, namely by Becher [3.245] at elevated temperatures in compression, and by Hockey [3.217] at room temperature during indentation and abrasion. Basal twinning, on the other hand, was also observed both in single crystals and polycrystals. Stofel and Conrad [3.90] and Conrad et al. [3.241] found basal twinning in sapphire during compression tests at elevated temperatures. In polycrystals, basal twinning was observed under compressive loading between 1200 and 1800°C by Becher [3.245] and at room temperature in regions around microhardness indentations by Hockey [3.221]. This compilation shows that twinning is a very important feature in the mechanical behavior of alumina, since it occurs during most types of mechanical loading.

As compared to plastic deformation by dislocation glide, twinning is favored by low temperatures and high strain rates. It has been pointed out that twinning may be the primary mode of plastic deformation at room temperature in alumina single crystals subjected to compressive loading [3.95]. High-strain rate experiments on polycrystalline alumina between room temperature and about 500°C clearly showed twinning to be the predominant deformation mechanism [3.112]. The strain-rate and temperature dependence of the occurrance of twinning was studied in detail by Conrad et al. [3.241]. Figure 3.34, taken from their work, is a strain-rate/temperature map showing three predominant deformation modes: at high temperatures and small strain rates, (0001) slip prevails, while rhombohedral twinning predominates at low temperatures and high strain rates, and the range of (0001) twinning is limited to a very narrow band of experimental parameters.

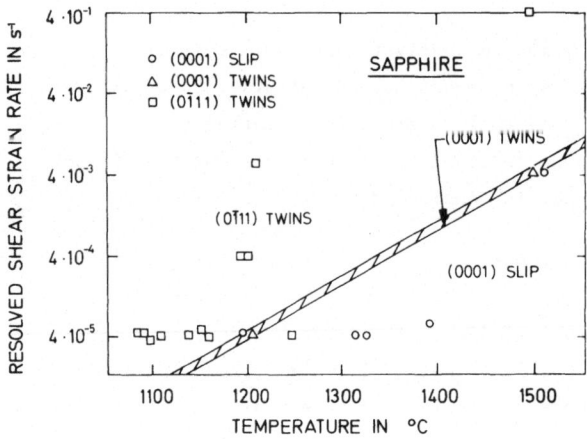

Fig. 3.34. Occurrence of twinning and slip as a function of strain rate and temperature. - From Conrad et al. [3.241]

The importance of rhombohedral twinning for the plastic deformation of sapphire is illustrated by Fig. 3.35, which is from the work of Bertolotti and Scott [3.232].

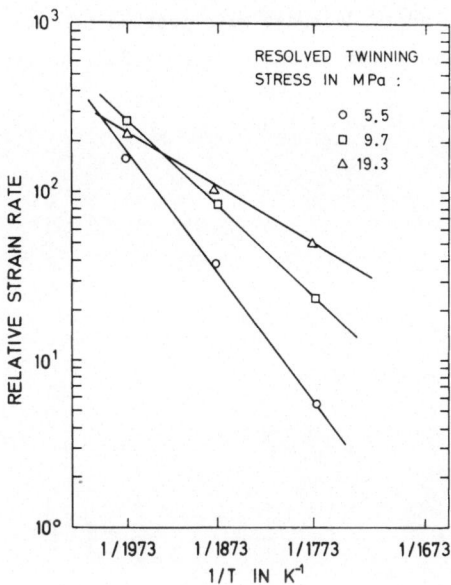

Fig. 3.35. Effect of temperature and resolved twinning stress on rate of creep by rhombohedral twinning. - From Bertolotti and Scott [3.232]

These authors separated the total strain rate obtained in compressive creep into a slip-induced and a twinning-induced fraction. Figure 3.35 is an Arrhenius plot of the twinning-induced strain-rate; it shows that it increases both with temperature and stress, and that rhombohedral twin growth is a thermally activated process, whose activation energy decreases with increasing stress.

An atomistic model of rhombohedral twinning in aluminum oxide was presented by Scott [3.246] which accounts for the faulted twin boundaries and their high interfacial energy observed experimentally [3.243].

3.5.3 Hardness

Aluminum oxide is one of the hardest materials known. Its high hardness promotes a series of applications in mechanical engineering, such as bearings and seals (Section 5.2.4). In order to understand the factors which influence and possibly improve the hardness, a number of studies has been published on the subject. In the following paragraphs, hardness of alumina will be discussed as a function of temperature, composition, and grain size.

The hardness of a material is a somewhat complex and not well defined quantity which reflects properties such as the elastic modulus and the resistance to plastic deformation and cracking. Clearly, hardness is expected to decrease with increasing temperature, and this dependence is also obtained in aluminum oxide. Figure 3.36 shows

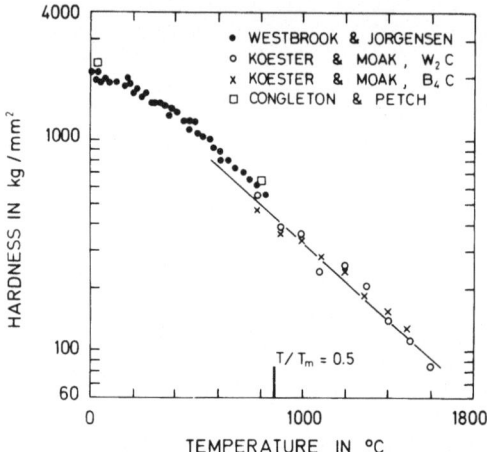

Fig. 3.36. Effect of temperature on microhardness of polycrystalline alumina.
Data points from Congleton and Petch [3.106], Westbrook and Jorgensen [3.247],
and Koester and Moak [3.248]

the variation of hardness of pure, polycrystalline alumina with temperature from various sources [3.106,3.247,3.248]. When the mean logarithmic changes of the hardness and of the tensile flow stress with temperature, $\Delta \ln H / \Delta T$ and $\Delta \ln \sigma / \Delta T$, are calculated from Figs. 3.36 and 3.33, respectively, a value of $9.1 \cdot 10^{-4} K^{-1}$ is obtained for the hardness and of $10.3 \cdot 10^{-4} K^{-1}$ for the flow stress of the prismatic and pyramidal slip system between 600 and $1600^{o}C$. While these two values lie close together, the change in the flow stress of the basal slip system, however, is nearly twice as much, namely $20.0 \cdot 10^{-4} K^{-1}$. These figures lend support to the idea that the hardness of aluminum oxide is controlled by the flow properties of the prismatic and pyramidal slip system families, which must both be activated to attain the five independent slip systems, as has been pointed out Section 3.5.1.2. The operation of the basal slip system alone obviously is not sufficient to accomplish the plastic flow necessary for a permanent hardness impression to be formed.

The effect of alloying Cr_2O_3 on the room temperature hardness of polycrystalline alumina was studied by Bradt [3.249], Belon et al. [3.250], Ghate [3.251], and Shinozaki et al. [3.252]. An increase in hardness was found in all studies, and in some cases a maximum and a subsequent decrease were observed. However, the increment was not very pronounced, reaching between 10 and 20% of the hardness of the unalloyed alumina for chromia contents of 10 to 20 mol%. Softening of alumina by solid solution was found by Kennedy and Bradt [3.253]. Alloying 2 mol% of $MgO \cdot TiO_2$ to hot-pressed alumina resulted in a hardness loss of 8% as compared to the hardness of the pure hot-pressed material. The rate of softening was assessed to be about 5 times the rate of hardening by Cr_2O_3.

The effect of grain size on the hardness of fine-grained aluminum oxide was studied by Skrovanek and Bradt [3.254]. They found a two-branch curve when the hardness was plotted as a function of the inverse square root of the grain size, as is true for metals. At grain sizes from 2 to 5 µm, the hardness was essentially constant, but decreased with further increases in grain size. The behavior was explained by the size relation between the grain size or the subgrain size and the dislocation pile-up length, as in metals.

3.5.4 Abrasive Wear

Plastic deformation properties also control the abrasive wear behavior of brittle ceramics. Abrasive wear is the result of a mechanical machining operation such as sawing, grinding, lapping, and polishing. It may also occur during solid particle indentation, as in sanding. While the aspects of mechanical surface finishing that refer to machining as a shaping process in the production of alumina parts are described in Section 4.4, some fundamental principles of the abrasive wear process in alumina will be discussed in the following paragraphs. A very detailed introduction to phenomena and mechanisms occurring during machining of ceramics is the review article by Rice [3.82].

Grinding a ceramic surface is a complex process, since the workpiece interacts with numerous abrasive particles of the grinding wheel which slide across the surface at high velocity. Thus the grinding operation consists of many single scratching events, each of them contributing to the material removal. The abrasive particles, generally diamond, are small and irregular in shape so that a sharp corner of the particle is usually in contact with the ceramic [3.255]. This small contact area produces a high localized stress concentration, and the particle penetrates the surface. Moving the particle along the surface generates a long plastic groove and a complex crack pattern. The damages caused by different particles usually overlap.

To simplify the complex situation, abrasion phenomena have been studied by means of single indentation experiments. Both indentation by a static indenter and by a moving indenter (scratching experiments) has been employed. According to present understanding, the process of material removel involves both fracture and plastic deformation. Figure 3.37 which is from the work of Evans [3.256] is a schematic of the damage which is believed to be caused by the passage of a pointed indenter or a sharp abrasive particle. A zone of plastically deformed material is produced beneath the plastic groove. Lateral cracks start from the plastic zone and extend parallel with the surface. Material removal is thought to be achieved by the curving of these lateral cracks to the surface, thereby isolating small chips or flakes of material. Additionally, two different sets of flaws perpendicular to the surface are introduced (not shown in Fig. 3.37) which control the strength after grinding [3.257,3.258]: one set is parallel with the grinding grooves (also called median cracks), and the other perpendicular to the grinding grooves (radial cracks).

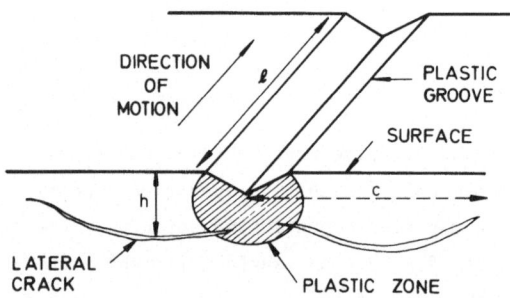

Fig. 3.37. Schematic of the damage caused by the passage of a pointed indenter or an abrasive particle. – From Evans [3.256]

Evidence for plastic deformation involved in the abrasion of single-crystal and polycrystalline alumina has been given by many authors. Using the etch-pit technique, Steijn [3.259] was one of the first to show that basal slip and prismatic slip is activated during abrasive wear of sapphire. Hockey [3.217] found high dislocation densities within a near-surface region in a TEM investigation of mechanically polished single-crystal and polycrystalline Al_2O_3. From X-ray diffraction line broadening observed in the stock removed during abrasion of polycrystalline alumina, Cutter and McPherson [3.200] inferred that heavy plastic deformation must have occurred. In a TEM study of the region about indentations produced by a quasi-static pointed indenter, dislocations and dislocation networks were observed by Hockey [3.221] and by Hockey and Lawn [3.260]. Becher [3.201] reported that grinding of sapphire caused a plastically deformed surface layer and that the basal slip depth was 30 µm. SEM examinations performed by Swain [3.258] of the surface region

148

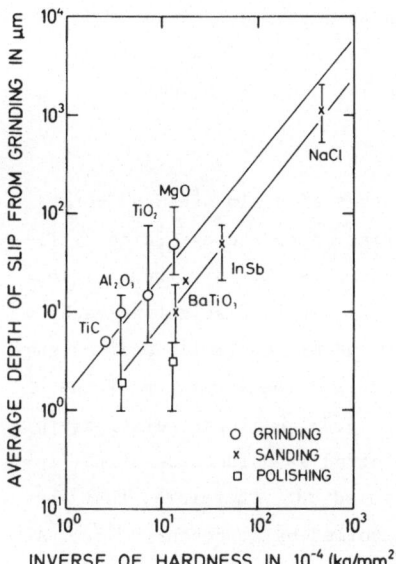

Fig. 3.38. Average slip depth caused by abrasion as a function of hardness for various ceramics. - From Rice [3.82]

adjacent to scratches in polycrystalline alumina revealed that slip bands were present within many grains, and that failure occurred where these slip bands intersected a grain boundary. On studying the wear behavior of alumina cutting tools, Ghate et al. [3.251] concluded that extensive temperature-enhanced plastic flow must have preceded crater wear at the cutting edge. In his review on machining behavior of ceramics, Rice [3.82] demonstrated that several abrasive processes such as grinding, sanding, and polishing, produce plastically deformed surface layers in a number of ceramics, among them alumina, and that the average slip depth correlates well with the inverse of hardness. His results are shown in Fig. 3.38. For alumina, the average slip depth is about 10 μm. Rice argued [3.82] that all mechanical finishing processes, including polishing, may leave a plastically deformed surface, the depth of which decreases with increasing hardness and fineness of operation.

Furthermore, in a detailed study on the machinability of several ceramics, including magnesia, zirconia, and alumina, Rice and Speronello [3.261] found that the machining difficulty which is defined as the inverse of the material removal rate increased linearly with hardness. They pointed out that a theory of the machinability of ceramics must especially include the effect of hardness. In a recent study, Lange et al. [3.262] determined compressive stresses induced by grinding in fine-grained polycrystalline alumina. From the reduction of these stresses due to surface layer removal they inferred that the majority of compressive surface stresses were within a surface layer of a depth ≤ 15 μm and argued that they were induced by the formation of a plastically deformed surface layer during grinding. All these studies demonstrate that plastic flow appears to be a major controlling mechanism in the abrasive wear of alumina.

Rice and Speronello [3.261] also investigated the effect of grain size and porosity on the machining difficulty of alumina. The variation of the machining difficulty with these two parameters was found to be the same as that of strength. That is, the effect of porosity was that given by Eq. (3.12), just substituting machining difficulty for strength. Even the numerical b value of Eq. (3.12) was the same as those of Table 3.5, i.e., b=4. The results obtained for the variation of the grinding difficulty with grain size are shown in Fig. 3.39. A linear relationship of the Petch type was obtained when the data were plotted as a function of the inverse square root of the grain size. The authors pointed out, therefore, that the abrasion behavior is, besides by hardness, also controlled by strength [3.261]. A similar effect of the grain size was found by Dawihl et al. [3.263]. They observed that the material removal rate increased by a factor of 2 when the mean grain size of the alumina was increased from 3 μm to 20 μm.

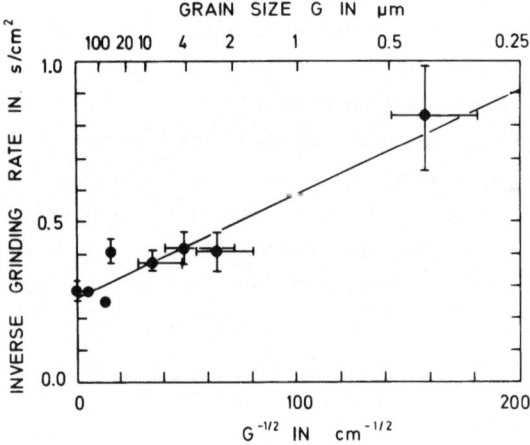

Fig. 3.39. Effect of grain size on grinding rate of alumina. – After Rice and Speronello [3.261]

Some information is also available on the environment-dependent abrasive behavior of alumina. Westwood et al. [3.264] studied the machinability of polycrystalline Al_2O_3 in water and n-alcohols with n varying from 1 to 12. They found that the drilling rate exhibited two maxima for n=5 and n=10, and that these maxima coincided reasonably well with the two maxima of the pendulum hardness found for single crystals and with the occurrance of a zero surface charge in n-alcohols. The drilling rate in pentyl alcohol (n=5) was found to be 6 times that in water. From the correlation between machinability, pendulum hardness, and surface charge it was concluded that the efficiency of drilling decreases when the energy of the drilling tool is expended in part to activate plastic flow instead of being used primarily to initiate fracture. Little is known, however, about the mechanism of how the chemical environment affects the near-surface plastic properties of a ceramic. The findings of Westwood et al. [3.264] agree well with those reported by Gruver and Kirchner [3.265]. Performing scratching experiments in n-alcohol environments on polycrystalline Al_2O_3 samples, these latter authors studied the penetration of subsurface damage and found out that the depth of damage (cracking) was greatest for butyl alcohol (n=4), which was the environment in which the surface charge was nearest zero. Swain et al. [3.266] reported that the material removal rate in n-alcohols was affected by the viscosity of the environment. They explained their results in terms of the environment-sensitive flow and fracture behavior of alumina.

Quantitative determinations of the depth of damage induced in alumina by scratching or grinding have been reported by several authors. The depth of median cracks produced by a sliding diamond pointer in air was measured by Gruver and Kirchner [3.265] as a function of the indenter load. A linear relationship was obtained between depth of damage and normal load for hot-pressed alumina, the median crack depths ranging from 10 to 30 μm. The authors pointed out that the depth of damage was at least an order of magnitude greater than the groove depth. Swain [3.258] derived a relationship based on indentation fracture mechanics between the normal load exerted by a sliding indenter, P_n, and the depth of median cracks, c, of the type

$$P_n / (\tan\psi \cdot \pi^{3/2}) \quad \propto \quad K_c \cdot c^{3/2}, \tag{3.23}$$

where ψ is the half angle of the indenter and K_c is the fracture toughness. On replotting the data of Gruver and Kirchner [3.265] according to Eq. (3.23), as well as other data on glass, Swain observed close agreement between prediction and experiment, as shown in Fig. 3.40. Swain also studied the nature of cracking about scratches in single-crystal and polycrystalline alumina. He observed that the crack pattern produced by a sliding indenter was very similar to that induced by a quasi-static pointed indenter. Lateral cracks were found to start from the region of the plastically deformed zone beneath the scratch, as depicted in Fig. 3.37. Rice [3.267] studied the effect of grinding direction on the strength of single-crystal

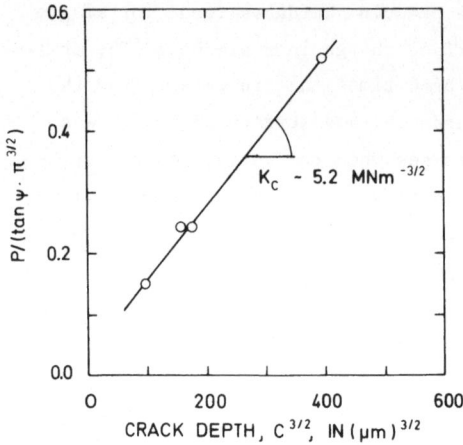

Fig. 3.40. Crack depth of median cracks induced by a sliding indenter in alumina as a function of normal indenter load. – After Swain [3.258]

alumina. Bend strength was found to be smaller for crystals oriented with their grinding direction perpendicular to the sample axis than for those having their grinding direction parallel with the sample axis. From this strength anisotropy it was concluded that the depth of flaws formed parallel with the grinding direction (median cracks) was greater than the depth of perpendicular flaws (radial cracks). This result was also confirmed for a number of other ceramics [3.257]. Some indirect conclusions concerning the effect of microstructure on the depth of damage induced by machining were drawn by Rice and Speronello [3.261] from their machining experiments. They emphasized that, for a given machining operation and a given ceramic material of approximately constant porosity and composition, but varying grain size, the depth of damage resulting from machining should remain nearly constant regardless of grain size.

A semi-quantitative model of abrasive wear of ceramics was presented by Evans [3.256]. In this model, material removal is supposed to occur by subsurface lateral cracking. According to the model, the volume of material that is removed by the passage of an abrasive particle is determined by the average depth h and average length c of lateral cracks that initiate at the plastic zone, as shown in Fig. 3.37. The depth h is assumed to be proportional to the extension of the plastic zone which is related to the plastic contact area between the ceramic and the abrasive particle and, hence, to the hardness H of the material and the normal load P_n. For the length of lateral cracks, a relationship very similar to that of Eq. (3.23) was established, thereby relating c to the fracture toughness of the material. The material removal rate V was shown to be

$$V \propto c \cdot h \propto P_n^{7/6}/(K_c^{2/3}H^{1/2}). \qquad (3.24)$$

Equation (3.24) relates the removal rate to fundamental material parameters and the machining conditions. It predicts that the removal rate increases as the hardness and the fracture toughness decrease. This is plausible since, for a given load and stress field, decreasing fracture toughness means longer lateral cracks, and decreasing hardness means an increase in the plastic zone size, both effects resulting in an increase in the removed material volume.

In Fig. 3.41 which is taken from [3.256], the inverse of the material removal rate of several ceramics is plotted against the parameter $K_c^{2/3}H^{1/2}$ according to Eq. (3.24). The experimental data were taken from the work of Rice and Speronello [3.261] who found a linear relationship between the inverse removal rate and hard-

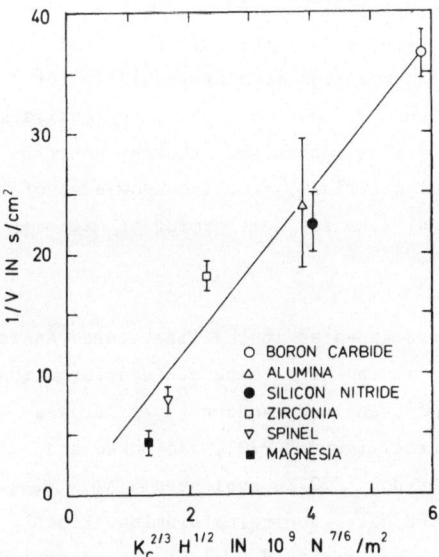

Fig. 3.41. Abrasive wear rate as a function of the material-dependent parameter $K_c^{2/3}H^{1/2}$. - From Evans [3.256]

ness. However, as shown by Fig. 3.41, Eq. (3.24) which predicts a variation of the inverse removal rate with the square root of hardness, is obeyed quite well. Thus, it was concluded [3.256] that reasonably good correlation exists of abrasive wear data and the prediction of the lateral cracking model.

On summarizing the abrasive wear behavior of alumina it may be stated that a number of experimental data are available which show that the material behaves very similar to other structural ceramics when subjected to abrasive wear. The material removal rate is found to increase with increasing grain size and porosity and with decreasing fracture toughness and hardness. It also depends strongly on the environment. Experimental evidence is given that material removal involves plastic flow and cracking. The material behavior can be understood at least qualitatively in terms of the lateral cracking mechanism which takes account for the occurrance of plastic deformation and fracture.

3.6 Creep

In a technical sense, creep means the slow and continuing deformation process which a material generally undergoes at elevated temperatures (normally above half the homologous temperature) and at low stresses. Creep phenomena are especially important for structural materials in engineering applications, since small deformations may affect the dimensional stability and structural integrity of service parts. Thus, knowledge of the creep behavior of a material is a prerequisite for safe design of high temperature service conditions. On the other hand, the analysis of creep data enables one to establish the underlying mechanisms, thereby contributing to the development of more creep-resistant materials. Finally, knowledge of creep behavior may help to find appropriate conditions for hot-forming of materials.

A series of reviews on the creep of ceramics have appeared in the literature. Apart from textbooks [3.1,3.4], the reader interested in the high temperature deformation behavior of ceramics may consult the articles of Evans and Langdon [3.2], Groves [3.268], Langdon et al. [3.269], Radford and Terwilliger [3.270], and Dokko and Pask [3.271]. Very recently, W.R. Cannon and Langdon [3.272] published a very detailed and comprehensive summary of ceramic creep data, including alumina. Finally, the work of R.M. Cannon and Coble [3.273] should be mentioned since it is concerned specifically with diffusional creep of alumina. In the following sections, we shall describe at first the basic creep mechanisms encountered in alumina. After that, creep data of pure and MgO-doped alumina will be reviewed, and then the effect of other dopants on the creep behavior of alumina will be discussed. Finally, high temperature failure mechanisms will be dealt with.

3.6.1 Basic Creep Mechanisms

On discussing the phenomena of slip in Section 3.5.1, it has been shown that gross plastic deformation of aluminum oxide can hardly be achieved by dislocation motion. This is due to the high stresses necessary to activate slip and to the pronounced slip anisotropy of the available slip systems. Thus, the more likely result of a high stress applied to an alumina polycrystal is the initiation of fracture rather than the activation of slip. Dislocation creep, if at all, will be reserved only for the regime of very high temperatures. However, at high temperatures and low strain rates alternative deformation mechanisms have been observed in many ceramics, as for example various kinds of diffusional creep and grain boundary sliding processes. These mechanisms will be discussed briefly in the following.

3.6.1.1 Creep Mechanisms at Small and Intermediate Stresses

It has been widely accepted that diffusional creep mechanisms are important deformation processes in those materials, and under such environmental conditions, where slip is difficult or impossible. This also holds for aluminum oxide at stresses which are too low to activate slip on five independent slip systems and at temperatures above about $1100^{o}C$ $(0.6\ T/T_m)$.

Diffusional creep models were developed more than 20 years ago. They mainly rest upon the idea that the deformation of a polycrystal is accomplished by a diffusional mass transport within the individual grains or along grain boundaries. The grain boundaries are assumed to act as perfect sources and sinks of point defects. The applied stress causes small deviations of the point defect concentration from the equilibrium concentration, the vacancy concentration at grain boundaries under a tensile stress being slightly increased and at grain boundaries under a compressive stress slightly decreased compared to the equilibrium value. The resulting point defect concentration gradient gives rise to a diffusional matter transport which results in an extension in the direction of the external tensile stress.

For mass transport by lattice diffusion, a steady-state creep equation has been presented by Nabarro [3.274] and by Herring [3.275], whereas the case of mass transport along grain boundaries has been treated by Coble [3.276]. These two kinds of diffusional creep are called briefly Nabarro-Herring and Coble creep, respectively. In an ionic compound like Al_2O_3, both the cations and the anions have to participate in the diffusional mass transport process in the stoichiometric ratio, but may diffuse along different paths, i.e., through the lattice or along grain boundaries. It has been shown by various authors [3.2,3.277,3.278] that in this case the total deformation can be composed by the sum of Nabarro-Herring type and Coble type creep, and that the creep rate $\dot{\varepsilon}$ is given by the relation

$$\dot{\varepsilon} = \frac{44}{\pi} \frac{\Omega_c \sigma D_{complex}}{kT(GS)^2} \tag{3.25}$$

where Ω_c is the volume of the Al_2O_3 molecule, σ is the applied stress, $D_{complex}$ is the effective diffusion coefficient of ambipolar diffusion given by Eq. (2.15), GS is the grain size and kT has its usual meaning. As can be seen from Eq. (2.15), $D_{complex}$ is composed of both the lattice and the grain boundary diffusivities of the two diffusing species, the aluminum ion and the oxygen ion.

It was already shown in Section 2.3.4 that in the case of Al_2O_3 the relationship for $D_{complex}$ can be simplified significantly. Since the grain-boundary diffusion of oxygen in aluminum oxide is very rapid compared to both the grain-boundary and the lattice diffusion of Al, Eq. (2.15) reduces to Eq. (2.16), which is

$$D_{complex} = (D_{Al}^{\ell} + \pi \delta_{Al} D_{Al}^{b}/GS)/\alpha \tag{3.26}$$

where the symbols have been defined in Section 2.3.4. Inserting Eq. (3.26) into Eq. (3.27) gives

$$\dot{\varepsilon} = \frac{44}{\pi} \frac{(\Omega_c/\alpha)\sigma D_{Al}^{\ell}}{kT(GS)^2} + \frac{44(\Omega_c/\alpha)\sigma \delta_{Al} D_{Al}^{b}}{kT(GS)^3} \tag{3.27}$$

Equation (3.27) is the diffusional creep rate relation modified for cation diffusion control in Al_2O_3. The first term reflects Nabarro-Herring type creep, whereas the second term describes the contribution of Coble type creep to the total deformation rate, both mechanisms operating in parallel, i.e., simultaneously and independent of each other. Furthermore, it should be noted that Eq. (3.27) predicts a viscous deformation behavior characterized by a linear dependence of strain rate on stress, and a strong grain size dependence of the deformation rate. For creep dominated by lattice diffusion a grain size exponent m equal to 2 is expected and for grain-boundary diffusion control it should be 3. For cases where both Nabarro-Herring and Coble creep contribute to the total deformation m should lie between these extremes.

156

3.6.1.2 Creep Mechanisms at High Stresses

When the stress is increased to a high level, or at large grain sizes, a transition from viscous creep behavior (stress exponent n=1) to non-viscous creep behavior (n>1) is observed experimentally. For aluminum oxide in the high-stress regime it has been proved repeatedly [3.272,3.279] that stress exponents are close to or slightly smaller than 3. There are several creep mechanisms known in the literature which predict a third-power stress dependence of the creep rate and which have been proposed to account for the experimental behavior of aluminum oxide.

A mechanism often invoked to explain non-viscous creep behavior of alumina is Nabarro climb. According to this mechanism proposed originally by Nabarro [3.235], the creep strain results from climb of edge dislocations without conservative dislocation glide being necessary to cause deformation. In other words, Nabarro climb is a modified diffusional creep process where point defects are created and annihilated at dislocations instead of grain boundaries. The creep rate of the Nabarro climb mechanisms was calculated by Weertman [3.280] as

$$\dot{\varepsilon} = \frac{\pi \zeta^2 \Omega_c \sigma^3 D_{complex}}{10 b^2 G^2 kT} .$$

(3.28)

where ζ is a constant of the order of 1, b is the length of the Burgers vector, G is the shear modulus, and $D_{complex}$ is the diffusion coefficient of ambipolar diffusion from Eq. (2.15). Excluding contributions from grain-boundary diffusion to intragranular climb processes and taking $D_0^\ell \ll D_{Al}^\ell$, Eq. (2.15) becomes $D_{complex} = D_0^\ell/\beta$, where β is a stoichiometric factor (β=3 for Al_2O_3). Thus, Eq. (3.28) predicts that (1) creep due to Nabarro climb is a non-viscous process having a third-power stress dependence, (2) the creep rate is controlled by the slow oxygen lattice diffusion, and (3) the creep rate is independent of grain size.

Aside from diffusional deformation processes, slip mechanisms have also been proposed to account for non-viscous deformation behavior of ceramics at high stresses and elevated temperatures. There are several dislocation creep processes with a third-power stress dependence known in the literature [3.272] which have been quoted to explain experimental creep data. Weertman [3.281] has proposed a deformation mechanism where the glide velocity of dislocations is the rate-controlling step. It has been argued [3.279] that the n=3 behavior of a series of nonmetals, including alumina, is due to the interaction of charged dislocations with other charged point defects during gliding along the glide plane, just in the sense of Weertman's microcreep model.

Another creep model with n=3 is the jog-dragging screw dislocation mechanism by Barrett and Nix [3.282]. In this model the strain is achieved by dislocation glide, while the deformation rate is controlled by the non-conservative motion of jogs in the screw dislocations. The creep rate is given by

$$\dot{\varepsilon} = A\rho_s \sigma D^{\ell}/kT \qquad (3.29)$$

where ρ_s is the density of mobile screw dislocations, and D^{ℓ} is the lattice diffusivity of the slower moving species. Following [3.272], use of the Taylor relation between stress and the density of mobile screw dislocations, $\sigma = \alpha_o Gb\rho_s^{1/2}$, gives

$$\dot{\varepsilon} = B\sigma^3 D^{\ell}/kT \qquad (3.30)$$

where A, B, and α_o are constants.

Finally, we would like to mention the dislocation creep model of Evans and Knowles [3.283]. In this model, the glide of dislocation loops yields most of the strain, whereas control of the process is due to the climb of dislocation links in a three-dimensional dislocation network. Climb results in a re-arrangement of the dislocation network by releasing individual loops which then can glide to a new pinning position. At high temperatures, where dislocation pipe diffusion can be neglected with respect to lattice diffusion, the creep rate is given by

$$\dot{\varepsilon} = f \frac{b\sigma^3 D^{\ell}}{G^2 kT} \qquad (3.31)$$

where f is a numerical factor.

The creep mechanisms so far discussed for high stresses have the common characteristic that the creep rate is independent of the grain size and varies with the applied stress raised to the third power. Since diffusional creep and dislocation creep are independent processes, it is expected that the latter mechanisms become increasingly important and gradually predominate over the former at increasing grain sizes and increasing stress levels, as in fact is observed experimentally.

However, it has been pointed out [3.284] that the precise nature of the dislocation process that controls creep at high stresses in alumina is not yet known. In any case, it seems to be certain that the most frequently quoted creep model by Weertman [3.234] in which the creep strain is achieved by dislocation glide, while the deformation rate is controlled by dislocation climb, does not apply to the creep behavior of polycrystalline alumina, since it predicts a stress exponent of n=4.5.

3.6.1.3 Creep Mechanisms at Low Stresses and Small Grain Sizes

It has been observed experimentally [3.104,3.285,3.286] that, at very low stresses and in very fine-grained materials, the creep rate of alumina deviates from that predicted by the diffusional creep model, Eq. (3.27), in that it becomes non-viscous and the grain-size exponent tends to unity. Grain-size exponents of unity indicate the operation of some type of interfacial process, which is rate-limiting, because the total interfacial area varies with the inverse of grain size.

Diffusional deformation of the Nabarro-Herring type and Coble type is based on the assumptions that point defects can be created and annihilated easily at grain boundaries and that grain-boundary sliding necessary to satisfy accommodation at triple points occurs easily. If one of these assumptions does not apply, diffusional creep becomes rate-limited by an interface reaction, as has been proposed by Ashby [3.287]. Such reaction may be the creation and/or annihilation of point defects or the sliding of grain boundaries. Several mechanisms have been considered which possibly may account for the observed creep behavior at very low stresses. Three of them will now be discussed briefly.

Interface control may be due to grain-boundary sliding. To maintain coherency across the grain boundaries, grain-boundary sliding must be accommodated by small plastic deformations at points of constraint, such as triple points and grain boundary ledges. Gifkins [3.288] has proposed a viscous grain boundary sliding mechanism in which accommodation occurs by stress-induced diffusional deformation. The creep rate calculated by Gifkins is

$$\dot{\varepsilon} = C_o \frac{\Omega_c \sigma \delta D^b}{kTGS} \qquad (3.32)$$

where C_o is a constant. The model predicts a viscous deformation mode and a grain-size exponent of unity.

Ikuma and Gordon [3.289] have shown that interface control of diffusional creep can be accounted for by the incorporation of an interfacial rate constant in the creep rate equation. This is done by substituting an effective interface-corrected diffusion coefficient D_{IF} for the ambipolar diffusion coefficient $D_{complex}$ in Eq. (3.25), which is given by

$$D_{IF} = \frac{D_{A1}^\ell K_{A1} GS/\alpha}{44 D_{A1}^\ell /\pi + K_{A1} GS} + \pi \delta_{A1} D_{A1}^b /(\alpha GS) \qquad (3.33)$$

where K_{A1} is the interfacial rate constant for aluminum point defect creation or annihilation at grain boundaries. In deriving Eq. (3.33) it was assumed that interfacial defect reactions and Al lattice diffusion are dependent processes which act only in series, and that grain-boundary diffusion occurs without the necessity of interfacial defect creation, thus acting in parallel. For a rapid defect reaction at boundaries, i.e. for $K_{A1} GS \gg 44 D_{A1}^\ell /\pi$, Eq. (3.33) reduces to Eq. (3.26), and the creep rate is still given by Eq. (3.27). However, if point defect creation and annihilation is the rate-limiting step, i.e., for $K_{A1} GS \ll 44 D_{A1}^\ell /\pi$, and if Al grain-boundary diffusion is slow, Eq. (3.33) reduces to $D_{IF} = (\pi/44\alpha) K_{A1} GS$, and Eq. (3.25) becomes

$$\dot\varepsilon = \frac{\Omega_c \sigma K_{A1}}{\alpha k TGS} \cdot \qquad (3.34)$$

It is noted that Eq. (3.34) predicts the same stress and grain-size dependence of the creep rate as Eq. (3.32), i.e., a viscous flow mode and a grain-size exponent $m=1$.

The last model introduced here to describe the creep rate in the stress regime of interface-controlled diffusional deformation is that by Ashby and Verrall [3.290]. In this model, creation and annihilation of point defects at grain boundaries are associated with the glide and climb processes of grain-boundary dislocations. Considering that the grain boundary dislocation density depends on the stress, the relationship

$$\dot\varepsilon = M\sigma^2 /GS \qquad (3.35)$$

was derived [3.290], where M is the mobility of the grain-boundary dislocations. The Ashby-Verrall model also gives a grain-size exponent of unity, but is the only one to predict a non-viscous deformation in the regime of very low stresses.

160

3.6.2 Creep of Pure and MgO-Doped Alumina

The creep behavior of polycrystalline alumina has been studied extensively. In the
review work of Cannon and Langdon [3.272], a total of 43 papers on the subject is
cited. It is beyond the scope of the book to present all of these studies in detail
However, the basic effects of temperature, porosity, grain size, and stress on the
creep behavior will be discussed, as well as the creep mechanisms in a range of
stresses and grain sizes. Since MgO is the most frequent sintering additive in
Al_2O_3, the discussion of MgO-doped alumina is separated from that of alumina containing other dopants.

Figure 3.42 is an example of a creep curve for polycrystalline alumina deforming
by diffusional creep at high temperature. It is taken from the work of Warshaw and
Norton [3.291], one of the first creep studies reported for this material, and

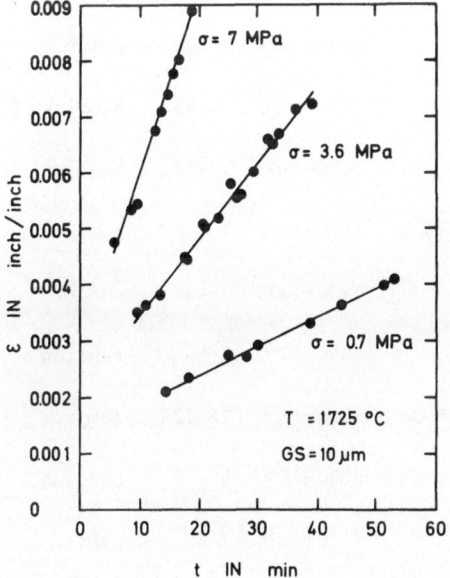

Fig. 3.42. Creep curve for polycrystalline alumina. - From Warshaw and Norton [3.291]

shows some features characteristic of the occurrence of diffusional creep: (1) the
deformation increases linearly with time; (2) there appears to be no transient
creep regime; and (3), when the slope of the creep curves is plotted against the
applied stress, a linear relationship between creep rate and stress is obtained.
It should also be noted that the material has high creep resistance even at a homologous temperature of T/T_m = 0.86.

Table 3.10. Creep of pure and MgO-doped polycrystalline alumina

Authors	Ref.	Material			Experimental Parameters		
	Year	Grain size μm	Density %	MgO content ppm	Tempe- rature °C	Stress σ MPa	Test method
Coble and Kingery	3.8 1956	23	50-95	-	1275	2-28	torsion
Chang	3.292 1959	25-30	95-97	-	1510-1570	-	tension
Folweiler	3.293 1961	7-34	>99	yes	1500-1800	0.7-180	bending
Warshaw and Norton	3.291 1962	3-100	97-100	yes	1600-1800	0.7-14	bending
Dawihl and Klingler	3.294 1965	7	99.5	2500	1150-1250	74-294	compr.
Passmore and Vasilos	3.295 1966	2	99.5	-	1357-1497	7-62	bending
Sugita and Pask	3.296 1970	3-7	>99	2300	1300-1470	7-103	compr.
Heuer, Cannon and Tighe	3.104 1970	1.2-11	>99	2500	1100-1700	7-190	bending
Engelhardt and Thümmler	3.297 1970	11-14	95-98	1000	1480-1700	10-74	bending
Mocellin and Kingery	3.298 1971	30-66	>99	260-1300	1580-1800	7	compr.
Crosby and Evans	3.299 1973	15-45	96-99	-	1450-1800	7-46	compr.
Davies	3.300 1975	15-30	99.5	1000	1450-1750	5-50	tension
Cannon and Sherby	3.301 1977	14-65	99	100-5000	1600-1700	28-124	compr.
Lessing and Gordon	3.302 1977	9-72 76-306	>98.5	-	1350-1550	5	bending
Hou et al.	3.303 1979	1.8-100	>99.5	20	1450-1525	7-50	compr.
Cannon, Rhodes and Heuer	3.285 1980	1.2-15	>99	2500	1192-1750	2-200	bending
Porter et al.	3.304 1981	2	>99	2500	1273-1479	6-220	bending
El-Aiat et al.	3.305 1981	1-40	>99.3	65	1450-1500	5-50	compr.
Carry and Mocellin	3.306 1983	0.6	99.7	500 plus 500 Y_2O_3	1500	25-30	compr.

Table 3.10. (continued)

Results			Creep Mechanism
Activation energy kJ/mol	Stress exponent n	GS exponent m	
–	1	–	GBS
840	–	–	DFC-D$^{\ell}$
545	1	2	DFC-D$^{\ell}$
545	1	2	DFC-D$^{\ell}$ at GS=3–13μm
774	4	–	DLC at GS=50 and 100 μm
502	1	–	DFC-D$^{\ell}$
597	1	–	DFC-D$^{\ell}_{A1}$ for σ<14 MPa
	2	–	other for σ>14 MPa
–	1.1–1.3	0	Localized plastic deformation
485–569	1.1–1.3	2.5	DFC-δ$_{A1}$D$^{b}_{A1}$ at GS=4.3–11 μm
	1.5–1.7	2.5	GBS at GS=1.2 μm
586	1	–	DFC-D$^{\ell}_{A1}$ for σ<30 MPa
	4	–	DLC for σ>60 MPa
523	–	–	DFC-δ$_{0}$D$^{b}_{0}$ at GS>32 μm
410–625	1.3	2	GBS controlled by diffusion
	2.5	2.7	Localized propagation of microcracks
638	1	2	DFC-D$^{\ell}_{A1}$ for σ<20 MPa
	2	–	GBS accomodated by disl.motion for σ>20 MPa
595	1.2	2	DFC-D$^{\ell}_{A1}$ at GS=14–30 μm
611	2.6	–	DLC at GS=65 μm
251–419	1.3	1.8	DFC-D$^{\ell}_{A1}$
	1.8–2.9	0	DLC for T>1550°C and GS>70 μm
544–754	1.2–1.3	2	DFC-D$^{\ell}_{A1}$
	2.4–2.7	0	DLC for GS>30 μm
419	1.1–2.0	–	IF controlled DFC at GS=1.2 μm
587	1.1–1.3	–	DFC-D$^{\ell}_{A1}$ for GS=15 μm
460	1.8	–	IF control or GBS
543–844	1.2–1.3	2	DFC-D$^{\ell}_{A1}$
–	1.1–1.8	–	GBS controlled by grain growth, DFC excluded

Abbreviations: GS grain size; GBS grain-boundary sliding; DFC-D$^{\ell}_{i}$ diffusional creep controlled by lattice diffusion; DFC-D$^{b}_{i}$ diffusional creep controlled by grain-boundary diffusion, the rate-controlling ion is indicated as a subscript; DLC dislocation creep; IF interface.

Table 3.10 summarizes the results of creep studies on pure and MgO-doped polycrystalline alumina. Further studies not contained in the table were reported by
Coble and Guerard [3.307], Fryer and Roberts [3.308], and Hewson and Kingery
[3.309]. The investigations cover a wide range of grain size, stress, and temperature. Tests have been performed under various types of loading, i.e., tension,
compression, torsion, and bending. Stress exponents found at intermediate stresses
are equal to unity or are slightly greater than one, while grain size exponents obtained experimentally are equal to two or somewhat greater. Most authors explain
these findings by the occurrence of diffusional creep, which can be of the Nabarro-
Herring type or of the Coble type, or can be a combination of both, as indicated
in the table.

Direct evidence of diffusion-controlled high-temperature deformation can be obtained
by verifying the prediction of the diffusional creep model for alumina, Eq.
(3.27), quantitatively. This is usually done by calculating the diffusion coefficient from the measured creep rate and comparing it with diffusivity data obtained
by other methods. However, assumptions must be made as to the predominant diffusion
path (lattice or boundary), i.e., wether the first or the second term of Eq. (3.27)
is rate-controlling. As was discussed in Section 2.3.5.1, Cannon and Coble [3.273]
compared diffusion coefficients obtained in this way from ten different creep
studies on aluminum oxide. Assuming lattice diffusion control, the comparison
revealed that the diffusion data agreed within a factor of two with the relationship

$$D_{Al}^{\ell} = 1.36 \cdot 10^5 \exp(-Q^{\ell}/RT) \ cm^2/s \qquad (3.36)$$

with $Q^{\ell} = 577$ kJ/mol, for materials of intermediate grain size and temperatures
above about 1400°C. Equation (3.36) was obtained by Cannon et al. [3.285] by
analyzing their own creep data. It is plotted in Fig. 2.8 (curve 12) together
with other data of the review of Cannon and Coble (curves 6-11). The figure
demonstrates that diffusion coefficients from different creep studies coincide
well with each other and that Eq. (3.36) also agrees reasonably well with cation
lattice diffusivities obtained from tracer diffusion experiments (curve 5) and
from the analysis of sintering experiments (curves 14 and 15). Thus, it has been
concluded that diffusional creep of alumina at not too small a grain size and at
high temperatures is controlled by cation lattice diffusion [3.273, 3.285]. As already mentioned in Section 2.3.5.1, the good coincidence of the diffusivities obtained from creep experiments on materials containing different MgO and impurity
concentrations was explained by the fact that in all studies both the MgO content
was above the solubility limit (only 300 ppm at 1630°C) and the impurity level

even of nominally pure samples was so high that the point defect concentration was not intrinsic, but composition-controlled.

For very small grain sizes and temperatures below about 1400°C, good coincidence of the diffusion data of several creep investigations was obtained by Cannon and Coble [3.273] when cation boundary diffusion control was assumed. In this range of parameters, data agreed well with the relationship

$$\delta_{Al} D_{Al}^b = 8.60 \cdot 10^{-4} \exp(-Q^b/RT) \ \text{cm}^3/s \tag{3.37}$$

with Q^b = 419 kJ/mol. Equation (3.37) was obtained by Cannon et al. [3.285] from the analysis of their creep data on very fine-grained (1-2μm) alumina. It is shown in Fig. 2.9 as curve 12, together with the other diffusivity data from Ref. [3.273], (curves 6-11 and 13). From this evaluation it has been concluded that creep of very fine-grained alumina is a boundary-controlled process.

The determination of the experimental activation energy of the rate-controlling process may help to elucidate the deformation mechanism. Figure 3.43, which is taken from the work of Heuer et al. [3.104], is an example of the determination of

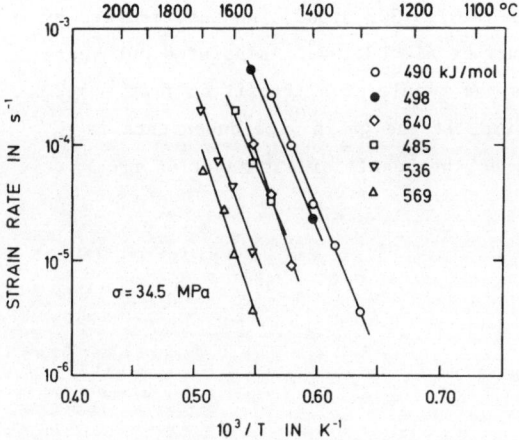

Fig. 3.43. Strain rate as a function of temperature for various grain sizes. From Heuer et al. [3.104]

activation energies from the slope of strain rate data in an Arrhenius plot. The spread of the slopes between 485 and 640 kJ/mol demonstrates, however, that activation energies cannot be determined very precisely. Furthermore, the mean activation energy of this very fine-grained material is much closer to Q^ℓ of Eq. (3.36) than to Q^b of Eq. (3.37), giving support to a deformation mechanism based on lattice diffusional control, in contrast to the conclusions drawn above for fine-grained alumina. Table 3.10 contains more activation energy data. Most of them, about 80 %, fall within a narrow range above and below Q^{ℓ} = 577 kJ/mol, while only a few data of those obtained from experiments on very fine-grained materials, [3.285,3.304,3.308] show a tendency to approach Q^{b} = 419 kJ/mol.

The effect of porosity P on the creep rate was studied by Coble and Kingery [3.8]. They observed an increase in the creep rate by nearly two orders of magnitude when the porosity was increased from a few per cent to 50 %. Spriggs and Vasilos [3.310] showed that the porosity dependence of these data could be linearized when plotted against $1/(1-P^{2/3})$, and suggested the incorporation of this term into the Nabarro-Herring creep equation to account for the porosity dependence of the creep rate.

The grain size exponent m was measured by several authors, as indicated in Table 3.10. Folweiler [3.293] determined m from the slope of straight lines in a double-logarithmic plot of stress vs. grain size at constant deformation rate and temperature. This plot is shown in Fig. 3.44. It demonstrates that, in the grain size range studied (7-34µm), the data lie very well on lines corresponding to a slope of m = 2, which is indicative of lattice diffusion control. Some other authors also reported m = 2 [3.291,3.299-3.101,3.303,3.305]. An intermediate grain size exponent, m = 2.5, was found by Heuer et al. [3.104]. As pointed out there, an intermediate value can be expected as the result of a transition from m = 2 to m = 3 behavior at decreasing grain size, if the grain size dependence is dominated by diffusional creep only, or as the result of a transition from a

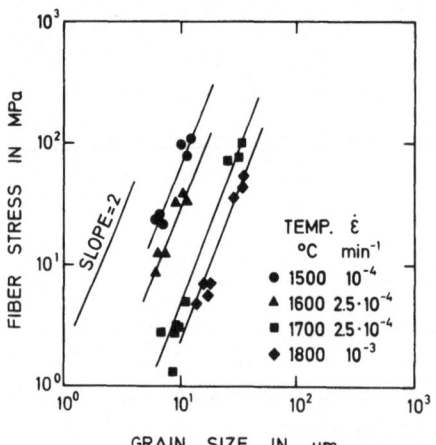

Fig. 3.44. Stress/grain-size relationship for diffusional creep. - From Folweiler [3.293]

boundary-controlled viscous diffusion mechanism (m = 3) to a non-viscous grain boundary sliding process (m = 1). Support was given to the latter suggestion [3.104].

From the foregoing discussion of diffusivities, acitvation energies, and grain size exponents, it seems to be a reasonable assumption that deformation of alumina at temperatures above 1400°C and a grain size above some microns is a lattice diffusional process. Thus, neglecting the boundary-diffusion term of Eq. (3.27), an expression for a temperature and grain size-compensated creep rate $\dot{y}$ can be formed which is given by

$$\dot{y} = \dot{\epsilon} kT(GS)^2 / (\Omega_c D^\ell_{Al} G) \tag{3.38}$$

where G is the shear modulus. According to Eq. (3.27), $\dot{y}$ should be equal to $(44/\alpha\pi)(\sigma/G)$ where σ/G is the relative stress. A plot of $\log\dot{y}$ vs. $\log(\sigma/G)$ can normalize creep data obtained on materials of different grain sizes and at different temperatures and should yield a sole "master curve", provided that there is no further effect of porosity and impurities. The slope of this curve is expected to be one for deformation dominated by viscous processes.

This plot is shown in Fig. 3.45. To generate the data points, D^ℓ_{Al} was taken from Eq. (3.36), G was taken from Fig. 3.1 and [3.12], and for Ω_c a value of $4.248 \cdot 10^{-29}$ m^3 was calculated from the lattice constants of Table 2.1. Figure

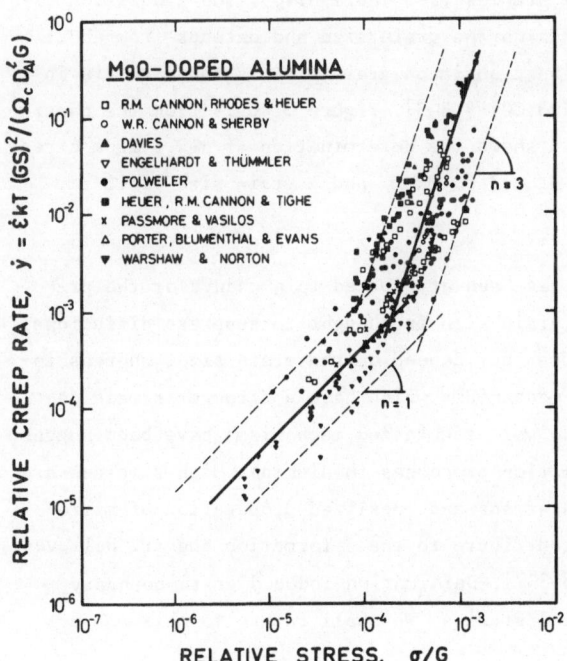

Fig. 3.45. Normalized creep rate vs. relative stress for MgO-doped alumina. Data points are from R.M. Cannon et al. [3.285], W.R. Cannon and Sherby [3.301], Davies [3.300], Engelhardt and Thümmler [3.297], Folweiler [3.293], Heuer et al. [3.104], Passmore and Vasilos [3.295], Porter et al. [3.304], and Warshaw and Norton [3.291]

3.45 contains 38 data sets from nine references covering a temperature range
from 1300 to 1800°C and a grain size range from 1.2 to 100 μm. Straight lines
and scatter bands of slope 1 and 3 were drawn tentatively. Some observations
can be noted from the figure: (1) The spread of the data is considerable, reaching
and order of magnitude along the stress axis. This spread must be due to composi-
tional and microstructural variations and to different impurity contents of different materials. Porosity as the origin of the spread can be excluded since seven of the nine references used materials of a density > 99%, and the largest porosity which occurred otherwise (5% in [3.297]) gives only a 16% increase in $\dot{\varepsilon}$
according to Spriggs and Vasilos porosity correction [3.310]. (2) A slope of
n = 1 is obeyed at low stresses. In a transition region of the relative stress
between about 10^{-4} and nearly 10^{-3} (corresponding to about 15–80 MN/m^2), the
slope increases to about 3, suggesting a change in the creep mechanism. (3) For
engineering design purposes, the scatter of the creep data is too large, per-
mitting only a rough estimate of the creep behavior of MgO-doped alumina in an
engineering application. Thus it still turns out to be an indispensable prerequisite to determine the creep properties of a specific material before putting
it into service when high-temperature requirements are to be met.

Figure 3.45 shows that the stress exponent at high relative stresses is about 3.
The increase to values greater than unity as the stress and/or the grain size
are increased was observed in several studies (see Table 3.10). The transition
stress is not well defined, but depends on the grain size and extends from about
20 to 60 MN/m^2 [3.297,3.299,3.300]. The transition grain size is found to lie in
the range between 30 and 70 μm [3.291,3.301-3.303]. Figure 3.46 is from the paper
by Engelhardt and Thümmler [3.297] and shows the determination of n≈2.5 from stress
change experiments at a stress level around 50 MN/m^2 and a grain size of 13 μm.

The deviation of n values from unity has been attributed to a change of the predominant creep mechanism. Increasing grain size is thought to suppress diffusional
creep in favor of a mechanism which does not depend on the grain size, whereas increasing stress should promote those mechanisms which have a stronger stress dependence than viscous deformation. Two well recognized mechanisms have been suggested to account for non-viscous deformation processes in alumina. High stresses are
thought to promote intergranular separations and localized propagation of microcracks. Such cracking phenomena can contribute to the deformation and are believed
to cause high n values [3.296,3.299,3.307]. Deformation-induced grain-boundary
separations were observed in many creep studies. We shall return to this subject
in Section 3.6.4.

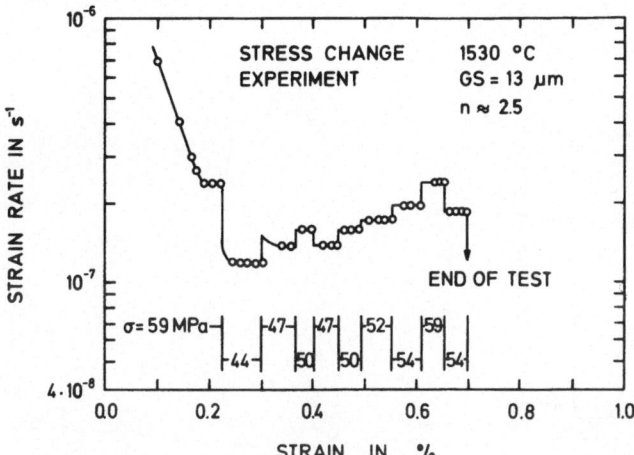

Fig. 3.46. Effect of stress changes on creep rate in the dislocation creep regime. - From Engelhardt and Thümmler [3.297]

The second mechanism suggested to account for non-viscous deformation behavior at high stresses is dislocation creep [3.291,3.297,3.300-3.303]. Since dislocation creep and diffusional creep are independent processes, the mechanism having the stronger stress dependence should dominate the deformation behavior at high stresses as depicted in Fig. 3.45. In fact, the occurrence of dislocation creep has been inferred principally from high stress exponents. Experimental evidence, however, for dislocation motion in polycrystalline alumina is scarce.

There is another argument that puts the operation of dislocation creep mechanisms in alumina in doubt. Plastic deformation of a polycrystal by dislocation motion requires five independent slip systems to be active. As discussed in Section 3.5.1.1, the fifth slip system must be of the family of the pyramidal slip systems which are only operative at very high stresses and temperatures, as can be seen in Fig. 3.33. Following a presentation originally used by Heuer et al. [3.311] to compare their diffusional creep data with flow stress curves of single-crystal alumina, Fig. 3.47 contains the flow stress curves of Fig. 3.33 together with stress/temperature data (which will be called "deformation curves" here) obtained from creep experiments on polycrystalline samples from five references. These deformation curves are crossplots of creep rate/stress plots at a constant creep rate of 10^{-5} s^{-1}. They show the stress necessary to maintain that creep rate as a function of temperature and at the grain size indicated. Looking at Fig. 3.47, there are two observations which exclude the deformation of polycrystalline alumina at high stresses to be due to dislocation motion: (1) Most of the deformation curves do not reach even the flow stress of the prismatic slip system, and none

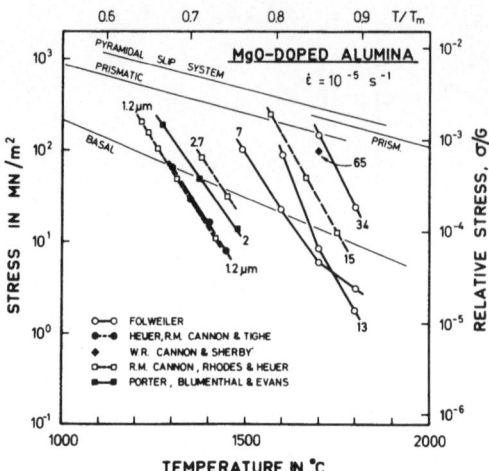

Fig. 3.47. Deformation curves of polycrystalline alumina of various grain sizes compared to flow stress curves of single crystals from Fig. 3.33. – Data points from Folweiler [3.293], Heuer et al. [3.104], W.R. Cannon and Sherby [3.301], R.M. Cannon et al. [3.285], and Porter et al. [3.304]

of them reaches the flow stress of the pyramidal slip system. At 1700°C, the stress to activate pyramidal slip is at least 300 MN/m². (2) The deformation curves depend on the grain size, an observation which contradicts all dislocation creep models. Rather, increasing the grain sizes obviously requires an increased stress and/or temperature to maintain a given deformation rate, as expected from the diffusional creep model, Eq. (3.27). It can be concluded, therefore, that, for grain sizes up to 65 μm and stresses up to about 300 MN/m², deformation of polycrystalline alumina due to diffusional mass transport is faster than due to dislocation motion. Thus, ambiguity still remains as to the dominant deformation mechanism in alumina at high stress and large grain size.

At very small grain sizes, in the range of 1–2 μm, it has been found that the deformation mode of alumina becomes non-viscous, particularly at small stresses. Stress exponents greater than one and less than two for this range of parameters were reported by several authors [3.104,3.285,3.303-3.306], as shown in Table 3.10. Figure 3.48, which is from Cannon et al. [3.285], demonstrates how the deformation rate falls behind that expected from an extrapolation of the diffusional creep regime for low stresses. Such non-viscous deformation phenomena have been interpreted repeatedly in terms of interface-controlled boundary kinetics [3.285,3.289,3.304]. In Section 3.6.1, three models were discussed which describe interface-controlled creep quantitatively. All three models, Eqs. (3.32), (3.34) and (3.35), predict a grain-size exponent m=1. However, experimental

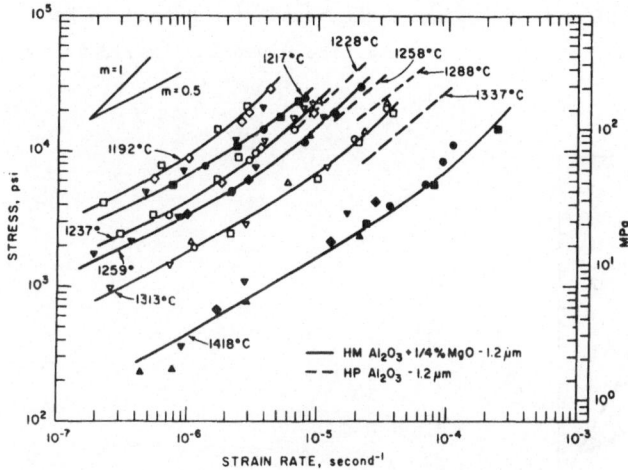

Fig. 3.48. Stress-strain rate curves for fine-grained MgO-doped alumina.
From Cannon et al. [3.285]. Reprinted with permission of the American Ceramic Society

evidence for an inverse grain size dependence of the creep rate of pure and MgO-doped aluminum oxide is still lacking. The only mechanism which predicts a non-viscous creep behavior is that by Ashby and Verrall, Eq. (3.35). Consequently, it was pointed out in [3.285] that the creep results are best interpreted in terms of this model which explains interface control as based on the stress-dependent density of grain-boundary dislocations [3.290].

A convenient way to illustrate which deformation mechanism is predominantly operative at given values of stress, temperature, and grain size is to present creep data in a deformation mechanism map. For an introduction into the construction and use of such maps, the reader is referred to the recent book by Frost and Ashby [3.312], where also maps of several ceramics including alumina may be found. For ceramic materials, principally two types of several possible maps have been used. In the first one, which has been suggested by Mohamed and Langdon [3.313], the grain size is plotted as a function of stress for a constant temperature. A deformation mechanism map of this kind for polycrystalline alumina was given by Heuer et al. [3.311] and is shown in Fig. 3.49. The second type of map, in which the grain size is plotted vs. the reciprocal temperature at a constant stress goes back to a suggestion made by Langdon and Mohamed [3.314], and such a map for alumina published by the same authors [3.284] is shown in Fig. 3.50. Both figures contain a number of fields, each of them characterized by the predominance of a certain deformation mechanism. Field boundaries are loci of equal contributions of neighbouring deformation mechanisms to the overall deformation rate. For the two types of maps shown here, the boundaries happen to be straight lines. Contours of constant strain rate are also given in Figs. 3.49 and 3.50.

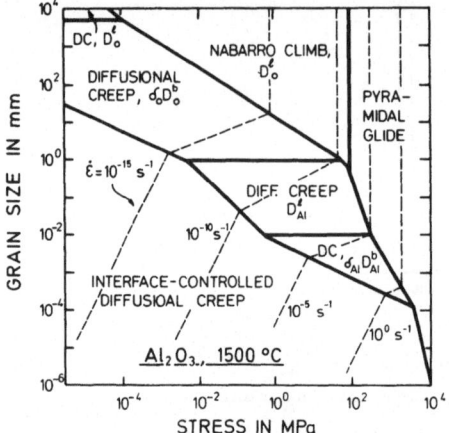

Fig. 3.49. Deformation mechanism map (grain size-stress) for MgO-doped alumina. - After Heuer et al. [3.311]

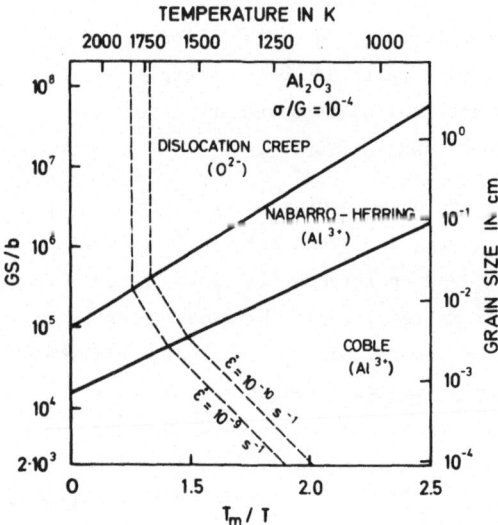

Fig. 3.50. Deformation mechanism map (grain size-inverse of temperature) for MgO-doped alumina. - After Langdon and Mohamed [3.284]

The deformation mechanism map of Fig. 3.49 essentially reproduces the sequence of deformation mechanisms discussed in the foregoing paragraphs. According to the figure, interface-controlled diffusional creep predominates at very low stresses and small grain sizes, followed by a regime of Coble creep and Nabarro-Herring creep controlled by the diffusion kinetics of the cation, when the grain size is increased. Anion diffusion kinetics should never be rate-controlling at grain sizes of practical importance, the same observation holding for creep dominated by Nabarro climb. Dislocation creep can only be expected at stresses of several hundred MPa, combined with an intermediate to large grain size. The same sequence of mechanisms is depicted by the deformation mechanism map of Fig. 3.50. At constant temperature and increasing grain size, the regimes of diffusional creep controlled by cation boundary and cation lattice kinetics, respectively, are passed through, before dislocation creep becomes rate-controlling. Creep kinetics controlled by oxygen diffusion does not appear in the map. Increasing temperature favors Nabarro-Herring creep at the expense of Coble creep, and dislocation creep at the expense of both kinds of diffusional creep.

Summarizing this review on creep of pure and MgO-doped alumina, it can be stated that sufficient creep data are available to establish deformation mechanism maps of the material which permit the determination of the predominant deformation mechanism for given service conditions (stress and temperature) and a given grain size. However, Fig. 3.45 demonstrates that the assessment of the deformation rate in a given high-temperature application is still an inaccurate matter. The transition from diffusional creep to dislocation creep at high stresses and large grain sizes needs further attention. Moreover, the occurrence of interface kinetics at low stresses and very small grain sizes, which is a relatively recent phenomenon in the creep of ceramics, must be elucidated further. In this context, the existence of a threshold stress must be clarified, since it is of utmost importance for long-term engineering applications.

3.6.3 Effect of Other Dopants

In Sections 2.3 to 2.5, the effect of a number of aliovalent dopants on self-diffusion was discussed and it was shown that such additives can increase aluminum diffusivities by increasing the point defect concentrations of V_{Al}''' or $Al_i^{...}$. Diffusion-dependent properties such as ionic conductivity and sintering rate proved to be dependent on the concentration and oxidation state of the impurity or additive present in solid solution. Similar phenomena may be expected for the effect of doping on the creep rate. Since oxygen grain-boundary diffusion is fast compared to both aluminum grain-boundary and lattice diffusion, only aluminum diffusion kinetics need to be considered in diffusional creep of polycrystalline alumina.

The effect of a number of both aliovalent and isovalent dopants on the creep behavior of alumina was studied in detail by Gordon and co-workers [3.277,3.278, 3.286,3.289,3.302,3.315,3.316] and by Kröger and co-workers [3.303,3.305]. Some of their results will be discussed in the following.

Agreement exists among these studies that aliovalent dopants accelerate the creep rate of aluminum oxide, while isovalent dopants do not. This general result coincides completely with results of sintering studies described in Section 2.5.3 on a number of aliovalent sintering additives, which were also observed to accelerate densification. Slightly non-viscous stress exponents were found in a wide range of temperatures, stresses, grain sizes, and dopant contents, most of them lying close to 1.30 [3.286,3.289,3.302,3.303,3.305,3.315,3.316]. Therefore, diffusional creep was indicated as the predominating deformation mechanism.

The influence of Fe and Ti additions on the creep rate of alumina polycrystals has been studied extensively by Lessing and Gordon [3.302,3.316] and Ikuma and Gordon [3.286,3.289] in a wide range of grain sizes. Their results are summarized in Fig. 3.51 and compared with creep data by Hou et al. [3.303] and Cannon and Coble [3.273] on pure and MgO-doped alumina, respectively. As shown in Fig. 3.51 (A), creep data obtained by different investigators on materials that are nomi-

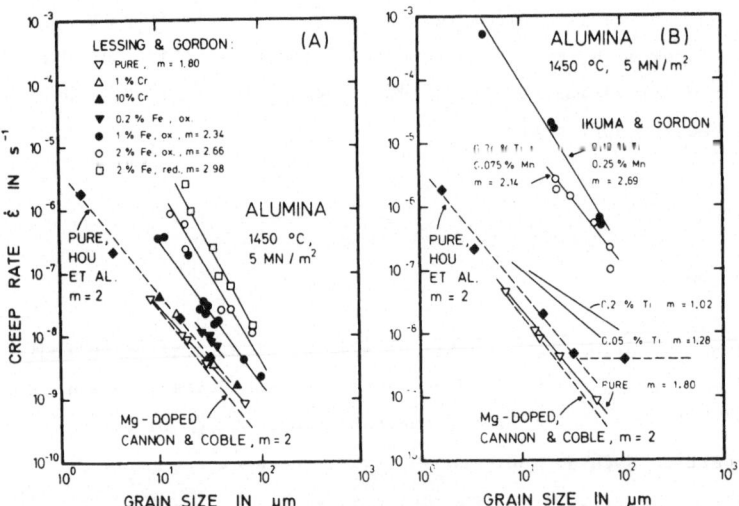

Fig. 3.51(A) and (B). Effect of various dopants on the creep rate of alumina polycrystals at 1450°C. - Data points from Lessing and Gordon [3.302], Ikuma and Gordon [3.289], and Hou et al. [3.303], the latter extrapolated from 1500°C using the activation energy given there; dashed line of Mg-doped material calculated from the diffusivity formula given by Cannon and Coble [3.273], Eq. (3.36), and using Eq. (3.27) without grain-boundary diffusion term

nally pure or contain small additions of Mg (0.25%) or even large additions of Cr (1 and 10%), agree very well among each other with respect to the magnitude of the deformation rate and the grain size exponent, which is close to 2. However, increasing additions of Fe in oxidizing atmosphere increase both $\dot{\varepsilon}$ and m, and a reducing atmosphere increases these two parameters even more. The explanation given later [3.286] to explain these findings was that divalent Fe enhances Al grain-boundary diffusion, i.e., it promotes Coble creep and increases the second term of the right-hand side of Eq. (3.27) or Eq. (3.33). Thus, a gradual transition from Nabarro-Herring to Coble creep occurs, until the inverse third-power grain size dependence finally dominates.

In the case of Ti additions, Fig. 3.51 (B), the creep rate is also enhanced above that of pure or MgO-doped alumina, but obviously reaches a saturation limit, particularly at small grain sizes. The interpretation of this behavior given by Ikuma and Gordon [3.289] was that an increasing amount of tetravalent Ti increases Al lattice diffusion to an extent that interfacial effects of point defect creation and annihilation become rate-controlling. On discussing the interface-control creep model of these authors in Section 3.6.1.3, it was shown that this situation leads to a creep rate given by Eq. (3.34). It predicts the inverse grain size dependence of the creep rate shown in Fig. 3.51 (B) for 0.2% Ti doping.

Figure 3.51 (B) also shows that surprisingly high creep rates can be obtained when alumina is co-doped with small amounts of a divalent and a tetravalent additive, here Mn^{2+} and Ti^{4+}. Instead of neutralizing each other, as reported previously [3.317], they intensify their effect on the creep rate, increasing it by more than three orders of magnitude compared to the undoped material. Additionally, an increase of m to about 2.7 is observed, suggesting increasing grain-boundary diffusional creep. According to Ikuma and Gordon [3.286,3.289], the divalent dopant, either Mn^{2+} or Fe^{2+}, is believed to enhance both $\delta_{Al}D^b_{Al}$ and K_{Al} of Eq. (3.33), thereby both removing the saturation limit caused by interface control and enabling increased matter transport along grain boundaries. Once the limiting effect of interface-controlled point defect creation-annihilation is removed, the effect of tetravalent titanium to increase Al lattice diffusion can now become fully active to enhance the creep rate even more.

From these results it is clear that the effect of aliovalent dopants unequivocally is to enhance the creep rate. The influence of the oxygen partial pressure p_{O2}, however, on the creep behavior is still somewhat ambiguous. An important effect of certain dopants on the creep rate is to make it sensitive to p_{O2}. Varying oxygen partial pressure can change the valence state of transition metal ions present in solid solution, as discussed in Sections 2.3 to 2.5, thereby causing a change of the charge-compensating native defect concentration.

The influence of p_{O2} on the Al lattice diffusivity calculated from diffusional creep data is shown in Fig. 3.52 for several dopants. For chromium and magnesium doping, no effect of p_{O2} is observed, and the magnitude of the diffusivity is not very far from that obtained by tracer diffusion studies [3.318]. This can easily be understood considering that Cr^{3+} and Mg^{2+} are ions whose valency does not change with p_{O2}. In contrast to that, doping with the transition metals titanium and iron leads to both a pronounced increase in the diffusivity and a strong p_{O2} dependence [3.315]. In the case of Fe doping, a ten-fold increase in diffusity is observed when the atmosphere is changed from oxidizing to reducing. The phenomenon can be explained in terms of Eqs. (2.8) to (2.12): Fe^{2+} causes $Al_i^{\cdots}$ to be formed. Since the concentration of divalent iron, $[Fe^{2+}]$, depends on p_{O2}, $[Al_i^{\cdots}]$ and D_{Al}^{ℓ} are also p_{O2} dependent. The variation of the creep rate of the Ti-doped material is the opposite of that observed in Fe-doped alumina. Again, the phenomenon can be understood by looking at the defect chemistry and using Eqs. (2.21) to (2.23): increas-

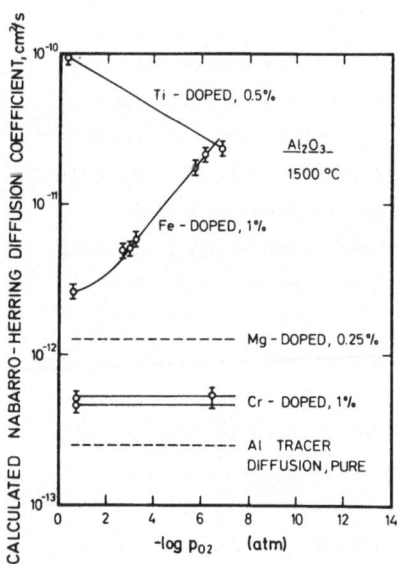

Fig. 3.52. Effect of oxygen partial pressure on Al lattice diffusivity. – Data points from Hollenberg and Gordon [3.315]; dashed line of Mg-doped material calculated from the diffusivity formula given by Cannon and Coble [3.273], Eq. (3.36); dashed line of Al tracer diffusion from Paladino and Kingery [3.318]

176

ing p_{O2} causes an increasing Ti^{4+} concentration, which is compensated by an increasing V_{Al}''' concentration. Increasing $[V_{Al}''']$, however, increases D_{Al}^{ℓ} according to Eq. (2.2).

A much weaker effect of p_{O2} on the creep behavior was reported by Hou et al. [3.303]. At temperatures around $1500^{\circ}C$, the creep rate of aluminum oxide, both undoped and doped with 500 ppm Fe, proved to be nearly independent of p_{O2}. Doping with 5000 ppm Fe resulted in some increase in the creep rate and a p_{O2} dependence still much weaker than that predicted by Eq. (2.12). It was argued that this was caused by the presence of a small amount (in the range of 50 ppm) of a fixed valency impurity such as Mg^{2+} or Ca^{2+}, which masked the effect of the Fe doping. Creep rates almost independent of p_{O2} were also found by El-Aiat et al. [3.305] for aluminum oxide polycrystals both doped with 0.5 and 3 w/o Fe, 500 ppm Ti, and undoped. No unequivocal explanation of this behavior could be given, but it was supposed once more that the presence of a fixed valency impurity (Mg^{2+}) in a concentration of some ten ppm prevented the expected strong p_{O2} dependence. It is worth mentioning in this context that no effect of p_{O2} on the sintering rate of MgO-doped alumina was detected by Coble [3.319].

The softening effect of oxygen partial pressure and of co-doping may present a means of improved processing of alumina at elevated temperatures. However, knowledge of co-dopant effects on the creep resistance is still rudimentary.

3.6.4 High-Temperature Failure Mechanisms

High-temperature fracture of ceramics is frequently a consequence of creep deformation. Formation of grain-boundary voids and cavities induced by stress or strain leads to tertiary creep and ultimate failure. The subject has gained growing attention during the last few years, because high-strength ceramic materials are being designed or used in an increasing number of applications and at increasing service temperatures.

High-temperature failure in ceramics has been reviewed recently by Evans and co-workers [3.320,3.321]. According to present understanding, the failure process involves a sequence of subprocesses, i.e., cavity nucleation, growth, and coalescence to microcracks, and subsequent growth of these cracks until a macrocrack of critical size is reached which leads to catastrophic failure [3.320]. Nucleation and growth of cavities in ceramics shows great similarity to corresponding processes occurring in metals under creep conditions. The two principal types of cavities described for metals [3.322], i.e., grain-boundary cracks and spherical voids formed on grain boundaries perpendicular to the stress direction, are also found in ceramics. They are shown schematicly in Fig. 3.53.

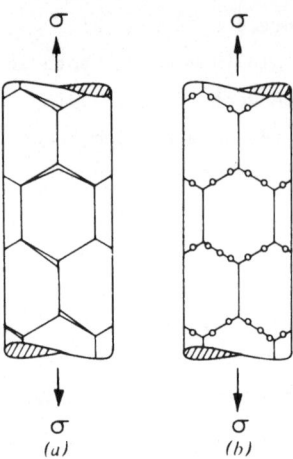

Fig. 3.53. Schematic of intergranular creep fracture: (a) triple point cracking, and (b) cavitation. - From Miller and Langdon [3.322]. Reprinted with permission of the Metallurgical Society of AIME

Cracks can nucleate at triple points or on grain boundaries due to unaccommodated grain-boundary sliding (Fig. 3.53a). Mechanisms for the generation of such cracks in ceramics were studied in detail by Vasilos and Passmore [3.323], who pointed out that there are numerous ways in which grain-boundary cracks may be nucleated by grain-boundary sliding. Experimental evidence of intergranular seperations in alumina was given by several investigators. Grain-boundary cavities, mainly along planes normal to the tensile stress and at high stresses, were reported by Folweiler [3.293], Warshaw and Norton [3.291], Coble and Guerard [3.307], and Carry and Mocellin [3.306]. Localized propagation of microcracks along grain boundaries after some creep strain was observed by Crosby and Evans [3.299] and Davies [3.300]. In a TEM study, Heuer et al. [3.104] observed triple-point voids and displaced triple points in deformed specimens, which were attributed to unaccommodated grain-boundary sliding. The same result was obtained ten years later by the same authors [3.311]. Additionally, a density decrease from 99 to 93% due to deformation was found. It was stated that the formation of triple point cavities contributed considerably to the total deformation, thereby representing a competitive deformation mechanism.

Failure at elevated temperature may also originate from pore-like voids situated at grain boundaries perpendicular to the tensile stress, Fig. 3.53 b. In single-phase materials, these cavities may grow and coalesce by diffusional mass transport along the grain boundaries. In ceramics containing a glassy phase on the grain boundaries, cavity growth can occur by the viscous flow of the intergranular phase [3.320]. The theory of diffusion growth of grain-boundary voids was first developed by Hull and Rimmer [3.324], and subsequently improved by Raj and Ashby [3.325]. In two more recent papers, Evans and Blumenthal [3.321] and Hsueh and Evans [3.326] analysed the formation of pore-like cavities in ceramics. They were thought to nucleate preferentially at three-grain junctions and in regions of microstructural inhomogeneities, which may be regions of a small dihedral angle (large grain-boundary energy), large grain size (small diffusional creep deformation and, hence, difficult stress relaxation), or a small ratio of the surface diffusivity to the grain-boundary diffusity, Δ. The analysis further assumed that triple point-nucleated voids can grow along the grain boundary by a combination of surface and grain-boundary diffusion. Coalescene of voids should occur when two extending voids meet each other in the middle between two opposite triple points to form a full-facet cavity. Depending on the stress level and the value of the diffusivity ratio Δ, the cavity shape should be equiaxed at small stresses (equilibrium contour) or crack-like at high stresses (non-equilibrium contour), when the surface diffusivity is too small to accommodate the driving force of the stress. Expressions for the growth rate of equiaxed cavities and crack-like cavities were developed by Hsueh and Evans [3.326] which permit the assessment of the failure time.

In a microscopic study of alumina deformed in creep, experimental evidence of all three types of cavities predicted in [3.326], i.e. equilibrium, crack-like, and full-facet cavities, was given by Porter et al. [3.304]. Pore-like cavities were also observed by Davies [3.300] at low stresses. A quantitative study of cavity formation in alumina was published recently by Page and Lankford [3.327]. Using the small-angle neutron scattering technique, these authors determined the number, size, and shape of cavities formed in creep deformation as a function of creep strain. Figure 3.54, which is taken from their work, shows the cavity volume fraction in two materials of different grain size. It increases linearly with creep strain, the coarse-grained material exhibiting the steeper slope. The number of cavities per unit volume was also found to increase with creep strain. At 8% strain, a two-fold increase had occurred. The cavity size, however, remained constant during deformation, the cavity diameter being about 100 nm. Thus, the increase in total cavity volume was attributed to the nucleation of new cavities rather than to cavity growth [3.327].

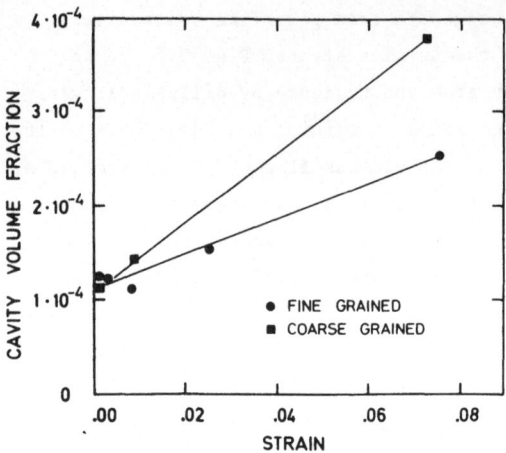

Fig. 3.54. Effect of creep strain on formation of cavity volume. - From Page and Lankford [3.327]

There are indications that cavitation does not occur in alumina if the diffusivity is raised sufficiently. In their studies of the effect of doping on diffusional creep, Gordon and co-workers [3.315,3.316] were unsuccessful in detecting any signs of cavities or intergranular separations. Although both cavity growth and creep are diffusion-controlled processes, an enhanced diffusivity seems to promote mainly creep deformation, thereby providing sufficient matter transport at critical locations along the grain boundaries to avoid any material separation.

Only little is known about the coalescence of cavities to microcracks and the extension of these cracks at elevated temperatures. Hsueh and Evans [3.326] presented a zone-spreading model and Raj and Baik [3.328] a diffusion model to describe the approximation of cavities to each other and their subsequent coalescence. However, experimental evidence of their predictions is still lacking. The final stage of high-temperature failure preceding catastrophic fracture, i.e., the propagation of microcracks, was analysed by Chuang [3.329]. His diffusive crack-growth model yields a power-law relationship between crack propagation rate and K_I, $\dot{a} \propto K_I^{12}$, which still needs to be verified experimentally.

Evidence of the importance of a grain boundary glassy phase in high-temperature fracture of alumina was given by Lankford [3.114], Evans and Rana [3.320], and Tree et al. [3.330]. The finger-like or globular shapes of the residues of the glassy phase on intergranular facets, which were found in SEM investigations of surfaces fractured at elevated temperatures, indicate that gross rearrangement of the amorphous film must have occurred before or during the propagation of the crack front. It is only recently,however,that such microstructural features of the fracture process attracted the attention of researchers, and the study of the details of high-temperature failure of structural ceramics is just at the beginning.

3.7 Strengthening Mechanisms

Many efforts have been made to increase the strength of ceramic materials. Strengthening mechanisms applied to polycrystalline alumina will be reviewed in this final sub-chapter on mechanical properties.

The strength of ceramics, S, is properly described by Eq. (3.5). Using Eq. (3.7) to substitute for γ_I and allowing for a term which considers the effect of a compressive surface stress, σ_c, this equation can also be written as

$$S = K_{Ic}/(YC^{1/2}) + \sigma_c, \tag{3.39}$$

where K_{Ic} is the fracture toughness, C is the flaw size, and Y is a geometrical factor of about 2. Equation (3.39) demonstrates that attempts to improve the strength of a ceramic effectively must aim at increasing either K_{Ic} or σ_c, or decreasing C. Consequently, the effect of strengthening mechanisms discussed in the following sections either consists of increasing K_{Ic}, as is the case for strengthening by second phases or by transformation toughening, or inducing compressive surface stresses by several methods. Other strengthening mechanisms such as solid solution hardening, precipitation hardening, and work hardening, were already treated in Section 3.5, and the effect of spontaneous microcracking on toughness was discussed in Section 3.2.2.

3.7.1 Second Phase Dispersions

Similarly to metals, attempts have been made to increase the strength of ceramics by the incorporation of second phase particles. Both metallic and ceramic phases have been used. The basic idea of strengthening ceramics by a second phase is to impede the propagation of the crack front by pinning it at the inclusions, rather than to impede the movement of dislocations, as in metals. Models have been presented by Lange [3.331] and by Evans [3.332] which permit the assessment of the increase in fracture energy or fracture toughness depending on the size, shape, and concentration of second phase particles. The strengthening effect is thought to be due to the bowing of the crack line between two pinning obstacles. Since the crack front can be treated as having a line energy per unit length, crack front bowing means consumption of additional energy, which must be supplied by the elastic stress field ahead of the crack front, thereby leading to an increased value of the fracture energy at the moment of breakaway [3.331,3.332]. The pinning capacity or impenetrability [3.332] of the obstacles plays a crucial role in this strengthening model.

Table 3.11. Strengthening alumina by second phase dispersions

Authors	Ref. Year	Second phase characteristics Type	Size μm	Volume fraction f_v, %	Mechanical properties[4] K_{Ic} MN/m$^{3/2}$	γ_{WOF} J/m^2	Strength MN/m^2
		Metallic Phases					
McHugh et al.	3.333 1966	Mo particles	≈2	0-16	–	–	370 (0) 690 (5-6)
Rankin et al.	3.334 1971	Mo particles	diam. 80	0-5	5.26 (0)[1] 6.73 (5)	–	–
Simpson & Wasylyshin	3.335 1971	Mo wires	diam. 50	0-12	–	20 (0) 4,500 (12)	135 (2-8)
Lloyd & Tangri	3.336 1974	Mo wires	4		–	4.8 fold increase[2]	–
Claussen	3.337 1973	(Cr,Al)$_2$O$_3$– Cr eutectic		10^3	2.6-3.6	200	–
		Ceramic Phases					
Rasmussen et al.	3.338 1965	Fe$_2$O$_3$ particles TiO$_2$ particles		0-20 0-5	– –	– –	no effect no effect
Wahi & Hübner	3.339 1976	BaO·6Al$_2$O$_3$ p[5]		4-40	3.1 (4) 4.9 (40)	–	155 (4) 244 (40)
Wahi et al.	3.340 1976	BaO·6Al$_2$O$_3$ p		4-29	3.40 (4) 4.59 (29)	–	–
		MgAl$_2$O$_4$ p	4-6 1	9 12	4.40 (9) 4.03 (12)	– –	– –
		Fe$_2$O$_3$ p TiC p	4 3	9 33	4.78 (9) 5.58 (33)	– –	– –
Grellner	3.341 1978	TiC p	0.7 0.5-1	4 and 35	–	–	520 (4) 480 (35)
Grellner et al.	3.342 1980	TiC p	≈1	0-96	3.7 (4) 4.4 (35)	–	–
Wahi & Ilschner	3.343 1980	TiC p	≈1	0-35	4.55 (0) 4.99 (35)	19.4 (0) 16.5 (35)	305 (0) 670 (35)

[1] Value of f_v given in parenthesis
[2] as evaluated by the present authors
[3] f_v of metal phase in the metal/ceramic eutectic
[4] largest value given in the table is that of maximum effect achieved
[5] particles

Experimental work on strengthening alumina by second phase particles is summarized in Table 3.11. McHugh et al. [3.333] and Rankin et al. [3.334] reported some strength increase by a metallic dispersion, but they explained the results by the refinement in grain size which was obtained when Mo particles were added to the alumina powder before sintering. The strong effect found in [3.335,3.336] of Mo wires on the work of fracture was attributed to the additional energy consumption during cracking due to pull-out, necking, and rupture of the fibers. Considerable increase in the work of fracture was also found in metal-ceramic eutectics [3.337]. However, the whole research area of metal additions was not pursued further.

Some strength improvements have been achieved by the use of ceramic dispersions in alumina. Wahi and co-workers [3.339,3.340] studied the effect of a number of ceramic inclusions on K_{Ic}. Their results are shown in Fig. 3.55 where the fracture toughness of the composite related to the fracture toughness of the alumina matrix, K_{Ic}^M, is plotted vs. $\bar{R}/\bar{d}$, with $\bar{R}$ the mean particle radius and $2\bar{d}$ the mean particle spacing. As shown in the figure and in Table 3.11, the strengthening effect of the various dispersions is on the order of ten per cent. A comparison with the prediction of Evans' crack front pinning model [3.332] showed that the agreement between theory and experiment was only reasonable up to a $\bar{R}/\bar{d}$ ratio of about 0.35.

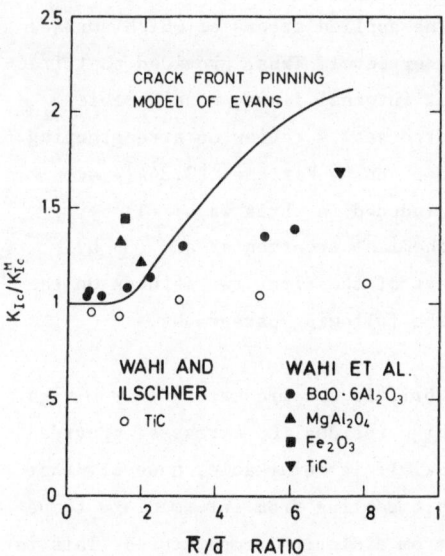

Fig. 3.55. Strengthening effect of various second phase dispersions. - Data from Wahi, Hübner, and Ilschner [3.340] and Wahi and Ilschner [3.343], pinning model from Evans [3.332], Eq. (19)

SEM examination of fractured surfaces showed fracture steps behind the second phase particles, thus giving evidence of the blocking action of the obstacles [3.340]. Investigations of the effect of titanium carbide dispersions [3.341-3.343] showed only a minor influence on K_{Ic}, but a distinct increase in strength, which is probably due to a grain size refinement in the presence of TiC particles during fabrication. Hot-pressed Al_2O_3-TiC composites are widely used as high-strength cutting tools (Section 5.2.5). K_{Ic} data of Al_2O_3-TiC composites [3.343] plotted in Fig. 3.55 demonstrate the weak influence of the carbide content. This result was confirmed by fractographic examinations which revealed transgranular fracture of the TiC particles and the absence of fracture steps around them, thus excluding that crack front pinning had occurred [3.343]. Concluding the discussion of strengthening alumina by second phase dispersions it should be noted that no breakthrough has been achieved on this field and that research activities drifted to other, more promising areas.

3.7.2 Compressive Surface Stresses

Compressive surface (CS) stresses can lead to a substantial strength increase in materials which are susceptible to tensile failure, such as glass and ceramics. Residual CS stresses and applied tensile stresses may cancel each other due to linear stress superposition, as is reflected by Eq. (3.39). In the presence of residual CS stresses, stress intensity factors acting at potential failure-initiating flaws are substantially reduced, and the applied stress at which surface flaws act to cause failure is shifted to a higher level. Thus, provided that fracture originates at surface flaws rather than at internal flaws, considerable strengthening can be achieved by residual CS stresses. A review on strengthening of structural ceramics by CS stresses has been given by Kirchner [3.344]. According to this reference, CS stresses can be induced in three ways: (1) by quenching from elevated temperatures; (2) by chemical treatments; and (3) by phase transformations in the surface. The effect of the first two methods on the strengthening of alumina will be reviewed in the following paragraphs.

It has been shown repeatedly that quenching a brittle ceramic material from high temperatures (around 1400-1700°C) can induce high residual CS stresses, provided that the samples do not fail from thermal shock. It is remarkable, however, that thermal shock failure occurs more often on rapid cooling from intermediate temperatures (about 300-1200°C) than on quenching from sintering temperatures. This is due to the details of cooling, thermal contraction, and the generation and relaxation of thermal stresses. On cooling from high temperature, tensile stresses in the surface and compressive stresses in the interior of the body are built up due

184

to the non-uniform temperature distribution across the sample section. In a material like alumina, where some limited plasticity is possible at elevated temperatures, these stresses are thought to be relieved by relaxation. On further cooling below 1100-1200°C, however, plastic deformation ceases, and thermal stresses can no longer be relieved. At this moment, the interior is still at higher temperature than the surface, and further cooling causes a larger thermal contraction of the interior than of the surface, thereby generating compressive stresses in the surface and tensile stresses in the interior [3.344].

Strengthening alumina by quenching has been studied extensively by Kirchner and co-workers. In one of the first papers on the subject [3.345], a strengthening effect of about 40% was reported for a 96% alumina containing a glassy phase, which was rapidly cooled from 1500°C to room temperature in forced air. The presence of residual CS stresses was demonstrated by the slotted-rod test. In a further study on the same material [3.346], a 2.1-fold strength increase due to quenching and a 2.3-fold increase due to glazing and quenching was reported, as compared to the as-received material. Figure 3.56 shows this tremendous effect in a Weibull plot of strength. The investigation also showed that a low expansion glaze was effective in inducing additional CS stresses and in improving the strength even more. The average strength obtained on quenched and on glazed plus quenched samples was 694 and 767 MN/m^2, respectively [3.346]. The investigation of the time-dependent strength of 96% alumina revealed that the lifetime of quenched samples in water increased by many orders of magnitude, compared to as-received samples [3.150]. The indentation strength of the material also increased after

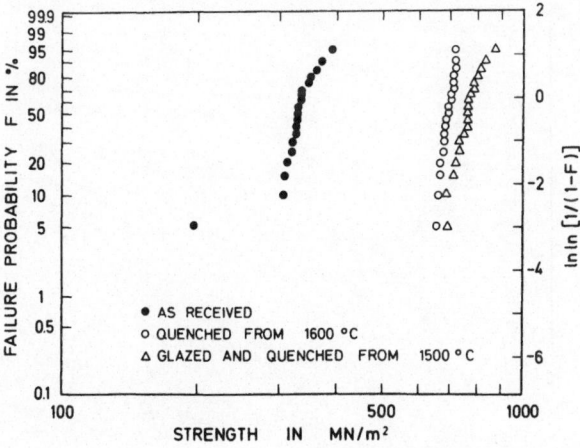

Fig. 3.56. Effect of quenching and of glazing and quenching on the strength of a 96 % alumina. − Data points from Kirchner, Gruver, and Walker [3.346]

quenching [3.347], and the determination of the temperature dependence of strength showed that a strength increase due to CS stresses could be retained at least up to 1000-1100°C [3.348]. Kirchner et al. [3.349] were also able to demonstrate that hot-pressed, single-phase aluminum oxide can be strengthened by quenching. Their results, as shown in Fig. 3.57, indicate a substantial strength increase (in the order of 30-60%, depending on grain size and temperature of the quench) in a range of grain sizes between 1 and 5 µm.

A calculation of the residual stress profile in a quenched rod of 96% alumina was performed by Buessem and Gruver [3.350]. The analysis revealed CS stresses of up to 210 MN/m^2, depending on the temperature difference of the quench and on the magnitude of the heat tranfer coefficient between the alumina surface and the cooling medium. In a fractographic study, a maximum CS stress of 400 MN/m^2 was assessed by Kirchner and Gruver from fracture mirror size measurements [3.351]. It has been pointed out [3.344] that this number seems to be the maximum strength increment attainable by strengthening due to thermally-induced CS stresses.

Strengthening alumina by quenching was also found by other authors. Stolz and Varner [3.352] reported a strength increase from 400 to 600 MN/m^2 when 98% alumina rods were quenched from temperatures in the range 1500-1700°C.

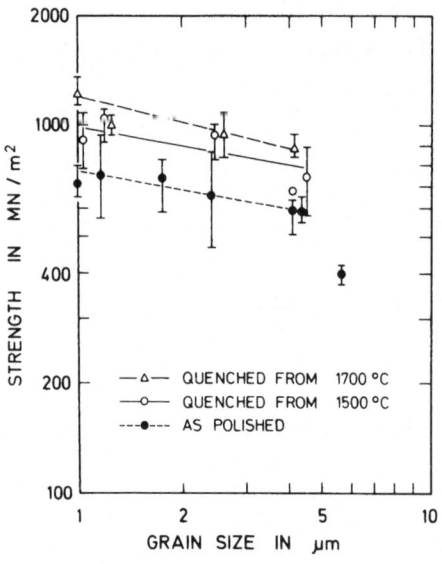

Fig. 3.57. Strength increase of hot-pressed, fine-grained aluminum oxide due to thermally-induced compressive surface stresses. - After Kirchner, Gruver, and Walker [3.349]

The generation of CS stresses by chemical treatment is achieved by changing the composition of the matrix material within a thin surface layer in such a way that the resulting compound has a lower thermal expansion coefficient than the matrix material. Thus, on cooling from the processing temperature the surface layer will go into compression and the interior into tension. Chemical strengthening of alumina was carried out successfully by Kirchner et al. [3.353] by packing alumina samples in Cr_2O_3 powder and annealing at temperatures between 1400 and 1600°C. During the heat treatment, Cr^{3+} diffused into the alumina surface and went into substitutional solid solution. The CS stress induced on cooling resulted in a strength change from 343 to 500 MN/m^2. The same annealing technique was applied by Frasier et al. [3.354]. The ceramic powders used were Cr_2O_3, Co_xO_y, Nb_2O_5, Fe_2O_3, and Cr_2O_3 + $CrCl_3$. All treatments except Fe_2O_3 and Nb_2O_5 resulted in strength increases of 20-30%. Single-crystal alumina also exhibited a significant strength increase when annealed in Cr_2O_3. A dramatic strength increase of sapphire rods was reported by Rahman and Jones [3.355]. Packing the samples in BaO paste and annealing at 1490°C resulted in a strength more than twice the initial strength, i.e., from 413 ± 52 to 1127 ± 95 MN/m^2. Finally, solid solution surface layer formation by an ion exchange reaction was achieved by Platts et al. [3.356]. Glazed alumina rods were heat treated in a molten salt bath containing KNO_3 at 400°C for 30 minutes. The exchange of Na contained in the glaze by K resulted in the formation of a surface layer of a lower expansion coefficient and, hence, in the creation of CS stresses, which caused a strength increase from 210 to 410-480 MN/m^2. Summarizing these results it is evident that strengthening by CS stresses is much more effective than strengthening due to the presence of second-phase dispersions.

3.7.3 Transformation Toughening

It is only a few years ago that it was discovered that the strength and fracture toughness of ceramics containing a finely-distributed ZrO_2 dispersion can be enhanced significantly by a phase transformation of these fine particles. Examples of such transformation toughening were first reported for zirconia and alumina in the years 1975 to 1977 [3.357-3.359]. On cooling from elevated temperatures, ZrO_2 undergoes a phase transformation from the tetragonal to the monoclinic structure at 1170°C, which is accompanied by a volume increase of about 3%. The monoclinic phase is the stable phase at room-temperature. The phase transformation from tetragonal to monoclinic symmetry is a martensitic reaction, i.e., it is a diffusionless and athermal change of lattice symmetry which occurs instantly when the driving force (the change of thermodynamic free enthalpy) is sufficiently high, without involving long-range diffusion processes.

For tetragonal ZrO_2 particles embedded in a ceramic matrix, phase transformation takes place under elastic constraint. Under this condition, transformation can still occur spontaneously during cooling to room temperature when the particle size is large enough so that the volume expansion can overcome the elastic constraint exerted by the matrix. In this case, microcracks can form around many of the transformed ZrO_2 particles. These cracks can extend and absorb additional fracture energy when a macroscopic crack propagates through the matrix, thereby increasing the toughness of the material (transformation-induced microcracking, Claussen [3.358, 3.360]).

For tetragonal ZrO_2 particles below a critical size, no transformation occurs on cooling, and the metastable phase is retained down to room temperature. In the stress field of a propagation crack, however, the transformation of particles adjacent to the propagating crack can be induced by the large tensile stress ahead of the crack tip. The energy required to build up the stress field around the transformed particles is supplied by the external stress. This additional energy consumption leads to an increase in the crack resistance and the fracture toughness of the material (stress-induced transformation toughening, Heuer [3.361], Lange [3.362]).

Experimental and theoretical work in the area of transformation toughening has expanded extraordinarily rapidly during the last few years. For a more detailed introduction to the field, the reader is referred to a series of excellent review articles published recently [3.361-3.365]. In the following paragraphs, the effect of transformation toughening on the mechanical properties of polycrystalline alumina will be reviewed. Transformation toughening has also been applied to a number of other ceramics, for example zirconia [3.359,3.366], boron nitride [3.367], silicon nitride [3.368], glass ceramics [3.369], β-alumina [3.370], and mullite [3.371]. Theoretical analysis of transformation toughening phenomena are also available in the literature [3.361,3.362,3.372,3.373].

Strengthening alumina by transformation toughening has been studied primarily by Claussen and co-workers [3.358,3.360,3.364,3.374-3.377]. Figure 3.58 shows the effect of the volume fraction of a ZrO_2 dispersion on the fracture toughness and strength from one of the first publications on the subject [3.374]. The greatest effect on K_{Ic}, namely an increase by a factor of 2, was achieved by adding 15 vol% of ZrO_2 particles with 1.25 μm diameter. However, no increase in strength was observed. On the contrary, the more the volume content or the size of the ZrO_2 particles increased, the more the strength decreased. The increase in K_{Ic} was explained by additional energy absorption in an extended microcrack zone ahead of

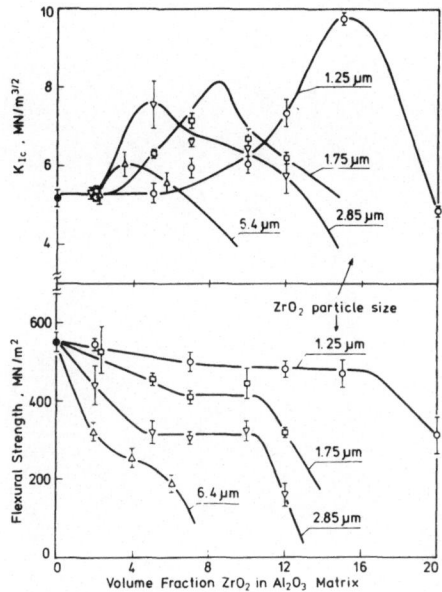

Fig. 3.58. Toughening effect of a ZrO_2 dispersion in alumina. - From Claussen, Steeb, and Pabst [3.374]. Reprinted with permission of the American Ceramic Society

the notch root of notched samples. The decrease of strength was attributed to an increased flaw size as a result of an increased number and size of microcracks, assuming that microcrack linking and joining had occurred on loading [3.374].

Yet, the basic concept of transformation toughening is not undisputed in the literature. The question being discussed is whether the toughening effect is caused by energy dissipation in a microcracked zone, as proposed above, or by stress-induced particle transformation in a limited region adjacent to the propagating crack and subsequent energy absorption in the local stress fields around the transformed particles. The latter concept has been advanced, among others, by Lange [3.362].

The fraction of retained tetragonal ZrO_2 particles can be controlled by the particle size and by the amount of Y_2O_3 dissolved in the ZrO_2 particles by extending the stability range of tetragonal ZrO_2 to lower temperatures. In an experimental study performed by Lange [3.362] of the Al_2O_3-ZrO_2 system over the whole range of compositions from one end member to the other, 2 mol % Y_2O_3 was added to ZrO_2 to allow complete retention of tetragonal ZrO_2 up to a volume fraction of 60%. The fracture toughness of the composite is seen in Fig. 3.59 to increase from 4.8

189

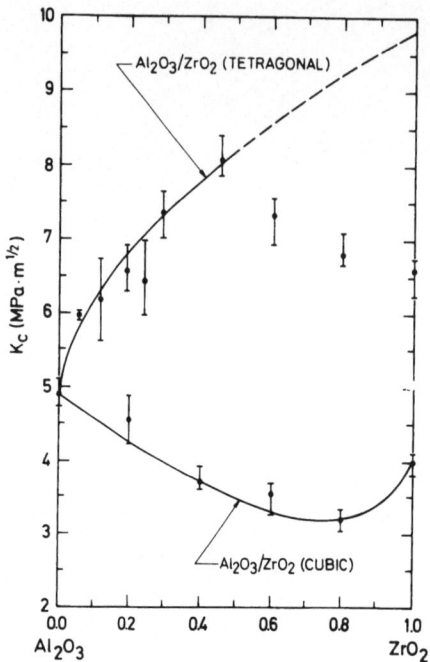

Fig. 3.59. Fracture toughness of Al_2O_3-ZrO_2 composites. - From Lange [3.362].
Reprinted with permission of Plenum Publishing Corporation

to over 8.1 $MN/m^{3/2}$ as the volume fraction of tetragonal ZrO_2 is increased. The
addition of fully stabilized, cubic ZrO_2 particles, on the other hand, continually
decreases the fracture toughness of the composites. This result may be taken as
evidence that toughening of alumina can also be achieved by the stress-induced
transformation of completely retained tetragonal ZrO_2 particles.

A further study of the effect of Y_2O_3 stabilizer concentrations on the transformation behavior of tetragonal ZrO_2 and the fracture toughness of Al_2O_3-ZrO_2 composites was reported by Becher and Tennery [3.378]. By optimizing both the ZrO_2 and
the Y_2O_3 content with respect to maximum fracture toughness, a maximum K_{Ic} of
8.5 $MN/m^{3/2}$ was obtained for 20 vol% ZrO_2 and 1 mol% Y_2O_3.

Detailed studies of the transformation kinetics of tetragonal ZrO_2 particles in
alumina matrices were published recently [3.379, 3.380]. Green [3.379] was able to
show that the transformation from tetragonal to monoclinic could be induced by
heat treating, and that there was a minimum ZrO_2 size required for the particles
to transform. This size was shown to decrease with increasing ZrO_2 content. Furthermore, microcracking was observed to occur simultaneously with the transforma-

tion. In a TEM study, Heuer et al. [3.380] determined quantitatively the fraction
of transformed particles as a function of particle size and temperature. From these
measurements, the relationship between the critical particle size for transformation and the transformation temperature was established. Their result shows that
the particle size necessary to completely retain the tetragonal phase to room temperature is 0.6 µm.

Finally, the work of Claussen, Cox, and Wallace [3.360] provides evidence for
toughening alumina by transformation-induced microcracking. An alumina material
containing 15 vol% ZrO_2 was studied whose particle size was adjusted to 1-2 µm so
that the transformation temperature was $\approx 600^{\circ}C$. At room temperature, only monoclinic particles were presumed to be present. As shown in Fig. 3.60, the fracture
toughness measured on cooling exhibited a steady increase from the matrix value
$(4.5$ MN/m$^{3/2})$ at $1200^{\circ}C$ up to 11 MN/m$^{3/2}$ at room temperature. On heating, K_{Ic}
remained constant, being equal to that of the matrix. The variation of the K_{Ic}
increase on heating and cooling correlated well with the onset of the formation
of monoclinic particles according to the upper part of Fig. 3.60. Furthermore, it
was observed that both Young's modulus and K_{Ic} of rapidly cooled samples showed
a time-dependent decrease at room temperature. The increase in toughness was explained by the increasing number of microcracks around transformed particles,
whereas the toughness loss at room temperature was attributed to the subcritical
growth of these microcracks to a size at which they are no longer efficient in
dissipating fracture energy [3.360].

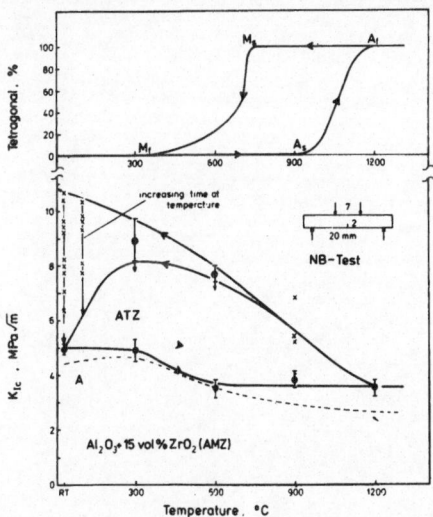

Fig. 3.60. Fracture toughness and tetragonal content of 15 vol% ZrO_2-containing
alumina as a function of temperature. - From Claussen, Cox, and Wallace [3.360].
Reprinted with permission of the American Ceramic Society

On closing the discussion of strengthening mechanisms in alumina it should be noted that toughening by the transformation effect of a finely-dispersed ZrO_2 phase obviously is the most promising approach developed so far. The 2-3 fold increase in fracture toughness achieved by transformation toughening represents an enormous success in material development. Toughness values similar to those of steels seem to be attainable also in ceramic materials. Thus, the prospect of a tough as well as strong ceramic no longer seems to be a contradiction.

4 Fabrication

The manufacture of high alumina ceramics is normally carried out according to powder metallurgical procedures rather than using conventional ceramic fabrication
methods. It involves the application of powder metallurgical methods to oxide powders. A high precompression and a sintering process without a liquid phase are
the main characteristics of the production process of high alumina ceramics.

Compacting methods, such as slip casting, in general use in the manufacture of
almost all other ceramic materials, for example porcelain, are normally not ap-
lied to high alumina ceramics because of insufficient precompaction. Instead of
slip casting other compacting methods such as dry pressing, hydrostatic molding,
extrusion, injection molding, and hot pressing have proved to be practicable and
successful in compacting the extremely fine-grained alumina powders having no
plasticity at room temperature.

Figure 4.1 gives a detailed review of all fabrication processes commonly used
for the production of high alumina ceramics. Dry pressing is the most economical

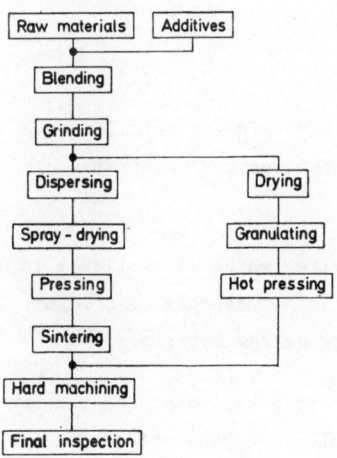

Fig. 4.1. Fabrication process diagram

compacting process. It is normally applied to small parts with simple shapes as well as to large precompacted blanks designated for subsequent machining in an unfired condition in order to obtain the intended final shape. Parts with larger dimensions which are designated for critical applications, or cylindrically shaped parts as cylinders, rods, or tubes with a length-to-diameter ratio of more than 4 to 1 are preferably manufactured by hydrostatic molding. Extrusion is restricted to long rods and tubes with small diameters but various profiles, as well as to flat ribbons for special applications. The injection molding technique is suitable for use in the fabrication of small parts with complicated shapes. Hot pressing, the most expensive powder compacting process, is reserved for relatively small parts for special applications with extreme requirements.

In contrast to other products fabricated by means of powder metallurgical processes, high alumina ceramics allow sintering in any atmosphere. The choice of a suitable furnace depends mainly on the dimensions of the products and on the sintering cycle required. During the sintering process the body obtains its final properties, characterized by its considerable intrinsic hardness. In this condition machining can be done by means of diamond tools only. Due to the total absence of plasticity at temperatures used in the most applications, high alumina ceramics cannot be deformed plastically. Above $1200^{\circ}C$ a limited plasticity allows limited deformations. But an increase in strength by deformations, as it is known from metals, does not take place. Most of the methods which are known from metal materials, e.g., transformation of microstructure, hardening, quenching, and annealing in order to vary the properties, are not applicable to high alumina ceramics. This material is naturally hard. In the following sections the various fabrication steps will be described in detail.

4.1 Preparation of Powders

The first precondition for successful fabrication of high alumina ceramics is a thorough preparation of the powders which depends on the various requirements of the different forming procedures.

The principal source of high alumina ceramics is bauxite, which is available in almost unlimited quantities all over the world. This is an essential fact, particularly at a time of general shortage of most important raw materials.

The extraction of bauxite is performed by means of a wet alkaline process, known as the Bayer process [4.1,4.2]. It is the most economical current method of disintegrating bauxite. The process produces different alumina hydrates with various amounts of crystal water, depending on the mineral deposites of the bauxite. During dehydration of the alumina hydrates by means of thermal treatment, a number of transition modifications are passed through, all ending in α-alumina [4.3]. Figure 4.2 illustrates the dehydration sequence, as a function of the temperature. In spite of its age, almost 100 years, the Bayer process is still the principal extraction method for bauxite. Its major output, however, is intended for the production of metallic aluminum [4.4].

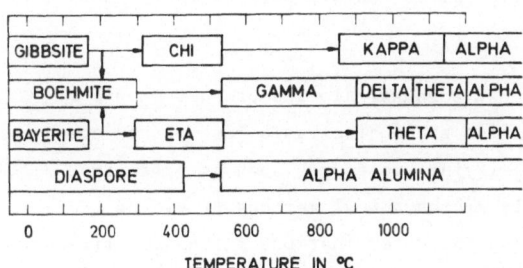

Fig. 4.2. Dehydration sequence of alumina hydrates in air (from Gitzen)

Generally, α-alumina powder, developed by the Bayer process with subsequent heat treatment, is normally the starting material for all fabrication processes of high alumina ceramics. For special reasons α-alumina powder, which is melted in an electric arc furnace with graphite electrodes followed by an extensive grinding process, is used as well [4.5]. Due to the relatively high Na_2O content, introduced during the Bayer process, in most cases the purity is not sufficient for the application of this powder to high-purity alumina ceramic products. This holds for the molten material corundum as well, since the melting process has no purifying effect. For enhanced requirements with respect to purity, other raw materials belonging to the group of ammonias and sulphates of alumina have to be taken into consideration, as, for example, ammonia-alum.

Effective crushing and grinding is an important prerequisite which provides the powder with the required small grain size and with the necessary surface activity for assisting the sintering process. For purity reasons, grinding is done preferably with grinding balls made out of the same material. Coarser fractions are crushed with metal impact crushers and ground with metal grinding balls, which are more effective because of their higher specific gravity. In this case a subsequent purification in a suitable acid is necessary in order to quantitatively remove the metal debris.

195

Basically, all well-known grinding facilities such as ball mills, vibration mills, impact mills, as well as devices for jet pulverization and autogeneous grinding can be applied to produce fine-grained alumina powders. The grinding processes mentioned above are described in detail in [4.6].

In addition to comminution, the grinding process is the most effective method of achieving a homogeneous distribution between the alumina powder and the additives. Alumina powders which are intended for high alumina ceramics are normally combined with natural, i.e., mineral additives, in most cases silicates, while alumina powders intended for high-purity alumina ceramics contain pure synthetic additives such as magnesium oxide.

The fineness of the powders required for an adequate alumina product should be of the order of 1 μm. In this extremely fine grain size range most of the commonly used grain size measuring and classifying procedures such as screening, windsifting, and sedimentation fail. Only microscopical methods providing additional information on the grain shapes and the determination of the specific surface, for example by means of nitrogen adsorption [4.7], have turned out to be useful in practice. The procedure is called the BET-method after the initials of the inventors. The specific surfaces of α-alumina powders can be as large as 20 m^2/g. Higher figures indicate the presence of γ-alumina. Pure γ-alumina shows a specific surface between 100 and 200 m^2/g.

The grain size distribution is also important with respect to an optimal particle packing, contributing to the desired compressibility [4.8]. Figure 4.3 shows a typical grain size distribution graph of an alumina powder which is usually used for most high alumina ceramic products.

In order to obtain a homogeneous bulk density distribution within the compacted body, the friction of the powders which have no plasticity, as well as the wall

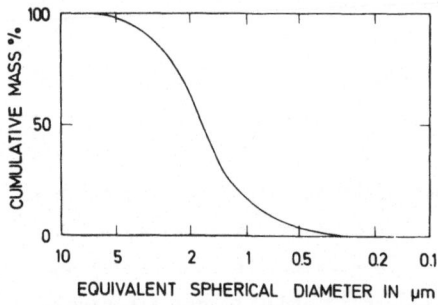

Fig. 4.3. Grain size distribution of an alumina powder

friction, have to be reduced by addition of about 1% of a suitable plasticizer, for example, carbon wax, stearic acids, or polyvinyl alcohols. A uniform distribution of the plasticizer can be obtained with the help of a spray drier, which also yields the free-flowing properties [4.9] which are required for automatic or semi-automatic compaction. The preparation procedure described here is applied to alumina powders intended for dry pressing and hydrostatic molding.

Extrusion and injection molding require a considerably greater quantity of plasticizers, of the order of 10% to 30% which are normally thermoplastics and resins. The choice of the organic components depends on the rheological requirements of the injection molding process [4.10]. Fatty acids, being very strongly adsorbed on the surface of the alumina powder, considerably improve the compaction [4.11]. The admixing of the plasticizers is carried out by means of a kneader in the temperature range of 150 to 200°C, depending on the amount and the composition of the organic binders, which have to be mixed very thoroughly with the alumina powder.

Due to the increasing plasticity of alumina with temperature above 1200°C, the powders for hot pressing do not require a plasticizer. Only the particle size distribution is of importance for this process.

4.2 Forming

For economic reasons, the raw material should be compacted and formed at the same time, i.e., during the same process. Compacting is necessary for bringing the particles of the powder as close together as possible in order to close the residual porosity by the surface tension during sintering. The choice of the forming process depends on the dimensions and the shape of the part to be fabricated, on the quantity of parts, and on the requirements of the final product.

The forming pressure to be applied should be approximately 100 MN/m^2, but is usually adapted to the forming process. Dry pressing and injection molding require a higher forming pressure than does hot pressing.

The green bulk density of the compacted body, as well as the final density after sintering, increases with increasing pressure within a certain range (Fig. 4.4). If the forming pressure is too low the products will not achieve the full final density, while excessive pressure, which has the same effect as insufficient plastification or inhomogeneous distribution of the plasticizer, can lead to defects in the compacted bodies such as flaws and cracks.

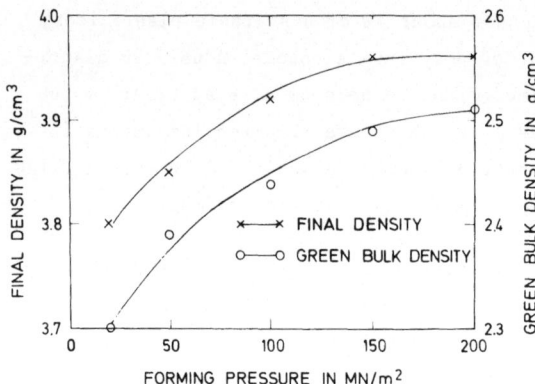

Fig. 4.4. Correlation between green bulk density, final density, and forming pressure of high-purity alumina ceramics

4.2.1 Dry Pressing

The dry pressing technique is the most common and most economical compaction process for the fabrication of high alumina ceramics. It is restricted, however, to parts with simple shapes and to wall thicknesses greater than 1 mm. Dry pressing works unidirectionally.

Hydraulic and mechanical presses in semi-automatic or fully-automatic operation allow an applied pressure of up to 200 MN/m^2 and achieve an output up to 30 parts per minute.

The punches in the die of a hydraulic press are operated by hydraulic pressure and compact the material at every stroke with the same pressure, independent of the quantity of powder in the die. The parts which are fabricated with a hydraulic press show very constant compaction but slight differences in height, due to filling differences.

In contrast to this, the punches of a mechanical press are operated by a rotating excenter. The pressure applied to the parts to be fabricated depends on the grade of filling of the die. The parts show constant dimensions but differences in compaction, depending on variations of filling. This results in differences in shrinkage upon sintering.

Figure 4.5 illustrates the dry pressing operation of a simple ring, as well as a ring with a flange, using a pressure of 150 MN/m^2. The procedure is carried out in three basic steps. It should meet the following requirements in order to pre-

vent different shrinkage and warping during sintering, which results in cracks
and in the development of internal stresses.

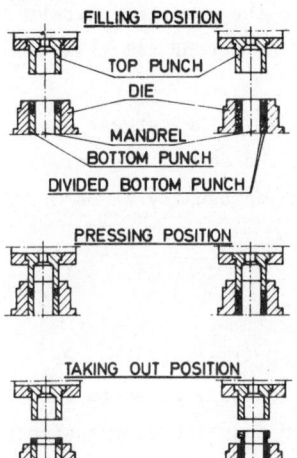

Fig. 4.5. Steps of dry pressing operation

(1) It is important that a homogeneous distribution of the powdered particles
should be used when the die is filled. This can be achieved by using a suit-
ably prepared free-flowing alumina powder and by filling the die evenly, for
example, by means of a fill shoe.

(2) In the pressing position a homogeneous compaction of the powder within the
desired shape should be insured.

(3) Removing the part from the die should be simple and without any risk of
damage.

The favorable economy of the dry pressing process can be obtained only if the
requirements described above are fulfilled.

Due to the substantial tool wear caused by the extremely abrasive alumina powder,
all parts of the dies which are exposed to wear, such as mandrels and punches,
are preferably made of cemented carbides. A lifetime of these tools of up to
200 000 cycles can be achieved. In spite of their considerable cost, the long
lifetime of cemented carbides results in economical operation.

Besides the fabrication of large series of small simple parts, the dry pressing
technique is also applied to the manufacture of blanks intended for subsequent

machining. In an unfired condition the pressed bodies are relatively soft. They can be machined like chalk. All kinds of machining, such as turning, milling, drilling, cutting, or grinding can be applied to this material. This process, however, is economical only for relatively large parts having complicated designs with no possibility of directly pressing the desired final shape, such as, for example undercuts, threads, or borings perpendicular to the pressing direction, as well as for small runs where the tool costs would be out of all reasonable proportion.

The dry pressing technique allows the manufacture of parts with the largest diameter at a reasonable cost. Depending on the dimensions of the facilities for pressing and sintering, parts with a diameter up to 500 mm can usually be obtained.

4.2.2. Hydrostatic Molding

Due to the internal friction of the powder material and lack of plasticity, dry pressing, used unidirectionally, allows a uniform compression only close to the top punch and bottom punch within the die. Inside the green body the compression decreases towards the center. In order to overcome these difficulties, a pressing technique has been developed, called hydrostatic molding, which ensures a very uniform pressure distribution within the pressed body [4.12,4.13].

A rubber mold is filled with the powder to be compacted, preferably by means of vibratory precompaction, inserted in a liquid environment, and exposed to a high pressure inside the liquid (Fig. 4.6). A uniform compression can be obtained, re-

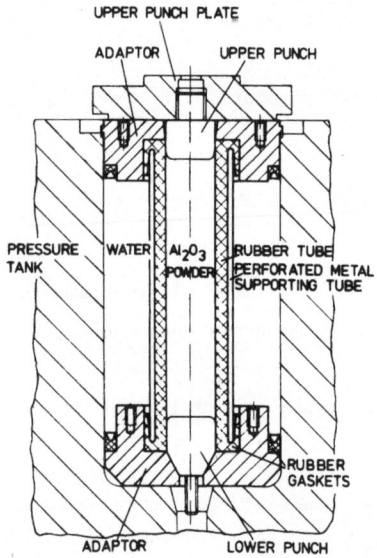

Fig. 4.6. Schematic description of hydrostatic molding

sulting in a symmetrical shrinkage. Because of the sealing problems related to
the high pressure, the maximum pressure to be applied should not exceed 80 MN/m^2.
Bore holes and inside threads can be fabricated by inserting suitable mandrels.
Due to the yielding of the rubber mold, the outer surface of the pressed body is
irregular. A perfect surface can be obtained by a subsequent machining in green,
i.e., in an unfired condition.

Hydrostatic molding is preferably applied to the manufacture of parts with a
length-to-diameter ratio of more than 4 to 1, such as long rods and tubes, particularly with small wall thicknesses, where problems of wall friction and irregular compacting are relevant. It is also used for compacting parts with larger
dimensions which are designated for critical applications with high requirements.
Hydrostatic molding is, however, much less economical than dry pressing, due to
the expensive facilities and the considerably longer time required for one pressing cycle.

4.2.3 Extrusion

Extrusion is one of the oldest ceramic forming processes, with a long history in
the field of refractories or other conventional ceramic materials [4.14]. The application of the extrusion process to high alumina ceramics is, however, not particular common, with one exception.

In contrast to conventional ceramic materials containing minerals of higher plasticity, for example clay, the extrusion of high alumina ceramic powders with no
plasticity of their own requires a considerable quantity of organic plasticizers.
As described in Section 4.1, a careful procedure is necessary in order to homogeneously distribute the organic binder throughout the alumina powder.

Extrusion is the only continuously working ceramic forming process which is usually carried out in two steps:

(1) Forcing a plasticized ceramic material under pressure through a suitable die
    with a shaped orifice.

(2) Cutting the unfired string into required length.

Extrusion of high alumina ceramics is preferably applied to the forming of long
tubes and rods with arbitrary profiles. It is restricted to a diameter below
30 mm and a wall thickness between 0.6 and 5 mm. The string proceeds with a velocity of about 10 mm/s.

The economy of this process is remarkable. The tool costs are relatively low.
Only the orifice has to be periodically replaced. In the case of orifices made
of cemented carbides a long service life can be expected. For this reason extrusion became a preferred procedure for microelectronic applications, where large
production quantities are involved. The forming of thin ribbons (with the wall
thickness of the order of 0.5 to 1 mm) by extrusion with subsequent punching in
an unfired condition turned out to be the most economical process of fabrication
high alumina ceramic substrates for microelectronics.

4.2.4 Injection Molding

The injection molding technique, well-proven in the field of plastics fabrication,
was first suggested for application to ceramic powders, particularly to powders
of high alumina ceramics, by Klingler in 1936 [4.15].

The economical application of injection molding, which is more expensive than
dry pressing, is obtained only in the mass production of parts with complicated
shapes, which do not meet the requirements of dry pressing on account of such
difficulties as undercuts, threads, borings, or breaks perpendicular to the pressing direction. Typical examples are thread guides for textile manufacture.

The procedure is similar to the injection molding of plastics; it is almost the
same as using plastic materials which have 70% to 90% of a mineral filler. Hydraulically operated presses with a load capacity between 0.1 and 1 MN have
proved good for the injection molding process. The compacting of the material,
which has to be preheated up to about $150^{\circ}C$, is carried out either by a piston
or by a worm thread which provides the required pressure of about 100 MN/m^2 for
the injection molding operation. In contrast to extrusion, the injection molding
process works discontinuously, producing parts with predetermined volume, at
an output of about three parts per minute for each die insert.

A press used for injection molding is illustrated in Fig. 4.7. The material is
transported to the injection tank by means of a dosage worm thread, heated up in
the heater, and injected into the water-cooled mold with the pressure built up
by the piston. After ejecting the parts by separating the two halves of the mold
the plasticizer has to be removed by a thermal treatment up to $400^{\circ}C$ in an air
atmosphere, which cracks or oxidizes the hydrocarbons of the organic binder. This
drying process has to be performed very slowly to prevent cracks. The heating-up
period takes five to ten days depending on the wall thickness.

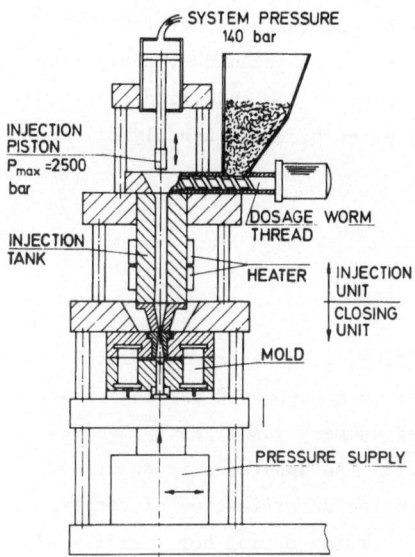

Fig. 4.7. Schematic description of a hydraulically operated press for injection molding

In applying injection molding to metal powders, carbide powders, or powders of other materials which are sensitive to oxidation, the drying procedure is the most critical part of the whole process because it is very difficult to remove the organic plasticizers in a protective, i.e., non-oxidizing atmosphere. Therefore, injection molding has so far not become a standard process in the field of powder metallurgy.

After drying, a deburring process is necessary in order to remove the burrs caused by an incomplete closing of the mold. Due to the high sensitivity to fracture of the parts in a dried condition prior to sintering, this procedure has to be carried out very carefully by hand, using soft tools of cotton or felt. An automation of the deburring is not possible which results in an expensive process.

The injection molding technique is restricted to parts with a maximum length of 100 mm, a wall thickness between 0.8 and 5 mm, and a maximum weight of 50 g. Sharp edges, abrupt changes of diameter, and high orders of accumulation of material should be avoided. The flow properties of the material and the design of the parts to be injection molded are the decisive operating conditions affecting the economy of the process. Inadequate operation conditions as well as incorrect positions of the injection channel or incorrect tool separation produce defects such as blisters, cracks, and isolated pores, which can be detected only in the sintered condition, that is at a higher level with respect to the production costs.

Due to the lubrication effect of the considerably higher quantity of plasticizers relative to the raw material prepared for the dry pressing technique, tools made of hardened steel offer a reasonable lifetime of about 25 000 injection cycles. Therefore, the application of tools made of cemented carbides is usually not necessary and is reserved for special cases only.

4.2.5 Hot Pressing

Hot pressing, a combination of pressing and sintering, is a well-known process in powder metallurgy, particularly in the fabrication of cemented carbides [4.16]. It is also known as pressure sintering. A detailed summary was given by Goetzel [4.17] in 1950, and in a later review by Rice [4.18]. An apparatus is described in [4.19]. Hot pressing plays an important role in the densification of ceramic-metal combinations [4.20]. The basic mechanisms occurring during hot pressing of alumina were described in Section 2.6.

Regarding high alumina ceramics, hot pressing is the most expensive fabrication process, due to the low tool life. Its application is, therefore, restricted to special cases of extreme requirements of the mechanical properties, such as high strength or high impact resistance.

The combination of pressing and sintering allows a reduction in temperature and pressure, while achieving the same densification as obtained for normally required fabrication conditions. Hot pressing results in a significantly finer grain size and consequently in higher mechanical strength.

The material of the tools such as dies and punches has to withstand high mechanical loads at high temperatures. Up to now graphite has proved to be the only material which meets these requirements. Additionally, graphite offers a relatively high electric conductivity, thus allowing direct electric heating either by direct current or by induction heating. An extremely high electric current of the order of 10 000 amperes at a low voltage of about 6 volts permits heating-up within seconds. The mechanical strength of the graphite dies withstands the required pressure of 40 MN/m^2 or more [4.21].

Normally the process is carried out at a temperature of about $1500^{o}C$ with a pressure of the order of 40 MN/m^2. Higher temperatures are not recommended. These enable reactions between alumina and graphite, which deteriorate the properties of the product, and reduce the tool life through a temperature-dependent decrease in the strength of the graphite.

For economic reasons hot pressing is usually carried out in an air atmosphere, sometimes in a flow of nitrogen in order to protect the graphite dies und punches from oxidation. A combination of hot pressing and high vacuum applied to high alumina ceramics as described by Rossi and Fulrath [4.22] will probably remain an exceptional procedure reserved only for laboratory uses and not for fabrication purposes. The same holds for the application of hot pressing to the joining of solid ceramic components [4.23]. For very special applications a hot isostatic pressing technique has been introduced into the field of powder metallurgy [4.24]. Due to the higher temperatures which are required for ceramic powders, this procedure has not proved to be economical.

4.3 Sintering

Sintering is the most important step of the fabrication process. It gives the high alumina ceramic body its final properties. During this procedure, which is usually carried out at 0.85 of the homologous temperature, thermal densification of the precompacted green bodies takes place. At this temperature there is sufficient atomic mobility. Under the influence of the surface tension, which is the driving force of this process, the green body consisting of a multitude of powder particles with a considerable internal surface, for energy reasons, tends to reduce its internal surface. As a result, the precompacted powder particles grow together, which leads to solid-state reactions such as recrystallization and grain growth [4.25]. The scientific basis of sintering phenomena was given in Section 2.5.

Densification, recrystallization, and grain growth occur in the same temperature range. Therefore, a strict control of the sintering process and of small additions of grain growth inhibitors, usually MgO, to the alumina powders are essential in order to achieve a fully dense sintered body with a fine-grained microstructure. In the course of sintering the density increases with the logarithm of time, and the grain size increases with the one-third power of time [4.26].

As illustrated in Fig. 4.8, the density grows linearly, according to the densification process during sintering, until it reaches saturation, representing the final density. A further increase in the sintering time does not improve either the density or the mechanical properties, on the contrary, it promotes grain growth which reduces the mechanical strength. For this reason the sintering process should be stopped as soon as the final density is obtained. Sintering in an air atmosphere achieves a maximum final density of about 99% of the theoretical density which corresponds to a minimum of residual porosity of about 1% due to

trapped air. Beyond saturation there is a decrease of density, which is very
slight in the case of high-purity alumina ceramics and larger in case of high
alumina ceramics due to volume-increasing reactions between the individual additives.

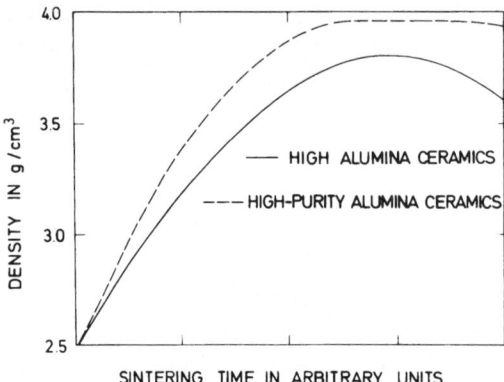

Fig. 4.8. Development of the density of alumina ceramics during sintering

The effects of the various additions to the alumina powder are very different.
In the case of high alumina ceramics, usually having silicate additions, a single addition shows only little effect on the properties. This effect is lost in
the total of all additions which first act to reduce the sintering temperature.

In contrast, the properties of high-purity alumina ceramics depend very strongly
on the various foreign oxides which are added in small amounts (less than 1%) to
the starting powder. As already outlined in Section 2.5.4, MgO is usually the
main additive used to promote sintering in terms of enhancing densification and
impeding grain growth. All other foreign oxides are present accidentally, depending on the purity of the alumina powder.

Basically the additives can promote or impede both densification and grain growth.
A combination of these possibilities leads to four different effects:

(1) Promoting densification and impeding grain growth;
    typical example: MgO.

(2) Promoting densification and promoting grain growth;
    typical example: TiO_2

(3) Impeding densification and impeding grain growth;
    typical example: Ni_2O_3.

206

(4) Impeding densification and promoting grain growth;
    typical example: SiO_2 [4.27].

Effect (1) should be the aim. It satisfies the basic requirements for a high density product with small grain size having the best mechanical properties. Effect (4) should be avoided because it leads to low strength and poor wear resistance. CaO impurities show less influence on sintering; they have, however, a long-term effect in reducing the mechanical strength due to calcium migration and segregation effects and reactions with the environment [4.28,4.29] (see Section 2.7).

The choice of the sintering temperature, which is usually between 1600 and 1800°C, depends on the surface energy, on the grain size distribution, and on the additives to the alumina powder. The sintering time, particularly the heating rate during the heating-up period, should be adapted to the size and the wall thickness of the body to be sintered. The same holds for the choice of a suitable furnace, such as tunnel kiln, box furnace, or special furnace. The furnaces are operated with gas, oil, or electrical energy. The energy consumption for the sintering of 1 kg of high alumina ceramics varies between 10 and 30 MJ.

Figure 4.9 illustrates a typical sintering cycle used for achieving saturation density in medium-sized parts with a diameter of the order of 50 mm. Larger parts require a longer sintering time, particularly a longer heating-up period. Smaller parts can be heated up more quickly, allowing a much shorter sintering time.

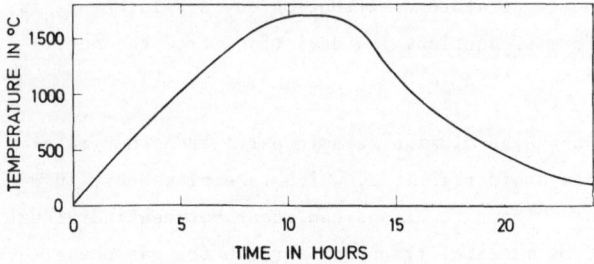

Fig. 4.9. Sintering cycle for medium sized parts

High alumina ceramics allow sintering in every atmosphere. For economic reasons, the sintering process is normally carried out in air. Hydrogen or high vacuum are applied if special properties such as light transmission are required. The use of inert gases is of no practical value.

Due to densification during sintering a linear shrinkage takes place which is of
the order of 20% or 60% shrinkage in volume, within a tolerance of about 1%. The
course of shrinkage occurs under maintenance of the geometric shape. Figure 4.10
illustrates the shrinkage of a complicated part. It can be seen that the shrinkage is always directed towards the so-called "neutral fiber".

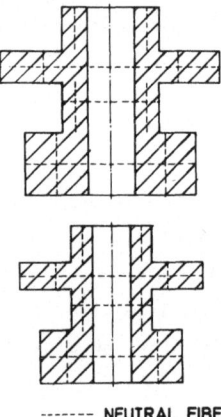

------- **NEUTRAL FIBER**   Fig. 4.10. Shrinkage

The shrinkage rate depends on the precompaction [4.8]. Higher precompaction means
lower shrinkage, and conversely. Non-uniform precompaction within the same body
leads to different shrinkage and consequently to warping, cracking, or to the development of internal stresses. The same results occur if the temperature distribution during sintering is not uniform. Therefore, a constant precompaction,
as well as a strict control of the temperature distribution during sintering, is
necessary in order to obtain uniform production. The deviations from the specified sintering temperature should be within 10°C.

During sintering all contact between high alumina ceramic parts and other materials must be prevented in order to avoid reactions, which take place very intensively in this temperature range. These reactions can occur between individual
bodies even without solid contact by material transport through the gas phase.
Therefore, all components of the furnace that are subjected to high temperatures,
as well as the base plates that the alumina parts are loaded onto during the sintering process, have to be made out of the same materials, usually out of porous
high alumina ceramics.

4.4 Hard Machining

A subsequent so-called hard machining after sintering is necessary if the required final dimensions cannot be attained by the forming and sintering process alone or if there are special surface requirements. Due to their extreme hardness in the sintered condition, high alumina ceramic products can be machined only with tools harder than alumina, such as diamond tools or diamond grit. In special cases boron carbide grit can also be used.

Hard machining of high alumina ceramics is classified into three different groups:

(1) grinding
(2) lapping
(3) polishing.

Grinding is usually applied for dimensional reasons. The grinding tools consist of cylindrical, spherical, or disc-shaped carriers which are covered with metal or plastic-bonded diamond grit of various grain sizes between 15 and 150 µm. According to the rate of material removal the operation is called rough or fine grinding, and requires tools with a coarser or a finer diamond grit respectively. The operation normally starts with a rough grinding process changing over to fine grinding step-by-step as the surface improves, depending on the dimensional and surface requirements.

The lapping process improves the flatness of a flat part. It is carried out using diamond grit or boron grit with a particle size between 1 and 10 µm as lapping agent. If boron carbide is used the process is restricted to the improvement of the flatness of the ceramic part to be machined. In order to obtain a smooth surface with a low roughness, it is necessary to use diamond grit. In contrast to grinding, where the hard particles are strongly bonded to the surface of the tool, lapping agents are loose powders usually suspended in oil, placed between the rotating ceramic work piece and a metal disc which rotates opposite to the ceramic part with a high contact pressure. Lapping can be applied to spherical parts as well. In this case the shape of the metal counterpart has to be well adapted to the shape of the ceramic work piece.

Polishing improves the surface finish by increasing the smoothness, thus reducing the surface roughness. Similarly to grinding operations, polishing starts with a coarser grit in the order of 10 µm, going down step-by-step to the submicron range. Diamond grit is the only effective polishing agent which offers a very low roughness in the range between 0.1 and 0.01 µm. The roughness is measured by means

Fig. 4.11.
High alumina ceramic surface, rough ground

Fig. 4.12.
High alumina ceramic surface, fine ground

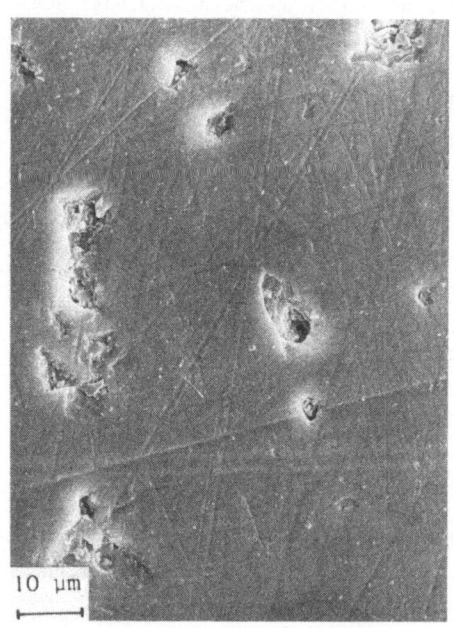

Fig. 4.13.
High alumina ceramic surface, lapped

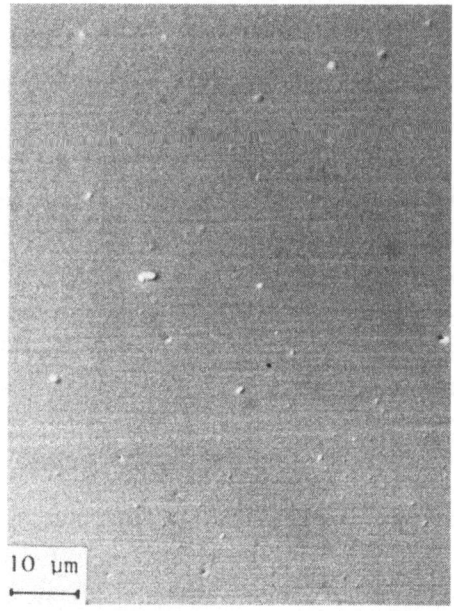

Fig. 4.14.
High-purity alumina ceramic surface, polished

210

of a surfindicator, scanning the surface with a diamond stylus. The figures mentioned here refer to the medium roughness, known under the term "center-line-average" (CLA). Such an extreme smoothness, as indicated above, can be achieved only with a high-purity alumina ceramic grade with full density and a fine-grained microstructure. Polishing can be carried out with hard carrier materials such as metals or with soft carrier materials such as felt, depending on the desired result. Hard carrier materials change the geometrical shape of the part to be machined, soft carrier materials do not.

The grinding, lapping, and polishing machines required for these operations are basically the same as used for the machining of metals and cemented carbides. All hard machining operations should be carried out using effective coolants in order to avoid temperature peaks, created by the high contact pressure, which lead to thermal-shock cracks. Diamond grinding and polishing tools such as grinding or cutting wheels have to be operated with water as coolant or an oil coolant. The same holds for the lapping or polishing grit of diamond or boron carbide which is normally suspended in oil or in a suitable paste.

During the rough grinding process the grains of the alumina microstructure are partly cut or broken out, depending on the grain size of the diamond tool which causes an increase in the surface roughness as compared to the as-fired surface. A reduction of the surface roughness can be achieved by a grinding process using tools with finer diamond grit, thus resulting in a finer cut of the alumina grains. Lapping and polishing operations cause a levelling of the surface. Figures 4.11 to 4.14 represent micrographs of rough and fine ground as well as lapped and polished surfaces of high alumina ceramics.

4.5 Quality Control

Any defect or imperfection of the product which arises during the fabrication process can be tolerated as long as its detection by the quality control is assured. Defects or imperfections of the product which are found only after delivery are not tolerable. They have to be avoided under all circumstances. Therefore quality control is one of the most important of fabrication steps.

Every part has to pass several quality control procedures, particularly crack inspection which is carried out first (immediately after sintering) by means of penetration methods. The parts to be inspected for cracks are dipped into a penetration fluid with extremely low surface tension and subsequently washed with water which dissolves the penetration fluid but has a higher surface tension, in order to remove the penetration fluid from the surface of the parts but not out of the cracks. This way the testing fluid which penetrates into even the smallest cracks or into small surface flaws can stay in the cracks without being dissolved by the water. The fluid either is of red color, offering a good contrast to the white ceramics, or contains fluorescent additives. In the latter case the subsequent visual crack inspection is carried out by means of ultraviolet light.

Cracks are usually defects which might significantly reduce the mechanical strength (Section 3.2). Therefore crack inspection is of the greatest consequence regarding the reliability of the product. It is repeated after all subsequent fabrication operations which are likely to produce cracks in the ceramic parts.

Other properties such as, for instance, density and grain size, which affect the mechanical strength are also determined at random for each production lot. For economic reasons the purity control is carried out with the raw material.

The most extensive control procedures are the dimensional inspections such as wall thickness, inner or outer diameter, radius sphericity, shape, and surface roughness, as well as the detection of surface defects. All of these inspections are performed to a standard of 100%. This also holds for the leak test of all hermetic metal-to-ceramic seals by means of a helium leak detector. Some of the control procedures are carried out during the fabrication, i.e., in the unfired state, in order to ensure the correct performance of the presses and the correct conditions of the plasticized raw material. In addition, all production facilities, particularly the furnaces, which usually have automatic temperature control, are inspected periodically.

4.6 Manufacturing Tolerances

The engineer involved in design problems is interested, first of all, in manufacturing tolerances. When dealing with high alumina ceramics he has to distinguish between two kinds of tolerances. One depends on the sintering and the other on the machining.

The dimensional deviations after sintering are due to irregular shrinkage. This has a number of causes, such as variations of the green density due to non-uniform precompaction which is caused in most cases by inhomogeneous distribution of the plasticizer as well as a non-uniform temperature distribution during drying or during sintering (Section 4.3). Both result in irregular shrinkage and thus in dimensional variations from part to part.

Usually the shrinkage S is related to the linear dimensions according to the following equation:

$$S = \frac{L_1 - L_2}{L_2} \quad , \tag{4.1}$$

where L_1 is the length before and L_2 the length after sintering. Equation (4.2) gives the relations between shrinkage S, green density ρ_g, final density after sintering ρ_f, and the content of plasticizers c in w/o [4.30]:

$$S = 1 - \sqrt{\frac{\rho_g (1 - \frac{c}{100})}{\rho_f}} \tag{4.2}$$

The tolerances due to sintering are normally ± 1.5%.

For most applications, the tolerances which can be achieved after sintering are too large. In these cases subsequent hard machining with diamond tools is necessary. By this means a tolerance of 10 μm can be achieved for relatively small parts. The final dimensions produced by machining alumina can be held within the same tolerance as those allowed for metal parts.

4.7 Principles of Design

In contrast to metals, high alumina ceramics offer their full advantage in those applications which require a combination of properties such as, for example, high mechanical strength combined with corrosion resistance and/or abrasion resistance or the maintenance of these properties up to elevated temperatures. In many of these cases metallic materials fail due to insufficient chemical inertness, poor wear properties, or plastic deformation. High alumina ceramics offer many advantages in these areas.

Frequently, however, the utilisation of high alumina ceramics is limited by the lack of tensile and impact strength. Therefore, one must try to compensate the disadvantages of the ceramic material by a suitable design. Ceramic parts should

be designed to operate under compressive loads. Similar principles could be transferred from the field of gray cast iron.

First of all, some principles have to be observed in order to prevent both notch
effects and the development of internal stresses during fabrication or application:

(1) No abrupt changes of diameter or wall thickness.
(2) All edges, breaks, and corners should be bevelled or rounded with as large a
    radius as possible.
(3) The load to be applied should be distributed over a large part of the surface.

These principles have to be met independently by the individual ceramic parts as
well as by the ceramic components of a compound design (Fig. 4.15).

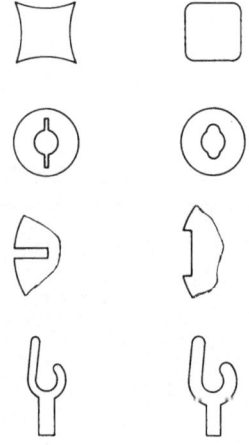

NON - APPROPRIATE
DESIGN

APPROPRIATE
DESIGN

Fig. 4.15. Examples of appropriate and non-appropriate design

One of the compound designs most frequently in use consists of ceramic and metal
components. Rubber and plastics are approved partner materials of high alumina
ceramics as well. Basically all kinds of connecting procedures can be applied
such as mechanical clamping, shrinking, soldering, brazing, and welding as well
as the use of adhesives. Mechanical clamping and shrinking are the only detachable
connecting methods. In those methods the high coefficient of friction between ceramics and metals in dry condition is utilized particularly to provide strong
pressure-fit connections. All other procedures mentioned above result in non-de-
tachable connections.

The choice of the connecting method depends on the operating conditions. Adhesives are used if no elevated operating temperatures are to be expected. Shrinkage has turned out to be a useful connection process for ceramic extrusion dies in a steel frame providing the necessary compressive strain. Soldering or brazing are normally applied to metal-to-ceramic seals in the field of electronic applications (Section 5.1.1) which require high temperature endurance and vacuum tightness. Welding still remains an exceptional procedure for very special laboratory applications.

The components of a compound design should be matched with respect to both thermal expansion and elastic properties in order to withstand elevated temperatures and deformations without damage.

5 Applications

Electronics and mechanical engineering are the main application fields of high alumina ceramics [5.1, 5.2]; two other fields have arisen that are of growing importance: medical and armor applications. Thermal and chemical applications are of minor importance so they are included in the Section "Other Applications".

This representation certainly does not claim to be complete. We have tried to restrict it to the most common and most important applications of high alumina ceramics. Numerous other less important applications, manufactured in small quantities or only in single pieces, had to be omitted out of regard for the volume of this book.

In the United States and Japan electronic applications are dominant, whereas in Europe the field of mechanical engineering comprises about 80% of all applications of high alumina ceramics (as illustrated in Figs. 1.1 to 1.3). Different properties of the material may be considered to be more important than others, depending on the nature of use and service conditions. Table 5.1 shows the characteristics of major interest in the various fields of application.

Electronic applications call for ceramic components with special electrical properties. In most cases high mechanical strength is not required, while in other fields of application it is of great importance. The wear behavior is of interest only in the fields of mechanical engineering and medicine, whereas high temperature properties are not relevant for armor and medical applications. Besides armor, a corrosive environment can be expected within the other fields. Therefore, corrosion resistance is required in many applications.

The alumina content of the high alumina ceramic grades being used in the application fields described here ranges from 95 to 99.9% Al_2O_3. The lower grades have a silicate binder, the higher ones - above 99%, designated here as high-purity alumina ceramics - have only small additions of MgO.

Table 5.1. Requirements of the various fields of application

Required properties	Mechanical engineering	Electronics	Medicine	Armor
Mechanical strength	x	(x)	x	x
Wear resistance	x	–	x	–
Corrosion resistance	x	(x)	x	–
Electrical resistance	–	x	–	–
High tempera- ture endurance	x	x	–	–

x yes   – no   (x) sometimes

Table 5.2 demonstrates the influence of the composition of two typical aluminas on their properties. Additions of silicates generally result in a decrease in mechanical strength, particularly at elevated temperatures, because they are main- ly present as a glassy phase at the grain boundaries. They thus facilitate grain boundary sliding and hence contribute to plastic flow. MgO, however, leads to an increase in the mechanical strength by accelerating the sintering process and in- hibiting discontinuous grain growth, as discussed in Section 2.5.

The variation of the properties with composition can be explained by differences in the microstructure [5.3]. Figure 5.1 represents the microstructure of alumina

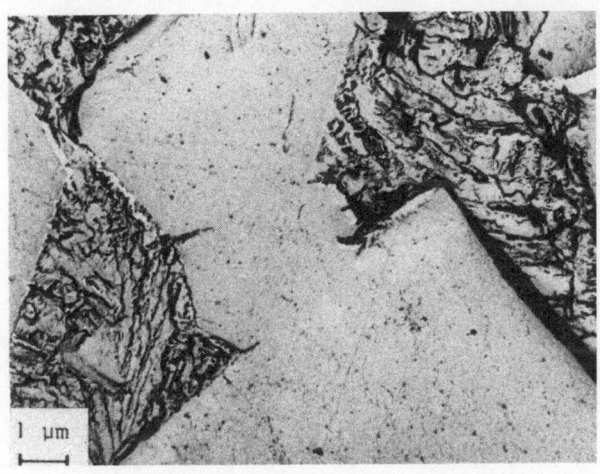

1 µm

Fig. 5.1. Microstructure of alumina ceramics with additions of 10% silicates

Table 5.2. Some characteristic properties of two typical high alumina ceramics of different composition

Property	Unit	Composition	
		99.7% Al_2O_3 +0.25% MgO	97% Al_2O_3 +Silicates
General properties			
Average grain size	μm	4	10
Density	g/cm^3	3.9	3.7
Residual porosity	%	0.2	7.2
Leak rate	mbl/s	$<10^{-12}$	$<10^{-12}$
Mechanical properties			
Compressive strength			
at 20°C	MN/m^2	5000	3000
at 1000°C	MN/m^2	2000	600
Flexural strength			
at 20°C	MN/m^2	500	300
at 1000°C	MN/m^2	400	200
Electrical properties			
Volume resistivity			
at 100°C	Ωcm	10^{14}	10^{13}
at 500°C	Ωcm	10^{12}	10^{11}
at 1000°C	Ωcm	10^7	10^6
Dielectric strength at 20°C	kV/mm	30	18
Loss coefficient (tanδ) at 20°C and 4000 megacycles	–	$1 \cdot 10^{-4}$	$14 \cdot 10^{-4}$
Thermal properties			
Thermal conductivity	J/cmsK	0.20	0.38
Thermal expansion	1/K		
between 0 and 300°C		$6.7 \cdot 10^{-6}$	
0 and 500°C		$7.3 \cdot 10^{-6}$	
0 and 1100°C		$9.5 \cdot 10^{-6}$	

ceramics with additions of 10% silicates. The alumina grains, which have relatively sharp edges, are embedded in a glassy silicate phase. They do not change their shape during sintering, which is an indication of low sintering temperatures. The silicate phase controls the strength properties. As the weakest link in

the chain of load-bearing microstructural features within the sintered body, it
is the reason for the considerably lower mechanical strength of silicate-contain-
ing alumina ceramics.

In cases of wear or corrosive attack the silicate binder, which has lower hardness,
lower wear resistance, and less chemical stability, is eroded or dissolved, leaving
behind isolated alumina grains that lose their adherence. This leads to a strength
loss of the whole mechanically attacked surface.

With respect to electronic applications, the silicate binder determines the
properties as well. As a consequence of the interconnected glassy phase a
decrease of volume resistivity and dielectric strength and a considerable
increase in the ionic conductivity of the system occur.

In contrast to this, the microstructure of a high-purity alumina ceramic
grade with an addition of 0.25% MgO shows an essentially single-phase material
consisting of equiaxed grains sintered together directly (Fig. 5.2). The absence of the silicate binder results in a considerable increase in wear resistance and corrosion resistance as well as in an improvement of the mechanical strength and of the electrical properties.

The surface grains which are shown in Fig. 5.2 have rounded edges due to the
effect of surface tension during sintering, which is an indication of high
sintering temperatures. When exposed to mechanical loads, fracture occurs at
considerably higher stresses compared to the silicate-bonded alumina material.
As observed in many cases, the fracture is transcrystalline if the microstructure
is coarse grained and intercrystalline if the microstructure is fine grained.

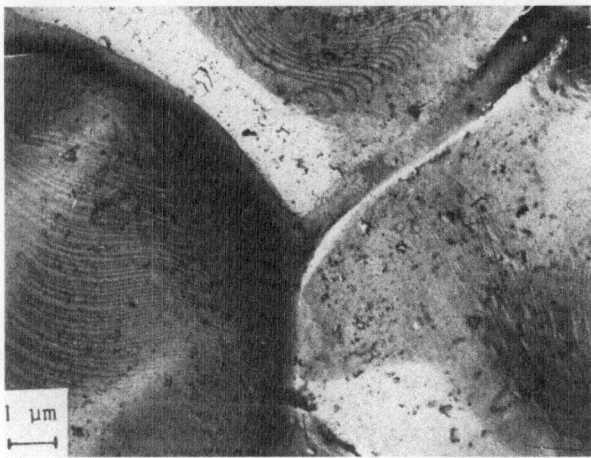

Fig. 5.2. Microstructure of high-purity alumina ceramics with a natural (as-fired) surface

The effect of grain size on the mechanical strength was discussed in detail in Section 3.2. The relationship between the average grain size and the compressive strength according to experimental results, as given in Fig. 5.3, shows that even a slight increase in grain size results in a significant decrease in the compressive strength [5.4].

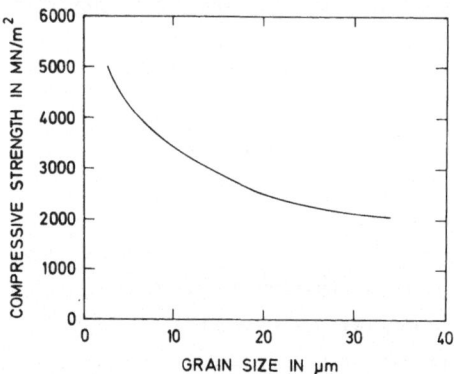

Fig. 5.3. Compressive strength of high-purity alumina ceramics vs. grain size

5.1 Electronic Applications

The applications in the field of electronics require an insulating material with relatively high mechanical strength for rectifier housings, with high volume resistivity at elevated temperatures (Fig. 5.4), and with low dielectric losses for transmitter tubes, with high density resulting in an increased thermal conductivity and in an excellent polishability for microelectronics, and with translucency for discharge lamps [5.3].

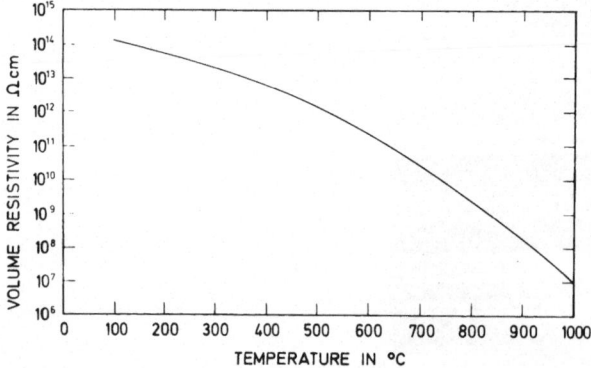

Fig. 5.4. Volume resistivity of high-purity alumina ceramics vs. temperature [5.5]

Most of these requirements can be met only by a high-purity alumina ceramic quality. A silicate content can only be tolerated if electrical resistance and dielectric properties are not specified, since it reduces both volume resistivity and dielectric strength and increases the dielectric losses.

5.1.1 Metal-to-Ceramic Bonding

Within the electronic field there are many applications which require hermetic metal-to-ceramic seals with a leak rate below 10^{-12} mbl/s and high bond strength at a service temperature of at least $600^{\circ}C$ [5.3].

In order to meet these requirements the molybdenum-manganese process, which will be described briefly in the sequel, is normally applied here [5.6]. Only in this special case is there a need for a certain amount of silicate phase. During this process a layer of a finely dispersed molybdenum and manganese suspension is deposited on the alumina surface by means of a brush or by printing. In a subsequent thermal treatment at temperatures in the range of $1500^{\circ}C$ in a hydrogen-nitrogen atmosphere with a dew point of about $25^{\circ}C$ the manganese oxidizes, thereby forming an aluminum-manganese-spinel. This spinel reacts with the silicates, giving the necessary bond strength at the interface between alumina and the metallizing layer [5.7,5.8]. Subsequently, the layer is covered with nickel in order to improve the wetability of the surface. The process described here allows brazing temperatures of up to $1000^{\circ}C$. Normally, a silver-copper eutectic alloy with a melting point of $780^{\circ}C$ is used as a braze.

The metal partner should be matched with respect to the thermal expansion in order to prevent excessive thermal stresses at the metal-ceramic interface. Alloys with a base of iron-nickel or iron-nickel-cobalt have proved satisfactory. The best correspondence in thermal expansion is given with tantalum, niobium, or a niobium alloy with 1% zirconium (Fig. 5.5). But tantalum and niobium cannot be handled in a hydrogen or nitrogen atmosphere. They need either a high vacuum or a special purified inert gas atmosphere, and this would increase the cost of the process significantly.

Another method of producing metal-to-ceramic seals is the so-called active-metal process. In this case brazing alloys with a base of titanium or zirconium are placed between the ceramic and the metal components. The procedure is carried out in a vacuum of about 10^{-7}b. Under these conditions titanium or zirconium oxidizes and reacts with alumina. The reaction products are very brittle. Therefore the bond strength and the vacuum tightness cannot be guaranteed for a long period of time. This is why the active-metal process did not succeed.

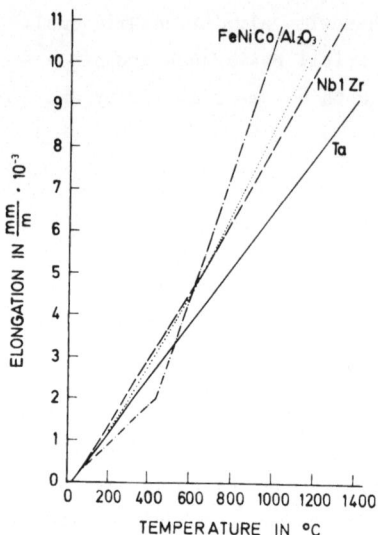

Fig. 5.5. Thermal expansion of high-purity alumina ceramics and of different metal partners, which are matched to alumina, vs. temperature

Due to the fact that the brazing temperature does not exceed 1000°C - this holds for the molybdenum-manganese process as well as for the active-metal process - the working temperature must be kept below 600°C. For higher operating temperatures other procedures had to be developed [5.9].

In the field of nuclear technology there are special applications for metal-to-ceramic seals with much higher operating temperatures, for example, in the case of thermionic convertors which require operating temperatures of at least 1000°C, as well as an extremely high corrosion resistance, particularly against cesium vapor, and which thus require extremely high-purity alumina ceramics. For this application a special metallizing process was developed [5.4]. Suspensions of tungsten powder with small additions of yttrium oxide were applied to the alumina surface using a firing temperature of 1900°C in a hydrogen-nitrogen atmosphere with a dew point of 18°C. The metallizing layers obtained with the process described here could be brazed in a temperature range of 1500°C with alloys based on palladium and vanadium using a niobium-zirconium alloy as metal partner. Those hermetic seals withstand operating temperatures up to 1200°C, meeting the requirements of thermionic converters.

A further increase in operating temperature can only be achieved by a process which does not comprise the metallizing and brazing technique, for above 1500°C the metallizing layers, which are only a few microns in thickness, would be dis-

222

solved by the brazing alloys [5.10]. In this case it is necessary to carry out a solid-state bonding without a liquid phase. This can be obtained by a combination of pressure and temperature below the melting point of both components, a kind of welding process [5.11,5.12]. Another possibility is the constitution of a laminated composite with metallic faces, a ceramic middle layer, and corresponding transition zones, as illustrated in Fig. 5.6.

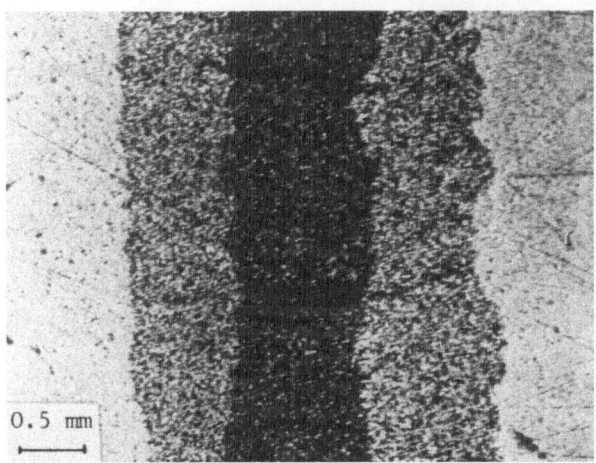

Fig. 5.6. Cross section of a laminated ceramic-metal composite [5.13]

Similar procedures have been used in order to join alumina layers to layers of tantalum and niobium [5.14]. But all these solid-state bonding processes have so far had no large scale application.

5.1.2 Spark-Plug Insulators

The first important practical application of high alumina ceramics in the field of electronics was the production of spark-plug insulators.

The requirements for the insulating material with respect to mechanical strength, electrical resistance, and dielectric strength at elevated temperatures grew with increasing engine power. In particular, the part of the insulator projecting into the engine is additionally exposed to thermal shock and chemical attack. Therefore it was necessary to choose a relatively pure alumina ceramic grade and to develop new production techniques – such as, for example, the injection molding technique – and an appropriate compound design for spark-plug insulators as well as for spark-plug suppressors [5.3]. Figure 5.7 shows spark-plugs (standard and interference-suppression types) and ceramic components belonging to them.

223

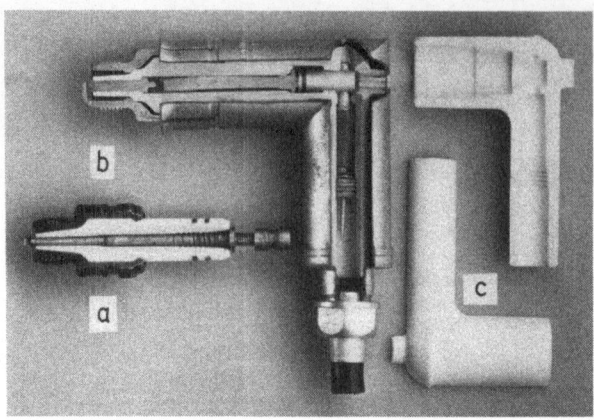

Fig. 5.7. Cross section of a standard spark-plug (a), an interference-suppresion spark-plug (b), and a spark-plug suppressor of high alumina ceramics (c)

Nowadays spark-plug insulators are normally made from ceramic materials containing 94% Al_2O_3. Only high power engines as for racing-cars or for propeller airplanes, require an insulating material with a higher alumina content, and this is a rather small portion of the production.

5.1.3 Components and Housings for Electron Tubes

Similar production techniques are applied to tube components with complicated shapes, such as they are used inside electron tubes (Fig. 5.8). These components require a high electrical resistance at elevated temperatures, whereas the strength properties are relatively unimportant. Since they are operated under high vacuum, the surfaces of the ceramic components must not emit gaseous substances, even under the influence of ionizing radiation.

Aluminas of less purity would deteriorate the vacuum because their additives, being silicate phases in most cases, have a higher vapor pressure compared to pure alumina. Furthermore, the additives used change their properties or transform into other phases under the effect of irradiation by electrons.

The trend to higher power, higher temperature, and higher frequency in the field of transmitter tubes has resulted in a replacement of glass by alumina ceramics, particularly for envelopes (Fig. 5.9) and for microwave windows. By using alumina instead of glass the operating temperature can be significantly increased. This permits either to enhance the electric power output at constant tube size or to reduce the size of the tubes at the same power output. Another advantage of using alumina consists in the fact that the degassing temperature can be raised and adsorbed gases can be evaporated more effectively. This is one of the reasons for the longer lifetime of ceramic tubes as compared to glass tubes.

Fig. 5.8. Components for the interior of electron tubes made of high alumina ceramics

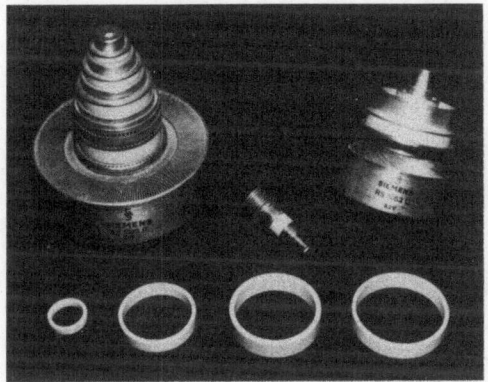

Fig. 5.9. High frequency transmitter tubes in metal-ceramic compound design for television and radio relay systems (back); high alumina ceramic components (front) (courtesy of Siemens)

A number of other favorable properties have further supported the use of alumina as a tube envelope material. Among these, we cite the good vacuum tightness (even at temperatures up to 1200°C), low dielectric losses, and the relatively high thermal conductivity which is particularly useful when effective cooling is needed [5.15].

Simple production, uncomplicated design, high power rating with respect to temperature, small space required, long lifetime, and high reliability are the determining factors for the application of high alumina ceramics in the field of space technology. The standards in this area are extreme and not comparable with the standards in other technical fields. Aside from the mechanical loading and vibrational stresses during the start of the rocket, completely new specifications arise because of weightlessness during the flight. This means that the inside of the tube must be absolutely free from production residues, microchips, or dust particles which in the absence of gravity could move into the electric fields between the narrow-spaced electrodes, leading to defects of serious consequence or short-circuits. Figure 5.10 shows as an example an electron tube based on alumina insulating parts. In 1965 this tube transmitted the first pictures of the Mars surface to earth over a distance of 220 million km.

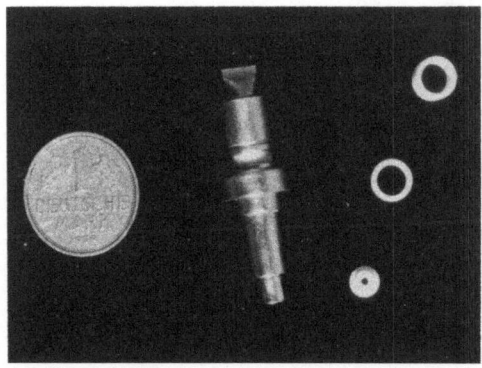

Fig. 5.10. Ultra-high frequency disc-triode in metal-ceramic compound design, prototype for Mars probe "Mariner IV" (middle); high alumina ceramic components of the tube (right) (courtesy of Siemens)

The most common application with the largest quantities described in this section are housings for rectifier tubes and thyristors (Fig. 5.11). They require hermetic metal-to-ceramic seals of high bond strength combined with high mechanical strength of the alumina parts. In most cases a very robust design is necessary for the assembling procedure of the rectifiers as well as for the operation. The electrical and thermal properties, however, are relatively unimportant. The operation temperature does not exceed 200°C [5.16]. In order to prevent a deposition of impurities on the relatively rough outer surface of the ceramic housing a glaze is usually applied, which is much smoother than a ground alumina ceramic surface.

Electrical terminals and bushings as feedthroughs for ultrahigh vacuum furnaces or for other special equipment which have to be heated at temperatures above 400°C also require replacement of the insulating material glass by alumina ceramics (Fig. 5.12) [5.17].

Fig. 5.11. Housings for silicon rectifiers and silicon thyristors in metal-ceramic compound design

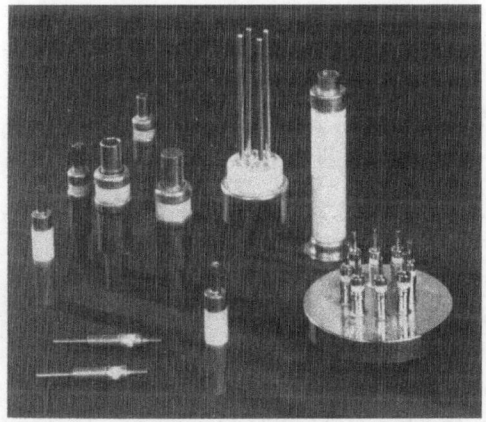

Fig. 5.12. Terminals for high temperature operation in metal-ceramic compound design

The application of high alumina ceramics of greatest dimensions up to now is that of vacuum chambers for particle accelerators. The problem of hermetically sealed accelerator chambers that are unaffected by high energy electron irradiation and therefore reliable in operation for a long period of time could be solved for the first time by the application of high alumina ceramics.

The accelerator consists of a circular tube with a diameter of about 100 m which is divided into single chambers. Each chamber is assembled by 15 elliptical alumina tubes of about 300 mm in length that are metallized and brazed to metal flanges at both ends (Fig. 5.13).

227

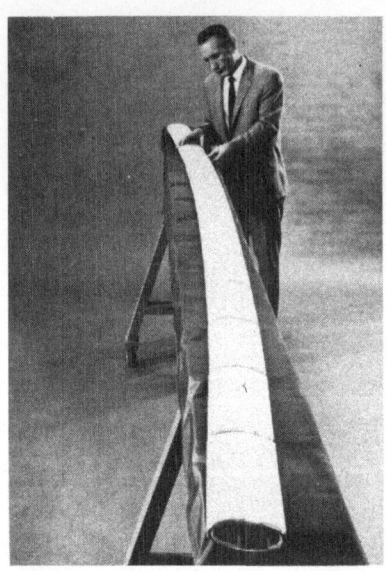

Fig. 5.13. Vaccuum chamber for an electron accelerator in metal-ceramic compound design

Due to the fact that the accelerator is exposed to an extremely strong magnetic field, in order to prevent highly undesirable eddy currents the flanges of the metal partner must not be made of a ferromagnetic alloy. The alloys based on iron-nickel or iron-nickel-cobalt which are normally matched to alumina ceramics with respect to the thermal expansion are ferromagnetic and unfortunately cannot be taken into consideration. Instead of these metals an alloy based on copper-nickel having a thermal expansion 50% in excess of alumina ceramics has to be used. Therefore a special design of a compression seal was chosen which takes advantage of the difference of expansion coefficients. During cooling down from the brazing temperature the metal flange shrinks onto the ceramic component, thus forming a precompressed seal with a considerably increased bond strength, as illustrated in Fig. 5.14.

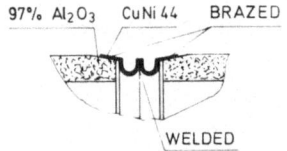

Fig. 5.14. Hermetic metal-to-ceramic connection (compression seal) [5.18]

Additionally, a conductive layer is deposited inside the ceramic tubes by metallizing in order to heat the whole system directly and to prevent electrostatic charges during operation when single high speed electrons collide with the inner wall.

The advantages of the ceramic design compared to previous conventional devices are obvious: better vacuum through higher degassing temperatures, fewer interruptions caused by leakage, and longer operation periods through higher reliability [5.3].

5.1.4 Discharge Lamps

Normally the light transmission of polycrystalline alumina ceramics is relatively small, in contrast to single-crystal alumina, which shows a complete transparency. This holds for the total transmission as well as for the in-line transmission. The reason for the small light transmission of polycrystalline alumina are grain boundaries and residual pores acting as scattering centers for the light. Thus, the translucency of a polycrystalline ceramic material can be increased by completely eliminating the residual porosity.

Residual pores are single and isolated voids caused by trapped air during sintering which is normally carried out in an air atmosphere. The sintering atmosphere plays an important role in obtaining a fully dense material and hence a high transmission coefficient. To eliminate the pores it is necessary to remove the gas from the pores before they become isolated. According to a basic study on the solubility of gases in alumina conducted by Coble [5.19] the alumina lattice shows only some limited solubility for pure hydrogen and pure oxygen whereas nitrogen and all inert gases are not soluble. From this it can be concluded that a sintering atmosphere containing nitrogen – like air, for example – or inert gases will never lead to a translucent alumina body. But when sintered in hydrogen – a sintering process in oxygen is not practicable – a remarkable total transmission of up to 96% can be obtained provided that the alumina is of high purity and the sintering time is long enough to allow the gas to diffuse through or into the alumina lattice. The disadvantage of this process, however, is an exaggerated grain growth because of the long sintering time, leading to a significant loss in strength compared to a fine-grained alumina body.

Therefore, a completely different production process of translucent alumina ceramics has been developed which prevents large grain size [5.20,5.21]. The basic idea is to remove the sintering atmosphere from the pores not by diffusion during sintering but rather by evacuation before the beginning of final-stage sintering. This idea can be realized by carrying out the final-stage sintering under high vacuum. In this case the green body of very pure alumina is prefired in air in order to remove all organic binders at a temperature low enough before the pores become isolated. Subsequently, the body is placed into a vacuum furnace and evacuated. During the heating-up period the pores, which are now free from any gas, can close completely at once; this permits the application of short sintering times, which leads to a sintered body with full density and a fine-grained microstructure. The total transmission turns out to be the same as that of alumina bodies sintered in hydrogen (Fig. 5.15).

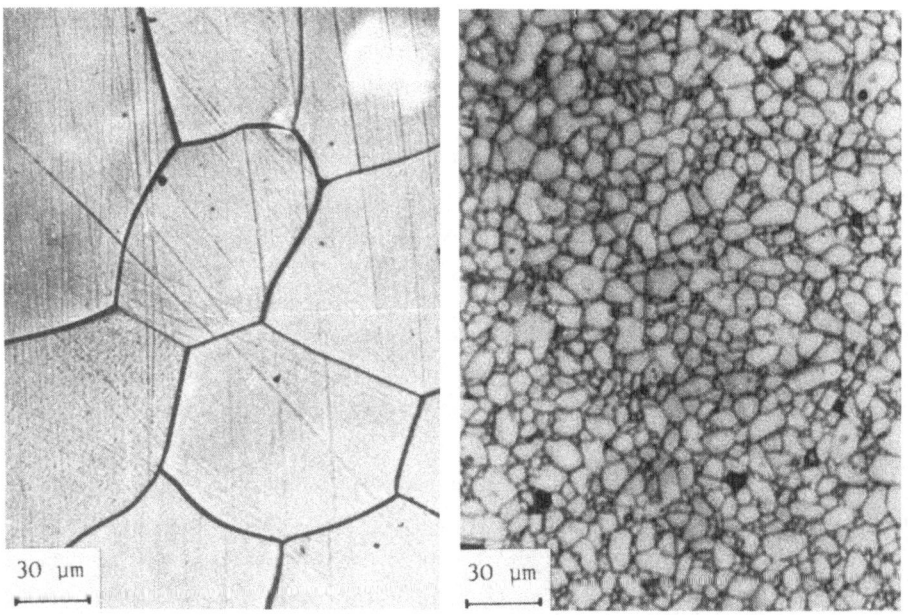

Fig. 5.15. Microstructure of translucent high-purity alumina ceramics having the same total transmission.
(a) Sintered in hydrogen; flexural strength: 350 MN/m^2;
average grain diameter: 50 μm.
(b) Sintered in high vacuum; flexural strength: 520 MN/m^2;
average grain diameter: 4 μm

The in-line transmission, however, is small in both cases. That means that the alumina body is not transparent like ordinary glass but only translucent like milky glass.

Besides hydrogen sintering and vacuum sintering, the hot pressing technique has also been suggested for manufacturing translucent alumina bodies [5.22]. This procedure, however, turns out to be very expensive if applied to shapes like tubes. Since the tube is almost the only shape used for translucent alumina, the hot pressing technique does not succeed in this application field.

The considerably high total transmission, up to 96%, offers a completely new application for this special alumina ceramic quality: envelopes for gas discharge lamps, particularly sodium vapor discharge lamps.

Using glass as the envelope material, these lamps can be operated at temperatures not exceeding 600°C. This limit is given by the restricted chemical resistance of glass. At higher temperatures the glass is attacked by the sodium vapor leading to a reduction of transparency or to a leakage. Contrary to glass pure alumina does not react with sodium vapor up to 1500°C. Therefore discharge lamps with envelopes made from translucent alumina ceramics can be operated up to that temperature [5.3].

A typical alumina-based high-temperature discharge lamp consists of a high-purity alumina ceramic tube sealed at both ends by niobium caps leading trough the tungsten electrodes. In order to protect the metal components from oxidation the interior is enclosed in a glass tube filled with an inert gas (Fig. 5.16).

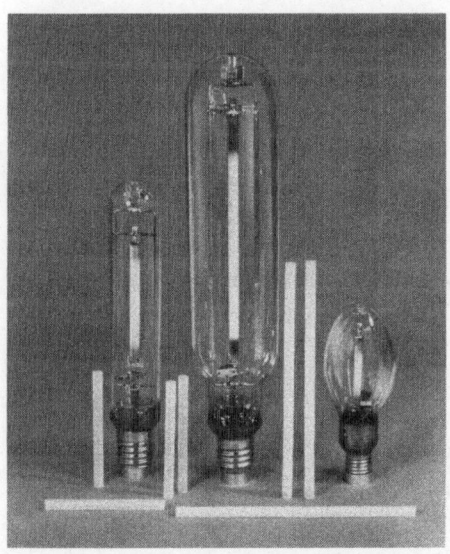

Fig. 5.16. Sodium vapor discharge lamps with tubes of translucent high-purity alumina ceramics for a lamp power from 100 to 400 W; ceramic tubes in front

Compared to conventional sodium vapor discharge lamps the light output of the ceramic lamp could be increased by 100% and the undesirable yellow color of the light could be changed almost to white due to the higher operating temperature. For the ceramic lamp a lifetime of 10 000 operating hours is guaranteed. The main application fields are lighting for airports, sport fields, golf courses, streets, and workshops (Fig. 5.17).

Fig. 5.17. Airport of Düsseldorf illuminated by sodium vapor discharge lamps with tubes of translucent high-purity alumina ceramics

5.1.5 Microelectronics

Microminiaturization of electronic circuits raises problems of heat transfer. Plastic substrates permit only low temperatures. Glass substrates have insufficient thermal conductivity. Therefore, these materials have had to be widely replaced by high alumina ceramics.

In addition, the automatic mass production of electronic circuits requires substrates with extreme precision with respect to outer and inner dimensions, flatness, as well as high mechanical strength, and with edges, which are insensitive to handling operations. Extreme precision is required by the tools that are necessary for the automatic assembly.

The surface has to be very smooth and free of pores in order to allow the application of extremly uniform metal layers by sputtering or vacuum deposition techniques. Deviations of the metal layer thickness as well as deviations from the values specified for the surface roughness of the substrates or the occurrence of surface pores will lead to an inhomogeneous distribution of the electrical properties within the same surface.

The extremely good surface finishing of alumina substrates, particularly those for the thin film technique, can be achieved by a thin glaze if no high thermal conductivity is required, or by a special polish. In this case a medium surface roughness of 0.02 μm (CLA) can be obtained if the alumina ceramics are of high purity and high density (Figs. 5.18 and 5.19). This procedure, however, is rather expensive and not suitable for large scale production. Therefore the development of high alumina substrates for thin film technique was directed towards the fabrication of smooth surfaces in the as-fired condition by considerably reducing the grain size [5.23 to 5.26].

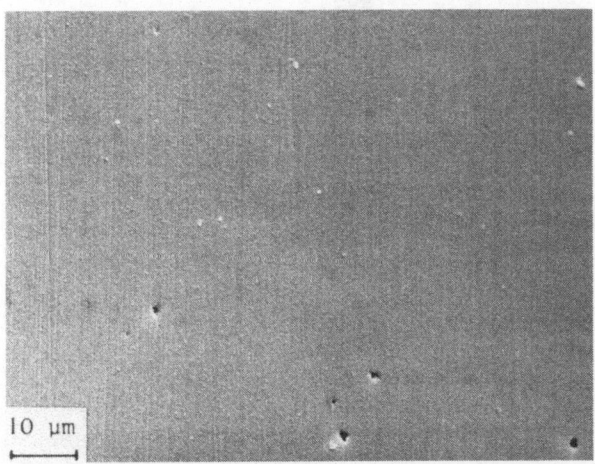

Fig. 5.18. Electron micrograph of a polished surface of high-purity alumina ceramics. Roughness: 0.02 μm (CLA)

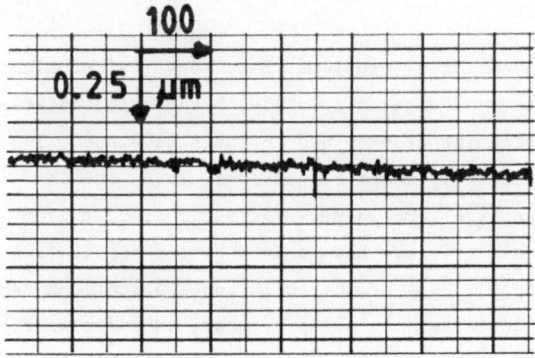

Fig. 5.19. Surface diagram of a polished surface of high-purity alumina ceramics. Roughness: 0.02 μm (CLA)

In the course of further miniaturization, the integrated circuits were embedded in the substrate material. This led to the so-called multilayer technique consisting of alternating layers of ceramics and metal. The printing of the metal paste onto the ceramic surface is carried out in the unfired state [5.27,5.28]. The firing requirements of ceramics and metal have to be coordinated, for the sintering process takes place for both materials at the same time. Figure 5.20 shows various types of layer packages which have been manufactured according to the production technique described above.

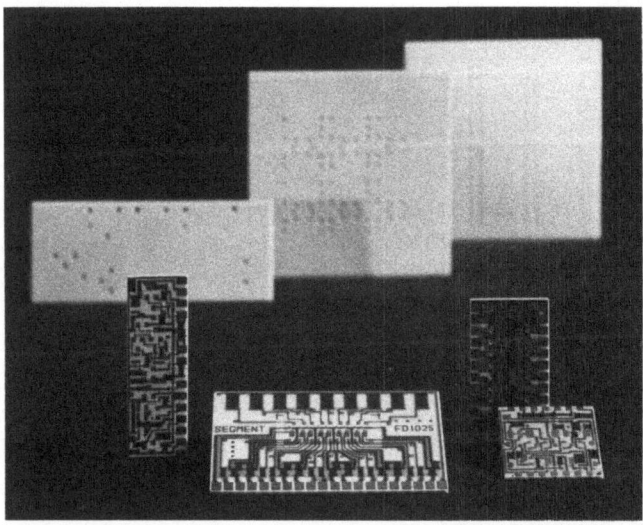

Fig. 5.20. Various types of layer packages (courtesy of Feldmühle Kyocera)

5.2 Mechanical Engineering Applications

Technical progress is often limited by material problems. Many technical problems had not been solved or had only been solved unsatisfactorily, because the materials with properties required by a particular problem had not been developed. In the field of mechanical engineering, numerous material problems, i.e., the increase of wear resistance, strength at high temperatures [5.29,5.30], corrosion resistance [5.31], or dimensional stability, have been solved by the application of high alumina ceramics [5.32,5.33,5.34]. Most of the following applications require a material with a considerable resistance to abrasive wear, adhesive wear, erosive wear, and fatigue wear [5.35]. Because of their extreme hardness, high alumina ceramics are well suited to meet these requirements. High hardness, while necessary, is not sufficient for the suitability of a material exposed to extreme abrasion, for instance. The relevant property is high wear resistance. However, it has to be considered that wear resistance cannot be described as an isolated material property. The behavior of the material depends on the existing operating conditions. A number of elementary wear processes, each of which depends differently on the physical properties, contribute to the totally observed wear [5.36, 5.37].

High alumina ceramics withstand wear and abrasion to a considerable extent. In applications where two materials slide upon each other, alumina can successfully be combined with materials of lower hardness. Due to gas adsorption phenomena at the surface, high alumina ceramics can even be combined with itself, provided that sufficient lubrication exists.

As described in Section 2.1, the atomic structure of the alumina crystal is characterized by hexagonal stacking of closely packed oxygen layers, with aluminum ions on octohedral sites. Based on this structure, a mechanism has been described in [5.38] to explain the superior wear and friction properties in terms of the adsorption behavior of the surface. A cross section through a sphere model of the surface is shown in Fig. 5.21. The outer ionic layer consists of oxygen ions rather than aluminum ions. Since the bonds are not saturated, a surface charge remains. Polarized molecules, such as water or long-chain carboxylic acids are attracted by this charge and become adsorbed due to van der Waals type bonding. This proceeds until a monolayer is formed. This process is known as chemisorption [5.38]. At higher concentrations of water vapor or carboxylic acids, additional layers are bonded physically. They are very stable up to a temperature of $200^{\circ}C$. By acting as a protective layer, the adsorbed molecules are expected to reduce the measureable wear to zero. In fact, this can be proved in practice under normal loading conditions (below 10 MN/m^2) provided that the articulating surfaces are sufficiently parallel [5.38].

5.2.1 Thread Guides for Textile Machines

One of the first large-scale applications for high alumina ceramic parts in the field of mechanical engineering was thread guides for textile machines. Increasing working speeds as well as the introduction of synthetic fibers called for deviation elements and guide elements with increased wear resistance.

Previously applied materials such as hardened steel, glass, or glazed porcelain were not suitable due to their poor abrasion stability. They wore out within a short period of time (from several hours up to several days) when the extremely abrasive synthetic threads were drawn across the surface with a speed in the or-

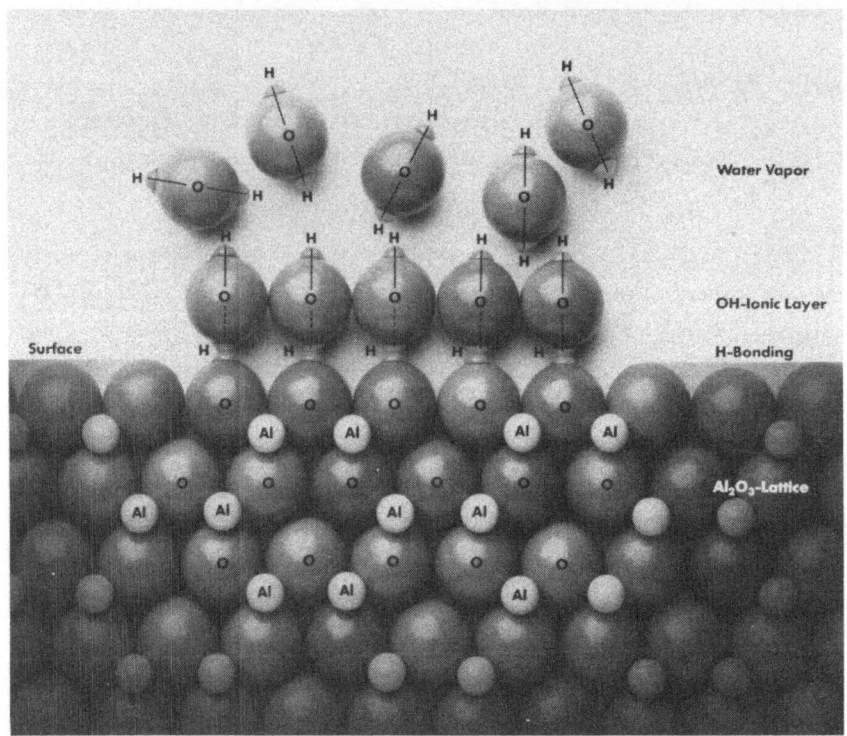

Fig. 5.21. Adsorption behavior of aluminum oxide

der of 1000 m/min. This is a typical example of solving a wear problem with high alumina ceramics, for ceramic thread guides withstand these conditions very successfully. The lifetime could be increased up to more than ten years [5.37,5.32, 5.33].

Besides wear resistance, other properties are of importance such as a constant low friction and a microscopically smooth surface, in order not to damage the very sensitive single filaments of a thread. All requirements for thread guides listed below can be fully met with high alumina ceramics [5.36].

Constant friction	Mechanical strength
Wear resistance	Dimensional stability
Corrosion resistance	No damage to the thread.

Figure 5.22 presents a collection of thread guides made of high alumina ceramics, such as deviation elements, stretch pins, and thread brakes. Figures 5.23 and 5.24 show thread guides in operation.

236

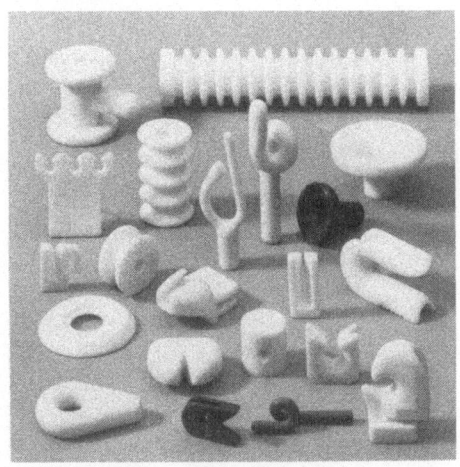

Fig. 5.22. Collection of thread guides made of high alumina ceramics

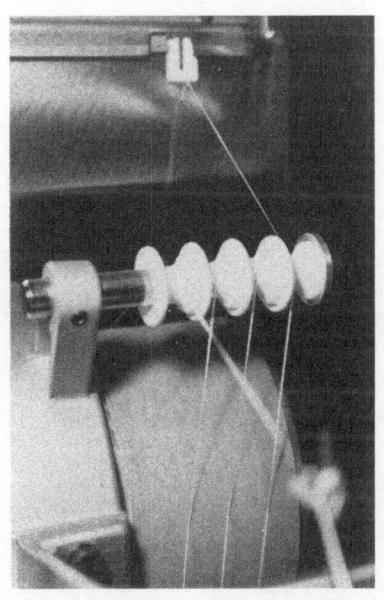

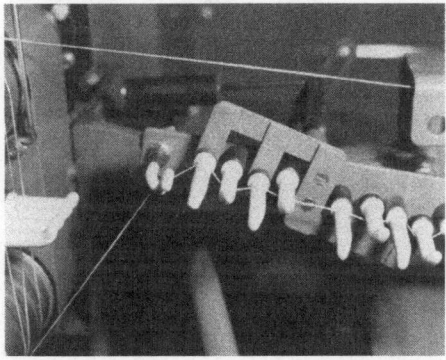

Figs. 5.23 and 5.24. Thread guides made of high alumina ceramics in operation

The most suitable surface to meet the requirements listed in the table can be obtained by a controlled sintering process. As seen in Fig. 5.2, the surface is
shaped by the surface tension when high sintering temperatures are applied. The
grains of the surface tend to form spherical shapes. Under the electron microscope the surface looks similar to an orange peel. All sharp edges, previously
present, have disappeared, and nothing is left to damage the thread which slides

smoothly across this curvature (Fig. 5.25). The grain size and the grain size distribution control the friction by determining the extent of the contact area.

If high friction is required, as for example in the case of a thread brake, the surface profile can be removed and leveled by a grinding and polishing process with diamond tools (Fig. 5.26). The area of contact between the thread and the thread guide is thus enlarged considerably, leading to a significant increase in friction.

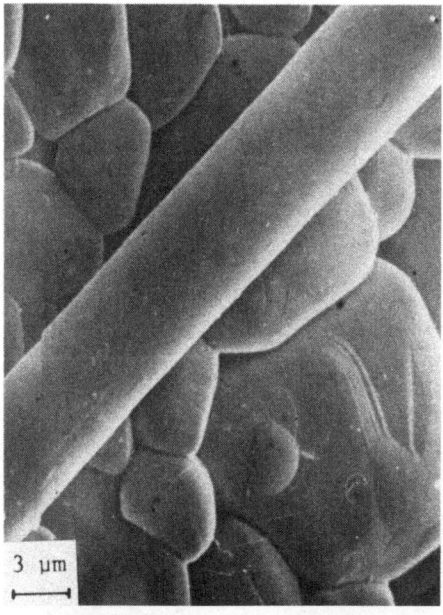

Fig. 5.25. Synthetic fiber sliding across the natural surface of a thread guide made of high purity alumina ceramics

As can be seen from Fig. 5.22, thread guides normally have a rather complicated shape. It was therefore necessary to develop a suitable technique for economical mass production. The injection molding process, developed in 1936 for spark-plug insulators, could be successfully transferred to the production of thread guides.

Problems of fixation can be basically solved with the aid of adhesives. But screwing and clamping also turned out to be appropriate methods. When applying these different kinds of fixation it has to be taken into consideration to expose ceramic materials to compression and not to tension or bending (Section 4.7).

An interesting exception to that rule can be found in the field of "false-twist-spinning" [5.39] by means of the so-called "diabolos", which are used in the

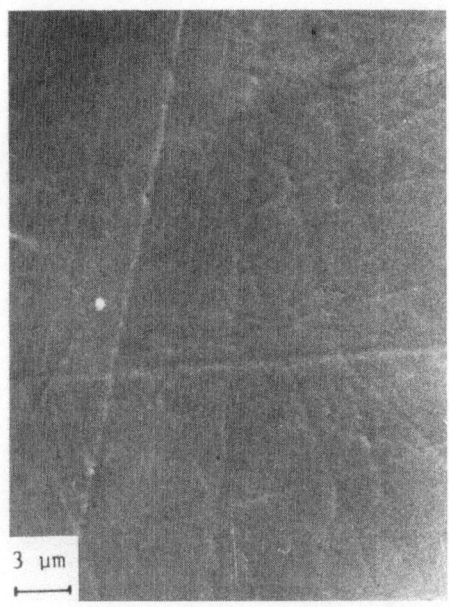

3 µm

Fig. 5.26. Polished surface of high purity alumina ceramics

fabrication of stretch yarn. In this process the smooth synthetic fibers are given an enlarged surface by curling in order to improve their thermal isolation as well as their elasticity. The "diabolo" (Fig. 5.27) - a ceramic pin with a length of a-bout 5 mm, a diameter of about 2 mm having a lace of about 0.7 mm around which the fiber is entwined by 360° - is inserted into a spindle rotating at 600 000 rpm in an air bearing (Fig. 5.28). The ceramic pin is exposed to tensile stress due to

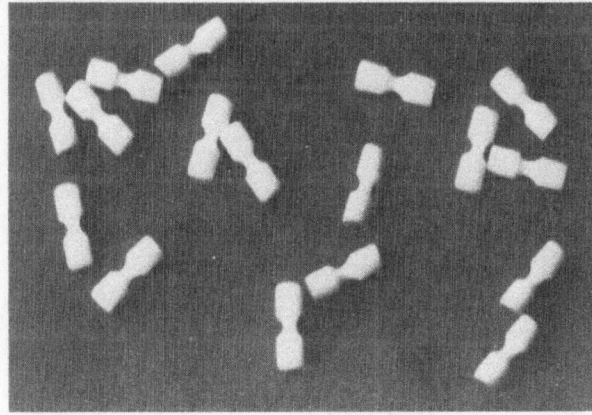

Fig. 5.27. "Diabolos" made of high-purity alumina ceramics for "false-twist-spinning" process

the extreme centrifugal force as well as to bending stress due to the entwined
fiber being drawn away with a speed of at least 1000 m/min. Therefore, only high-
purity alumina ceramic grades with high mechanical strength, high wear resistance, and a special surface structure (Fig. 5.25) keeping the friction low and
within a narrow range can meet the requirements [5.40].

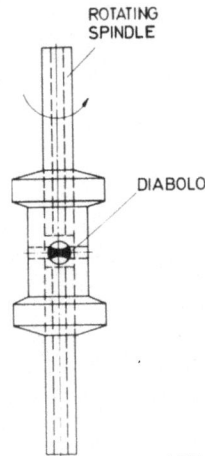

ROTATING
SPINDLE

DIABOLO

Fig. 5.28. Spindle for "false-twist-spinning" process with inserted "diabolo"

Since higher rotating speeds could not be realized with this process, it was
necessary to develop new procedures in order to increase efficiency and output.
The problem was solved by the so-called friction-texturing process [5.39]. The
curling of the synthetic fiber is carried out by rotating ceramic bushings or
discs, driving the fiber by frictional forces without damaging it (Fig. 5.29).

Fig. 5.29 Bushings and discs, for the friction-texturing process, made of high alumina
ceramics

240

High working speeds can cause electrostatic charges to be a considerable handi-
cap for an uninterrupted textile production, particularly if the charges are of
high voltage. The textile manufacturer therefore tries to prevent the generation
of high electrostatic charges as much as possible.

The generation of electrostatic charges depends on the speed and the preparation of the thread as well as on the electrical conductivity of the ambient atmosphere. Many attempts to solve this problem, basically by using conductive
thread guides made of metals or of titanium oxide ceramics, have failed. According to measurements of the electrostatic fields surrounding the thread, the
charges are generated predominantly by the friction between the thread and the
ambient air and not by the friction between the thread and the thread guide,
because the period of time during which a given element of the thread is in
contact with the air is much longer than the time the thread is in contact with
the thread guide. Besides this, metallic or titanium oxide thread guides have a
considerably shorter lifetime compared to thread guides made of high alumina ceramics.

Therefore it is more effective to shunt the charges via the air instead of shun-
ting them via the thread guide, and, indeed, it only needs an increase in electrical conductivity of the air, for example by increasing the humidity or by ionisa-
tion with the aid of a radioactive source, to reduce the electrostatic charges
of the thread to a minimum or to prevent their generation at all.

5.2.2 Wire Drawing Step Cones

Wire drawing operations, particularly the manufacture of steel and copper wire,
require wear resistant materials for drawing and diverting elements [5.1]. Therefore, step cones, capstans, pulleys, rollers, and other wire drawing elements
made of high alumina ceramics have proved quite adequate in this field for many
years.

During the wire drawing process the wire is guided over the diverting step cone,
through the wire drawing die to the drawing step cone, and from there to the next
step of the diverting cone, and so on. The reduction of the wire diameter takes
place in the die, normally made of diamond or cemented tungsten carbide, while
drawing the wire through the dies is carried out by the step cone (Fig. 5.30).
The drawing speed can come up to 3600 m/min when using modern drawing machines.
It is easy to understand that under these conditions extreme wear caused by slip
can occur. This is why metallic wear resistant materials have failed in many cases
and high alumina ceramics have solved most of the wear problems [5.41].

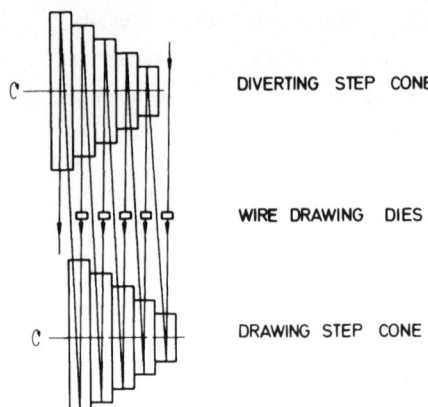

DIVERTING STEP CONE

WIRE DRAWING DIES

DRAWING STEP CONE

Fig. 5.30. Principle of wire drawing with step cones

In the field of fine wire drawing applications - in the diameter range below 0.1 mm - problems of friction play an important role. Here the sensitivity of the wire is comparable to a textile thread. Therefore, the experience in the textile field could be transferred to wire drawing, particularly those related to the smoothness of the ceramic surface being in contact with the wire. The wire surface has to be very smooth as well, in order to meet the requirements of the next operations, for example, those of a subsequent lacquer coating.

The static load on the drawing element grows with increasing wire diameter. In the field of intermediate wire drawing applications it is the ceramic drawing step cones that are primarily exposed to wear. More than 1000 tons of copper wire in the diameter range of 1 to 2 mm have to be manufactured without any sign of wear on the surface of the ceramic step cones. The lifetime of the ceramic elements depends on the wire diameter. It increases with decreasing diameter, starting with six or seven months in the intermediate range. In the field of fine wire drawing a lifetime of up to seven years can be achieved.

All ceramic wire drawing cones can be manufactured up to a diameter of 200 mm. Above this limit the process becomes uneconomical. This problem could be solved by the development of metal-ceramic combinations. They consist of metallic bodies, metallic flanges, and rings of high alumina ceramics comprising the wear resistant parts. The rings are held on their outside diameter by the metal flanges. The whole construction is screwed together (Fig. 5.31).

The tolerances of ceramic rings and metal flanges are adequate to ensure true concentric running and keep the wire on the track. The ceramic rings can be manufactured up to a diameter of 600 mm.

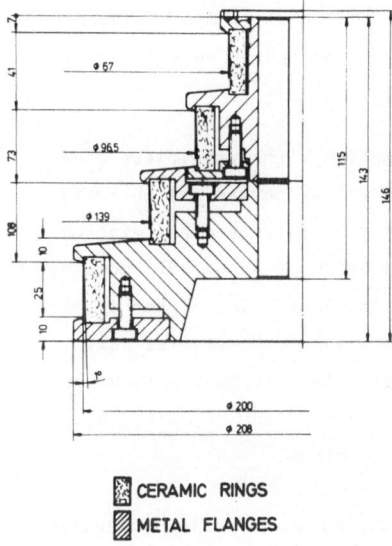

CERAMIC RINGS
METAL FLANGES

Fig. 5.31. Wire drawing step cone made of a metal-ceramic combination

The construction described here allows to replace individual rings as soon as signs of wear appear on the ceramic surface. Instead of replacing a worn cone, it is in principle, possible to regrind it. However, all the rings of the cone have to be ground as a system to ensure the correct ratio between the diameters from ring to ring. Since this procedure is rather expensive it is advisable to replace the individual rings in the case of a metal-ceramic composite construction.

Figure 5.32 shows various wire drawing step cones (from left to right): all ceramic, metal-ceramic, and plastic-ceramic combinations. Right in front there is a small solid ceramic cone consisting of three steps. These are the first steps

Fig. 5.32. Wire drawing step cones made of high alumina ceramics and of metal-ceramic and plastic-ceramic combinations

of a metal-ceramic or plastic-ceramic construction. This part is made in one piece for easier manufacturing. The first steps have the highest wear rate and they are, therefore, most frequently replaced.

The background (right) shows a plastic-ceramic composite with a stainless steel bushing on the bore. This construction has proved advantageous as a light-weight construction for fine-drawing applications. All parts of this cone not exposed to wear are replaced by plastics. With a weight reduction of about 30%, compared to solid ceramic cones, these plastic-ceramic constructions allow higher drawing speeds and offer better thermal shock resistance because of the smaller wall thickness of the ceramic rings. Additionally, the plastic component acts as a vibration absorber. The further development of the construction described here is a replacement of the plastic by the light metal aluminum.

Similar principles of design are also realized with wire guide pulleys, manufactured in aluminum-ceramic combination (Fig. 5.33). They can be applied to all current types of drawing machines leading the wire to and from the individual units of a drawing installation. Since pulleys are driven by the wire, a light-weight design with the lowest possible moment of inertia is required, particularly when superfine and fine wires are concerned [5.42].

Fig. 5.33. Wire guide pulleys of lightweight design made of aluminum-ceramic combination

Some wire drawing operations, particularly in the field of steel wire manufacturing, are carried out with step cones made of zirconia ceramics having a particular surface characteristic due to the extremely large grain size. In this case a special wear mechanism takes place where zirconia ceramics, which have a significant lower hardness and about 50% of the Young's modulus as compared to high alumina ceramics, turn out to be superior [5.43,5.44].

5.2.3 Paper Machine Covers

The efficiency and economy of paper manufacturing has improved considerably during the last 20 years by the application of high alumina ceramics [5.43]. The development of paper machines has taken place in two steps:

(1) Replacement of dynamic dewatering elements as table rolls by stationary elements [5.45].

(2) Increasing wear resistance of stationary elements by the replacement of high density polyethylene and wood by high alumina ceramics [5.46].

The most essential part of paper production is the so-called sheet forming operation. The paper suspension which consists of pulp, cellulose, and filler is injected homogeneously on an endless wire of bronze or plastics with a width of up to 10 m revolving with a speed of up to 1200 m/min. Within a distance of 10 to 20 m the paper suspension has to be dewatered until a sheet of paper with sufficient strength is formed. This happens in the so-called fourdrinier section. In this part of the paper machine the basic requirements for the product quality are established. Composition of paper suspension, wire, and covers are the three critical factors of each fourdrinier section, which have to be optimized by fitting them with each other (Fig. 5.34).

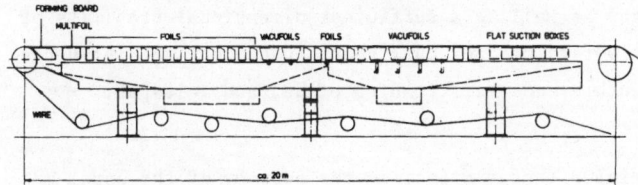

Fig. 5.34. Schematic view of a fourdrinier section of a modern paper machine

All stationary dewatering elements of a paper machine, being in contact with the wire, are exposed to considerable friction and wear. This is easy to understand, considering that each wire moves a distance in the order of 50 000 km. Previously used cover materials, such as polyethylene and wood, wear out with increasing wire speed and because of the abrasive effect of the filler. They show wear at the profile and at the skimming edge after only a few days, in many cases even after some hours. Increasing cover wear results in a nonuniform dewatering process over the whole width of the paper machine and in irregularities of the sheet formation [5.47].

All these disadvantages can be prevented by the application of wear resistant covers made of high alumina ceramics having the necessary dimensional stability. Figure 5.35 shows a fourdrinier section equipped with sheet formation and dewatering elements made of high alumina ceramics.

Fig. 5.35. Fourdrinier section equipped with covers made of high alumina ceramics for a wire width of 8 m

The uniformity of the dewatering process, a very important property which determines the paper quality, could be increased by using covers made of high alumina ceramics. According to their remarkable wear resistance, high alumina ceramics offer the required dimensional accuracy as well as a sufficient dimensional stability of the dewatering elements, which is retained for many years even under the influence of abrasive filler components such as quartz, titanium oxide, and feldspar.

The lifetime of the wire, a decisive factor affecting the economy of the paper manufacturing process, could be increased significantly by some 100% due to the hard, smooth, and compact surface of the ceramic material having an average roughness of less than 0.1 μm [5.43]. No abrasive particles can stick to this hard poreless surface. No wearing of the wire can occur. The special surface characteristic results in a decrease in friction and, hence, in a reduction of the electrical power consumption up to 45%.

Ceramic covers consist of individual segments. They are assembled according to different fixation methods, which in most cases are clamping strips and dovetail

246

guides. This holds for forming boards, supporting boards, and foils, as well as
for covers for felt suction boxes, for wet suction boxes, and for flat boxes.

The function of the forming board (Fig. 5.36) is to take up the paper suspension,
to skim the water, and to support the wire. Here exactly defined leading edges and
flat and smooth surface of the cover are of the utmost importance. By adjustment
variation of the forming board, the longitudinal and the transverse flow of the
paper suspension can be controlled. In this manner the preconditions of a uniform
sheet formation are established.

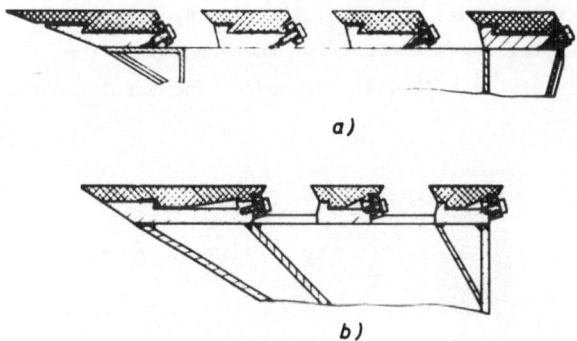

Fig. 5.36. Two different arrangements of forming boards made of high alumina ceramics

Supporting tables have similar functions. In combination with table rolls, they
are placed in such a way that the sag of the wire is reduced to a minimum. Worn-
out covers of supporting tables lead to a sagging wire, reducing dewatering efficiency.

Nowadays the most common dewatering and sheet formation elements are foils that
skim the water with their front edge from the bottom side of the wire (Fig. 5.37).

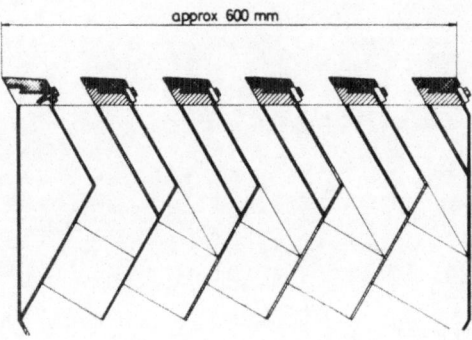

Fig. 5.37. Foils made of high alumina ceramics

247

The foil angle between foil blade and wire creates a certain underpressure causing a dewatering effect [5.48]. By a suitable arrangement of wide single foils or narrow multifoils the dewatering efficiency can be adjusted and controlled. Due to the dimensional stability of the ceramic foil edges and the constancy of the adjusted foil angle an increasing drainage, an improved sheet formation, and a better filler distribution can be obtained.

The main purpose of wet suction boxes (Fig. 5.38) is to fix the web under controlled conditions. Due to the relatively large open surface, from 40% to 60%, and due to the great number of narrow ceramic skimming edges, the dewatering effect of wet suction boxes is superior to that of foils. The dimensional stability of the wear resistant ceramic skimming edges account for the economy of the process. Wet suction boxes are preferably installed at the end of the forming zone.

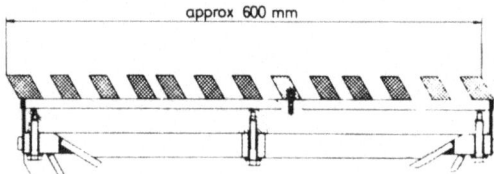

Fig. 5.38. Wet suction boxes with covers made of high alumina ceramics

Flat suction boxes (Fig. 5.39) are placed at the end of the fourdrinier with drilled or slotted covers. With a portion of 80% of the total wire contacting zone they represent an important factor of the wear mechanism defined by wire, cover, and filler. The cover material, the diameter of the holes, and the width of the slots, i.e. the open area of the covers, determine the wire wear to a great extent.

Fig. 5.39. High alumina ceramic cover for flat suction boxes

With increasing machine speed the cleaning and conditioning of the felt become a problem. Removing the impurites, which are abrasive residuals of the filler such as silicates, aluminates, and carbonates, cause cover wear leading to a damage of the felt. Therefore, it became necessary to replace conventional materials by high alumina ceramics for covers of felt suction boxes.

Summarizing the experiences with paper machine covers, three advantages have turned out to be decisive for the enhanced economy of the paper manufacturing process made possible by the exceptional hardness and wear resistance of high alumina ceramics:

(1) Improved productivity by higher machine speed and reduced down-time.
(2) Reduction of production cost by increasing the lifetime of the wire and felt as well as energy savings.
(3) Improvement of the paper quality.

5.2.4 Bearings

The term "bearings" shall be understood here in a broad sense, comprising components exposed to friction and wear by interaction with other components, such as seal rings, sleeve bearings, plain bearings, pump parts, and plungers (Fig. 5.40).

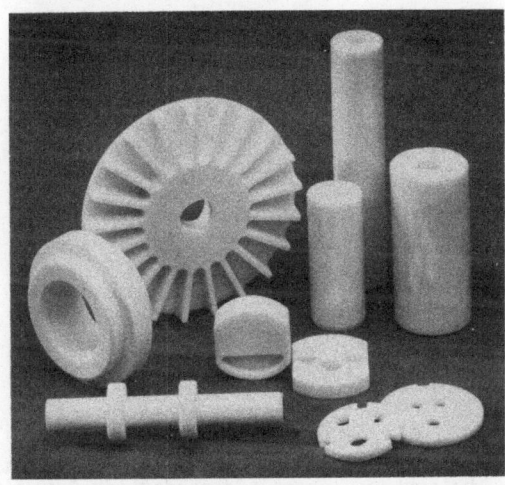

Fig. 5.40 Bearings and seal rings made of high alumina ceramics

Bearings are a typical example of a combined exposure to corrosion and wear [5.49]. In this special case, ceramic materials, particularly high alumina ceramics, meet these requirements to a much greater extent than metals [5.43]. Therefore, ceramic bearings are preferably applied to pumps for chemically aggressive fluids, sea water, or dirty water, as well as to high pressure pumps. The main field of application, however, are pumps for central heatings.

Within this group, mechanical seals with seal rings made of high alumina ceramics
have gained increasing importance. They have replaced the previously used stuffing-
box packings, particularly when applied to higher rotational speeds. Three reasons,
resulting from typical characteristics of high alumina ceramics were reponsible
for this replacement: longer lifetime, lower leakage losses, and absence of maintenance [5.36].

The mechanical seal consists basically of a stationary ring (1) which is called a
"seat", a rotating ring (2) which is called a "washer", a static seal (3) for the
seat and washer, respectively, a spring, and an anti-twisting device (Fig. 5.41).
The seat is preferably made of high alumina ceramics, the washer of carbon based
materials, teflon, stainless steel, or cemented carbide. In special cases when
both rings are made of high alumina ceramics (parts 1 and 2 of Fig. (5.41), effective lubrication has to be ensured because this material is sensitive to lubricant
loss when exposed to contact pressures higher than 10 MN/m^2 which are usually applied to bearings.

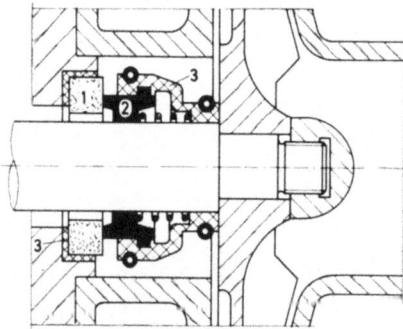

Fig. 5.41. Cross section of an axial seal ring packing

The condition of the sliding surfaces is important for the efficiency of the mechanical seal. The sliding behavior depends on the surface structure of both the
ceramic and the counterpart material. The tolerable deviation of the flatness is
restricted to 0.3 µm; the average surface roughness should not exceed 0.2 µm
(CLA). With respect to small sealing slits the leakage losses are proportional to
the square of the average roughness. These surface specifications, resulting from
many years of experience, can be met only with a high-density and small grain-
sized alumina grade allowing best polishing and lapping.

In many cases the surface can be characterized by the so-called "bearing area"
rather than by the surface roughness. When measured with a surfindicator the bearing area is the ratio of the bearing length to the measuring length of the roughness profile in a certain cutting depth. When determined by microscopy the bearing

area results from the ratio of the reflecting area to the total area. Best results can be obtained when the bearing area is in the range between 50% and 80%. A lower bearing area leads to increasing wear, a higher bearing area to rising temperatures. Both effects result in increased leakage, the latter causing damage to those counterpart materials which are less heat-resistant.

In order to meet the corrosion resistance requirements it is necessary to increase the purity of the high alumina ceramics above 99%. Additions of silicates reduce the corrosion stability [5.31].

Besides corrosion and wear resistance mechanical and thermal properties are also of essential importance. A high bearing pressure and elevated temperatures due to frictional heat can cause deformations of the seal ring leading to increasing leakage losses if the rigidity of the seal ring material is not sufficient. Therefore, a high resistance against deformations by mechanical or thermal stresses is required. In order to define this property a modulus of dimensional stability S was suggested:

$$S = Ek/\alpha \qquad (5.1)$$

where E is Young's modulus, k the thermal conductivity, and α the thermal expansion. Additionally, it is necessary to consider that local temperature peaks can cause cracks due to thermal stresses at the sliding surfaces. These cracks affect leakage losses and wear behavior. Extended surface zones can chip off, leading to complete damage of the sliding surface. Therefore, the resistance against thermal stress cracks also has to be sufficiently high. The definition for the thermal stress factor B was given in [5.50] by

$$B = \sigma(1 - \nu)k/E\alpha \qquad (5.2)$$

where σ is the fracture strength and ν Poisson's ratio. According to both equations, a suitable material for seal rings, that means a material with a high modolus of rigidity and with a high thermal stress factor, should have a small thermal expansion, but a high thermal conductivity, a high strength, and a medium Young's modulus because of the opposing influence of this property in the two equations.

The parameters which are relevant for seal ring materials are listed in Table 5.3 for a series of technically important materials. According to these specifications carbon-based materials are less rigid, have a low wear resistance, and are largely insensitive to thermal stresses. Stainless steel shows insufficient stability

and insufficient corrosion resistance. Cemented tungsten carbide meets all requirements with respect to dimensional stability and wear resistance but fails
when exposed to corrosion.

High alumina ceramics, however, seem to be an acceptable compromise with respect
to all required properties. With a dimensional stability of $650 \cdot 10^{-10}$ NJ/hmm^3,
a thermal stress factor of 11kJ/hm, a high wear resistance, and an excellent corrosion resistance, high alumina ceramics meet all requirements of a suitable material for seal rings.

Table 5.3. Properties of seal ring materials

Seal ring materials	Modulus of dimensional stability S $\cdot 10^{-10}$ NJ/hmm^3	Thermal stress factor B kJ/hm	Wear re- sistance	Corrosion resistance
Carbon based materials	4 to 75	1 to 170	bad	good
Stainless steel	85	21	minor	minor
Cemented tungsten carbide	2100	42	good	minor
High alumina ceramics	650	11	good	excellent

All considerations with respect to a comparison of requirements and properties of
seal ring materials can be transferred to sleeve bearings made of high alumina ceramics, which are used, for example, in circulating pumps for central heating systems (Fig. 5.42). Such bearings show a considerable life expectation of at least
10 years without any maintenance. No impurity of the water can cause wear indications, not even abrasives such as sand, rust, or limestone. Due to the high hardness of the ceramic material, solid particles entering the bearing cannot become
embedded between the sliding surfaces; they are crushed and completely ground
between the hard ceramic surfaces, thus providing an automatic cleaning of the
sealing faces.

Ceramic bearings can be operated with water lubrication. For this reason they are
gaining increasing interest because of the shortage of oil and, additionally,
they are of interest to the food industry.

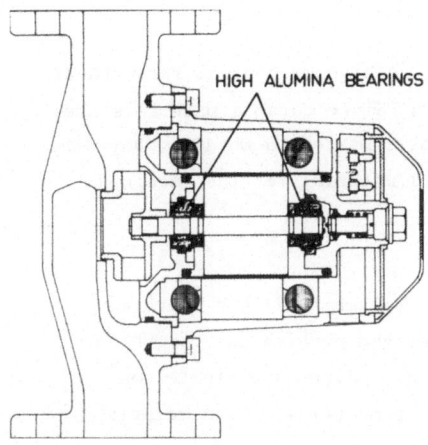

HIGH ALUMINA BEARINGS

Fig. 5.42. Circulating pump for central heating systems with bearings made of high alumina ceramics

One of the newest applications among the fields described here is seal discs for water taps (Fig. 5.43). The sealing function is carried out by two seal discs which are lapped plane. They allow more than one million opening and closing movements without wear. Abrasive residuals such as sand particles, dirt, rust, and limestone cannot scratch the sealing surfaces, due to the high hardness of this material. Ceramic materials of lower hardness, such as porcelain or low alumina ceramics would wear out within a short period of time. The complete fitting works without a moving rubber gasket, thus dispending with any maintenance. (Fig. 5.44).

Fig. 5.43. Seal discs for water taps made of high alumina ceramics

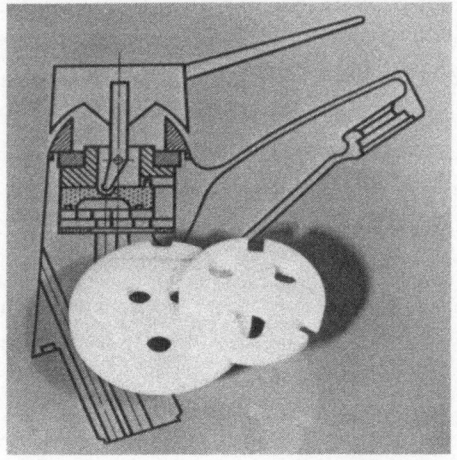

Fig. 5.44. Water tap with ceramic seal discs

253

5.2.5 Cutting Tools

Cutting tools require high wear resistance as well as high mechanical strength at elevated temperatures, particularly impact strength. Since ceramic materials are known to be brittle and have low impact strength as compared to metals, they were not expected formerly to be a suitable cutting tool material for the machining of metals.

After 1956, however, ceramic processing techniques became available which enabled the fabrication of alumina ceramics characterized by high density, high purity, and a small grain size on an industrial scale. Since a silicate content was known to deteriorate the high-temperature properties of the material considerably, high-purity alumina was thought to be the only material to ensure the high-temperature performance required in cutting operations. Based on the progress in fabrication technology, ceramic cutting tools were first developed for turning operations, and subsequently also for milling purposes.

No other application of high alumina ceramics needs a comparably good material performance. Specifically, a small grain size is of extraordinary importance in preventing catastrophic failure of the tools. High density as well as a fine-grained and uniform microstructure with an average grain size below 5 µm were essential for improving the impact strength [5.51,5.52]. Later on, this value was decreased gradually to 1.5 µm.

Many attempts were made to solve the problem of producing a fine-grained microstructure with the aid of silicate additions according to previous experience. But all these attempts failed, for the addition of silicates results in absolutely insufficient strength and insufficient wear resistance of the ceramic material. Only small additions of magnesium oxide can inhibit exaggerated grain growth without deteriorating the properties. Therefore all known ceramic cutting tool materials consist of high-purity alumina with only small additions of magnesium oxide in the order of some tenths of a percent (compare Sections 2.5 and 4.3).

Two exceptions turned out to be successful: additions of titanium carbide increase the thermal conductivity of aluminum oxide, resulting in an improvement in the thermal shock resistance and in the cutting edge stability of the ceramic tools, and additions of zirconium dioxide reduce the brittleness [5.53,5.54].

In the course of time the development of cutting tool materials has been directed toward increasing operation speed (Fig. 5.45). Beginning with carbon steels whose maximum cutting speed is in the range of 10 m/min, the first improvement was achieved by adding about 25% of a carbide phase, resulting in the so-called high-speed steels, which increased the allowable speed to 30 to 50 m/min. With stellites, the products of a parallel development with cobalt base alloys, cutting speeds up to 80 m/min could be achieved. Further progress was obtained by a considerable increase in the carbide content. This led to the development of cemented carbide materials which contain tungsten carbide, titanium carbide, or both up to 95% and more. Cemented carbides are the most important tool materials of today.

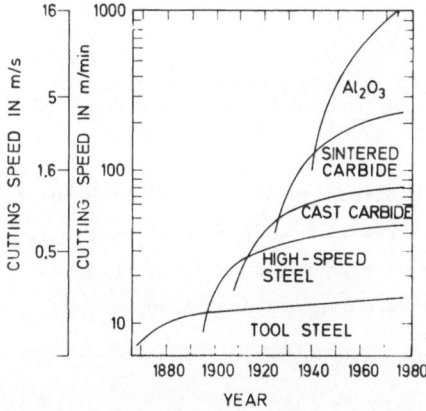

Fig. 5.45. Progression of cutting speed depending on the time (after H.Tully)

The manufacture of cemented carbides is carried out by powder metallurgical processes. In order to prevent exaggerated grain growth during sintering, cobalt is used as a binder for tungsten carbide/titanium carbide alloys; nickel and molybdenum are used as binders for titanium carbide base materials.

With decreasing binder content, cemented carbide tools can be exposed to increasing operation speeds up to 200 m/min. Speed and temperature are closely connected. The upper limit of the cutting speed of cemented carbides is given by the oxidation and the plastic deformation behavior of the metallic binder which produces a measurable overall deformation even at a temperature of 500°C. With rising temperatures, the plastic deformation increases. Figure 5.46 gives a comparison of the creep rates for various cemented carbides and aluminum oxides at a temperature of 1150°C under a compressive stress of 150 MN/m^2 [5.55].

255

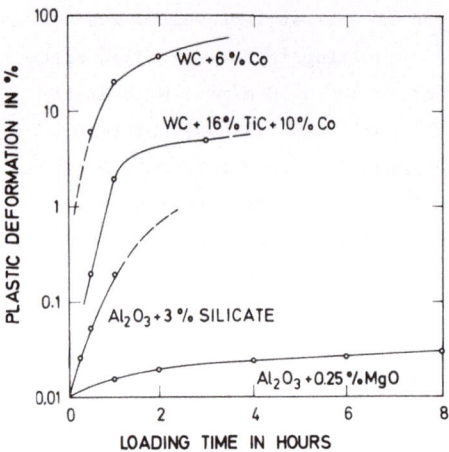

Fig. 5.46. Creep curves of cemented carbide and aluminum oxide of various compositions [5.55]

Figure 5.47 demonstrates the superiority of the compressive strength of high-purity alumina ceramics at elevated temperatures as compared to metallic tool materials. It should be noted that the compressive strength of high-purity alumina ceramics shows about the same value at 1100°C as steel at room temperature [5.56].

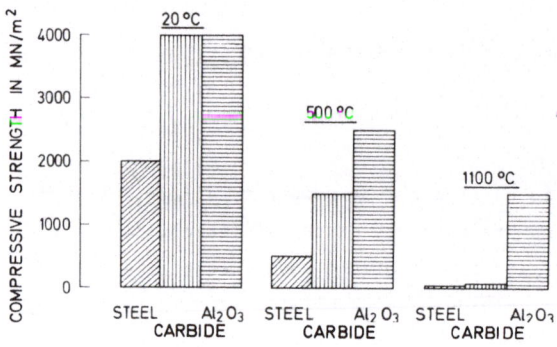

Fig. 5.47. Compressive strength of steel, cemented carbide, and high-purity alumina ceramics at various temperatures

Since inserts made of high-purity alumina ceramics do not contain a metallic binder, they can be used up to a cutting speed of 2000 m/min, depending on the material of the workpiece. On the other hand, the maximum cutting speed is limited by the motor power and by the diameter of the workpiece. The economic effect resulting from the ability of applying these extreme cutting speeds is the essen-

256

tial advantage of ceramic cutting tools compared to all other tool materials [5.57,5.58].

Additionally, metallic materials used as tools have a greater chemical affinity to the metallic workpiece compared to ceramic materials. Therefore, diffusion welding effects take place between the metallic workpiece and the metallic tool, increasing with increasing temperature, which is the essential reason for the wear of the metallic tool material. In contrast to this, high-purity alumina ceramics, which are basically different from metals with respect to this property, show no sign of welding with the metallic workpiece up to a temperature of 1000°C. As listed in Table 5.4 high-purity alumina ceramics show noticeable welding to steel at far higher temperatures, compared to cemented carbides according to bond strength measurements [5.55].

Table 5.4. Welding behavior of cemented carbide and high-purity alumina ceramics with respect to steel [5.55]

Temperature	Bond strength related to the actual contact area	
	Cemented carbide	Aluminum oxide
700°C	110 MN/m^2	1 MN/m^2
750°C	-	1 MN/m^2
900°C	-	1 MN/m^2
1300°C	-	100 MN/m^2

The welding behavior, indicated in Table 5.4, is the explanation of the lower fricton, of the lower cutting pressure, and therefore of the lower power input of ceramic cutting tools as compared to metallic tool materials within the proper speed range for ceramic tools [5.59].

According to Fig. 5.48 the wear of ceramic cutting tools does not depend on the cutting speed between 100 and 300 m/min, which is the range of lowest wear. Below this region, corresponding to a range of lower temperatures, the wear increases due to the higher sensitivity of the ceramic to vibrations because the ceramic material is not able to heal microcracks by diffusion at lower temperatures [5.60].

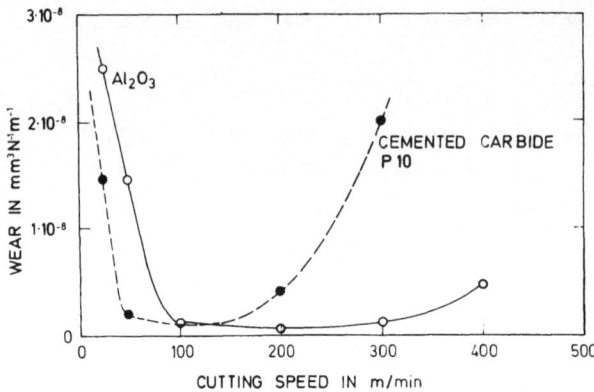

Fig. 5.48. Correlation between wear and cutting speed of cemented carbide and high-purity alumina ceramics [5.55]

Crack healing was already observed to occur at $600^{\circ}C$ [5.29]. In this study, with increasing cutting speed (above 300 m/min), corresponding to a range of higher temperatures, the wear increased as well, due to plastic deformation. Compared to ceramics the wear/cutting speed curve of cemented carbide is shifted toward lower cutting speeds. From Fig. 5.48 it can be concluded that high-purity alumina ceramics are clearly superior to cemented carbides at higher cutting speeds, leading to essential technical and economical advantages. The graph described here refers to the machining (turning) of steel. The same correlations were found with respect to the machining of grey cast iron.

The wear characteristics of high-purity alumina ceramics also differ from that of cemented carbides. Whereas a cratering takes place on the rake face of the carbide tools, which is a kind of eroding, ceramic tools show a flank wear on the cutting edge, which is a kind of abrasion (Fig. 5.49).

The question of which temperatures appear at the ceramic cutting edge was extensively investigated by means of thermocouples, which were inserted into the tool just below the cutting edge, as well as by pyrometrical measurements. The color of the metal chips can also be an indication of the temperature at the cutting edge. According to the results of these investigations, which were in good agreement, the temperatures rise as high as $1000^{\circ}C$ when steel is machined with a cutting speed of about 700 m/min. Machining of cast iron produces somewhat lower temperatures [5.61].

Ceramic tool tips are normally fixed by clamping combined with an adjustable chip breaker (Fig. 5.50). This permits a fast change-over of blunted edges. The cutting geometry of the tool tip given by the tool holder is designed in such a way that

258

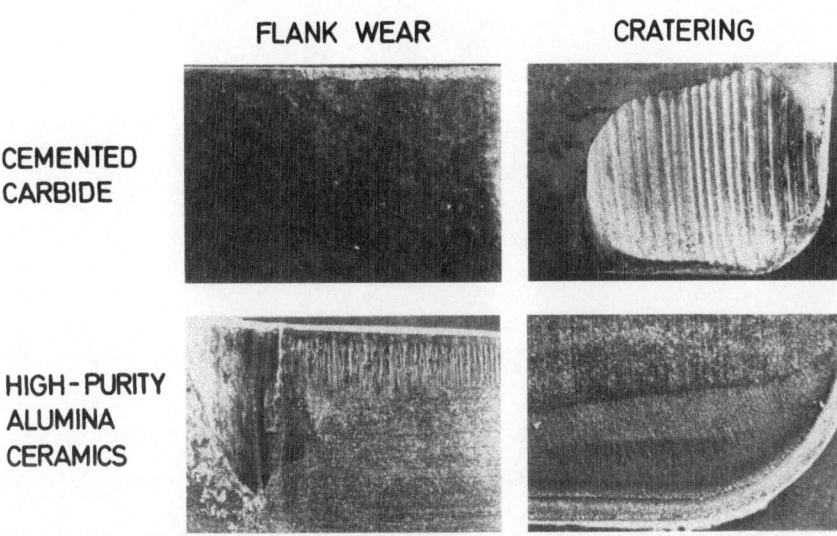

FLANK WEAR   CRATERING

CEMENTED
CARBIDE

HIGH-PURITY
ALUMINA
CERAMICS

Fig. 5.49. Tool wear of cemented carbide and high-purity alumina ceramics (8x)

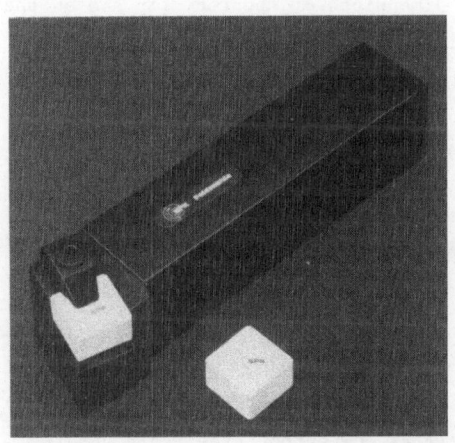

Fig. 5.50. Cutting tool consisting of tool holder and tool tips of high-purity alumina ceramics

the cutting edge is preferably loaded in compression. For that reason the rake angle is negative, having an average value of -6°. Due to the rectangular shape of the tool tip the clearance angle is 6° as well and the inclination angle 4°. A protective chamfer should be provided in order to protect the edge against vibrational and shock loads and to allow higher loads (Fig. 5.51). In manufacturing tool holders, special care has to be taken to ensure an exact tool-tip seating face [5.55].

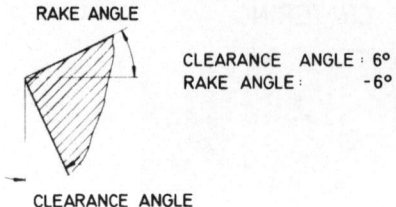

RAKE ANGLE

CLEARANCE ANGLE : 6°
RAKE ANGLE :   -6°

CLEARANCE ANGLE

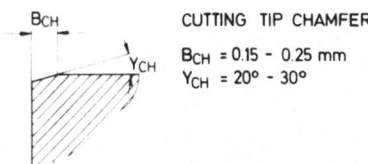

B_{CH}

Y_{CH}

CUTTING TIP CHAMFER

B_{CH} = 0.15 - 0.25 mm
Y_{CH} = 20° - 30°

Fig. 5.51. Cutting tool geometry

With regard to the cutting conditions, ceramic cutting tools are preferably operated with high cutting speeds, as indicated before. Cutting speeds up to 1200 m/min
are applied in practical operation. Basically a speed of 2000 m/min can be achieved. The maximum depth of cut depends on the dimensions of the tool tip and on the
power of the motor. In contrast to metallic cutting materials, which are insensitive to high feed rates, the maximum feed with ceramic tools was limited formerly
to about 0.4 mm/rev because of their lower shear strength. As the development of
ceramic materials proceeded the feed could be raised up to 0.8 mm/rev.

Ceramic cutting tools are normally used to machine cast iron. In this field ceramics can display their full superiority compared to metallic tool materials. About
80% of all cutting operations carried out with ceramic tools are involved in cutting cast iron, including chilled cast iron. The remaining 20% are split up into
the machining of high carbon steel, tempered steel, nitriding steel, and also
hardened steel. Tough metals producing long chips during the cut, such as steel
qualities alloyed with nickel or cobalt as well as low carbon steel, are unsui-
table for machining by ceramic tools. The same holds for aluminum, titanium, and
their alloys.

The next three figures illustrate three examples of machining operations with ceramic cutting tools. Figure 5.52 refers to the machining of a cylinder liner and
Fig. 5.53 to the machining of a brake disc, both made of gray cast iron; Fig. 5.54
refers to the machining of a drive shaft made of tempered 16MoCr5 steel. Tables
5.5 to 5.7 give the corresponding cutting conditions, respectively.

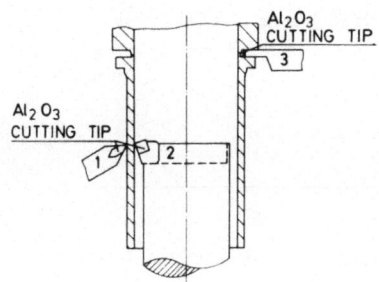

Fig. 5.52. Machining operation of a cylinder liner [5.55]

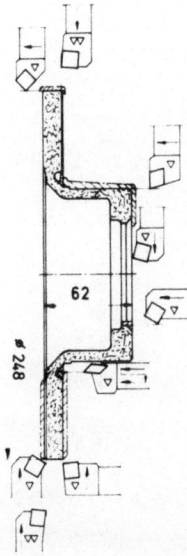

Fig. 5.53. Machining operation of a brake disc

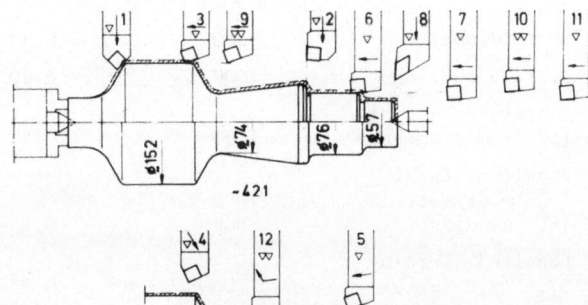

Fig. 5.54. Machining operation
of a drive shaft [5.62]

Table 5.5. Cutting conditions for machining a cylinder liner of gray cast iron

Tool No.	Machining process	Cutting speed m/min	Feed rate mm/rev	Depth of cut mm
1	Longitud. turning (outside)	315	0.45	2.5 to 3
2	Longitud. turning (inside)	245	0.45	3 to 4
3	Plunge-cut	315 to 345	0.20	8

Table 5.6. Cutting conditions for machining a brake disc of gray cast iron

Cutting speed	1000 to 1600 m/min
Feed rate	0.20 to 0.40 mm/rev
Depth of cut	2 to 3 mm
Machining time	0.55 min

Table 5.7. Cutting conditions for machining a drive shaft of 16MnCr5 steel

Tool No.	Machining process	Cutting speed m/min	Feed rate mm/rev	Depth of cut mm
1	Chamfering	300	0.25	2
2	Facing	400	0.30	2
3	Longitud. turning	600	0.30	3.5
4	Bevel turning 30°	700/350	0.30	3.5
5	Bevel turning 7°	600	0.30	3.5
6	Longitud. turning 77 mm diameter	600	0.30	3.5
7	Longitud. turning 58 mm diameter	600	0.30	3.5
8	Profile turning	1000	0.20	3.5
9	Longitud. turning 152.4 mm diameter	1400	0.25	0.5
10	Longitud. turning 100 mm diameter 76.4 mm diameter	1000 800	0.25	0.5
11	Longitud. turning 57.4 mm diameter	700	0.25	3.5
12	Profile turning 100/74 mm diameter 74/152.4 mm diameter	1100 1800	0.25	0.5

The cylinder liner illustrated in Fig. 5.52 is machined on a vertical copying lathe in a floating clamping fixture. Tool No.1 and tool No.2 operate at the same time and at the same position with respect to the height of the work piece, each performing about 60 pieces per tool edge. Tool No.3 subsequently cuts off the cylinder liner by plunging, producing about 200 pieces per tool edge.

The machining of brake discs, brake drums, and cylinder liners made of cast iron are the most successful applications for high-purity alumina ceramics with respect to the machining economy by reducing the machining time and, with respect to the tool economy, by increasing the number of pieces produced per cutting edge.

As compared to tools made of cemented carbide the total economizing factor varies between 50% and 250%.

Due to the considerably higher cutting speeds which can be realized with high-purity alumina ceramics, machining with these tools offers an excellent surface finish (Fig. 5.55). By this means the time required for subsequent grinding and honing can be eliminated completely, leading to a considerable reduction in the total machining time.

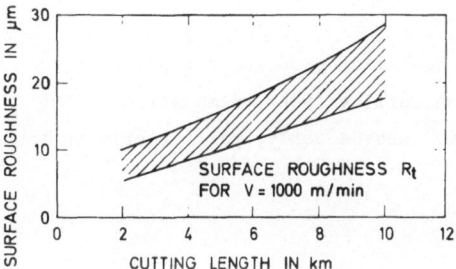

Fig. 5.55. Range of surface roughness vs. cutting length [5.55]

According to Fig. 5.56, the production costs (pce) of a turning operation when machining steel Ck 45 decrease rapidly up to the range of 600 m/min, with increasing cutting speed when using ceramic tools. Above this range the costs are nearly independent of the cutting speed. When using cemented carbide tools the production costs (pca) show a sharp pronounced minimum, rising steeply with increasing cutting speed above 200 m/min.

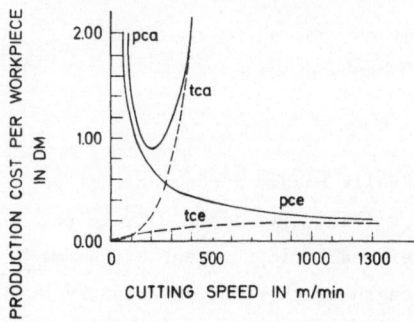

Fig. 5.56. Production costs depending on the cutting speed; pca: production costs carbides; tca: tool costs carbides; pce: production costs ceramics; tce: tool costs ceramics [5.63]

The curves of the production costs (pce and pca) are obtained from a combination of the curves of the machine costs and of the tool costs (tce and tca) referring to the machining of a certain workpiece represented in Fig. 5.56.

The machine costs are inversely proportional to the cutting speed; the higher the speed the lower the machine costs. Higher speed naturally means higher output in the same period of time.

Due to growing wear, the tool costs of ceramic cutting tools (tce), being extremely low in the lower cutting speed range, increase only slightly above this range. In order to minimize the production costs it is necessary to decrease the machine costs by reducing the machining time per workpiece. The most effective way to do this is to increase the cutting speed.

Figure 5.57 gives the correlation between the number of produced workpieces per tool life (nce and nca), the tool life, and the cutting speed, demonstrating the convincing economy of ceramic tools compared to carbide tools, particularly in the cutting speed range above 200 m/min.

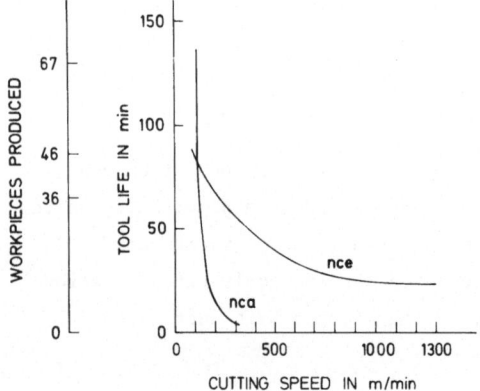

Fig. 5.57. Number of produced workpieces and tool life depending on the cutting speed; nca: number of workpieces per tool life (carbide tools); nce: number of workpieces per tool life (ceramic tools) [5.63]

Whereas turning operations with ceramic tools gradually became a conventional production technique, the solution of the problem of milling operations carried out with ceramic tools took a much longer time. A considerable improvement of the mechanical strength, particularly of the impact strength of the tool material by achieving a smaller grain size and a more uniform microstructure, was indispensable.

In contrast to turning operations, during a milling operation the cutting edge of the tool is exposed to extreme impacts. The impact loads during milling are not comparable with those occuring during an interrupted cut when turning, which are normally withstood by ceramic cutting tools.

264

Cutting tools made of cemented carbides applied to milling operations are normally used with a cutting speed up to 200 m/min and with a table feed rate up to 1000 mm/min. Under these cutting conditions the maximum surface roughness of the workpiece can be reduced to 8 μm. A further improvement of the surface roughness can be achieved only by grinding.

Ceramic cutting tools, however, permit a cutting speed of 1000 m/min and a table feed rate of 5000 mm/min, thus leading to a surface roughness of 4 μm and lower (Fig. 5.58). Due to this, a series of grinding operations can be performed more economically.

Fig. 5.58. Surface roughness of 2 μm, achieved by a single ceramic tool milling cutter

Ceramic milling tools are normally applied to finishing operations and roughing of grey cast iron. In order to obtain the required surface finish of the workpiece, an exact adjustment of the principal edge with respect to the working plane is important.

Grey cast iron and tempered and hardened steel, as well as chilled cast iron, are the favored materials for machining by milling operations with ceramic cutting tools.

5.3 Medical Applications

Due to their properties, the applications of high-purity alumina ceramics in the field of medicine are concentrated primarily to bone and joint replacement. In the course of the last nine years, this material has played an increasingly important role as a bone substitute for artifical joints and dental implants, as well as in the field of maxillary and oral surgery.

5.3.1 Artificial Joints

The requirements of artificial joints are similar to those of wear resistant parts in the field of mechanical engineering. Therefore it was obvious that one should make use of the advantage offered by the superior tribologic properties of the ceramic material and to apply it to artificial joints [5.64].

Generally, ceramic materials are known to be brittle, having poor flexural, tensile, and impact-strength properties. On the other side, however, they show an excellent biocompatibility and a superior tribologic behavior compared to metallic and plastic biomaterials. In order to meet the requirements of heavy load bearing artificial joints, it was necessary to select the most suitable ones from the large field of ceramic materials – and this was high-purity alumina ceramics – to improve their mechanical properties and to optimize the design [5.65,5.66].

5.3.1.1 Requirements

Artificial joints require complete body stability and biocompatibility. This holds for the implant itself as well as for the wear debris. A high mechanical strength is required, particularly fatigue strength. The loads to be withstood increase up to the sixfold body weight. For the articulating parts, high wear resistance and low friction are necessary [5.67]. The complete body stability and biocompatibility is ensured for high-purity alumina ceramics by means of animal tests and clinical experience for a long period of time [5.68-5.71]. The strength requirements can be met only by high-purity, high-density alumina ceramics with small grain size. Therefore, it is the aim of the production process to obtain a small-grained uniform microstructure. A high density is important as well, because each pore acts as a notch, reducing the strength and influencing the sensitivity to body fluids.

5.3.1.2 Mechanical Properties

The mechanical properties required by the ceramic components of heavy load bearing artificial joints are listed in Table 5.8. The figures in column 3 are minimum values, corresponding to national and international standards for ceramic implant materials [5.72]. The figures in column 4 correspond to the ceramic material BIO-LOX (product of Feldmühle) described here, which has higher purity and density, smaller grain size, and higher strength compared to the standard material. It is chosen for the following discussion because the majority of the ceramic-metal hip prostheses being implanted at present have components made of BIOLOX.

The properties listed in Table 5.8 refer to static strength values. Due to the fact that heavy load bearing joints like the hip joint, for example, are normally loaded dynamically with about 1 to 2 million load cycles per year, it was necessary to determine the dynamic properties of the ceramic material.

Table 5.8. Properties of ceramic implant materials (based on high alumina)

Property	Unit	Material according to standards	BIOLOX
Chemical composition	%	Al_2O_3 >99.5	Al_2O_3 >99.7
Density	g/cm^3	>3.90	3.94
Microstructure (average grain size)	µm	<7	4
Microhardness	MN/m^2	23 000	23 000
Compressive strength	MN/m^2	>4 000	5 000
Flexural strength	MN/m^2	>400	500
Young's modulus	MN/m^2	380 000	380 000
Impact strength	Ncm/cm^2	>40	50
Corrosion by organic acids	mg/m^2 per day	<0.1	<0.1

The dynamic fatigue strength was determined in the tensile and compressive zone between an upper and lower stress with a frequency of 10 cycles/s. Table 5.9 shows two results after $4 \cdot 10^6$ load cycles. The results plotted in a diagram appear similar to a so-called Wöhler-curve.

Table 5.9. Dynamic fatigue strength of high-purity alumina ceramics [5.73]

Tensile zone

Stress range	0 to 320 MN/m^2
Stress amplitude	± 160 MN/m^2
Mean stress	160 MN/m^2
Load cycles	$4 \cdot 10^6$

Compressive zone

Stress range	0 to 1600 MN/m^2
Stress amplitude	± 800 MN/m^2
Mean stress	800 MN/m^2
Load cycles	$4 \cdot 10^6$

The resistance against impact fatigue was determined in a pendulum testing machine by variation of the height of the pendulum weight. The criterion of evaluation was the impact energy corresponding to the appearance of the first cracks on the ceramic surface. It was found that with decreasing impact energy an increasing number of impacts was necessary for the formation of cracks. The results plotted in a diagram (Fig. 5.59) also show a shape similar to that of a Wöhler curve [5.73].

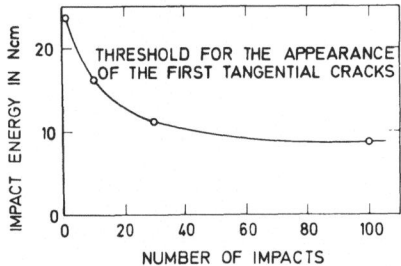

Fig. 5.59. Impact fatigue strength of high-purity alumina ceramics [5.73]

In the course of these tests a certain energy level was found at which no cracks can be observed even after 10 000 impacts. From this result it can be deduced that there is a lower limit of impact energy which can be endured for any long period of time.

5.3.1.3 Tribology

One of the most important properties of the articulating components of artifical joints is the wear and friction behavior, particularly the behavior over a long period of time [5.74].

In a hip joint simulator ceramic-to-ceramic joints were compared to previously known metal-to-UHMWPE (ultra high molecular weight polyethylene) joints. The load increased up to 5000 N, alternating with a frequency of one cycle/s. The test was carried out under Ringer's solution - a liquid simulating body fluids - with 10^7 load cycles, corresponding to a lifetime of about 10 years [5.67]. Figure 5.60 shows a comparison of the results. The initial state of the surface was the same for both joints. The articulating surfaces were highly polished, with an average roughness of less than 0.1 µm. Initially both joints showed the same friction, but after some hours the friction of the metal-to-UHMWPE joint increased, while the friction of the ceramic-to-ceramic joint decreased. The wear of both joints was very different when the test started, about 1:20 in favor of the ceramic joint. After some hours the wear of the metal-to-UHMWPE joint increased and the wear of the ceramic joint decreased, approaching zero.

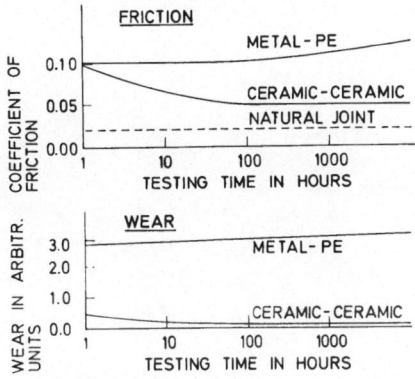

Fig. 5.60. Long-time behavior of high-purity alumina ceramics with respect to tribology effects [5.67]

These results can be explained by the surface change observed here. The surface roughness of the metal-to-UHMWPE joint increases, while the surface of the ceramic joint becomes smoother until, finally, no change is to be observed.

From these results it can be deduced that a constant improvement of the ceramic surface takes place, combined with a steady decrease of friction and wear. On the other hand, the surface of the metal-to-UHMWPE joint deteriorates, causing an increase in friction and therefore additional stresses to the bond between bone and implant.

Today the polishing technique of high-purity alumina ceramics has improved so far that it allows an average surface roughness of 0.02 μm (CLA). The result of this improvement is no detectable initial wear.

Following these tests the combination ceramic-UHMWPE was also investigated. Compared to the conventional metal-UHMWPE combination it turned out to be extremely superior. The wear of UHMWPE decreased by a factor up to 10. The ceramic heads, of course, did not show any trace of wear [5.75].

According to the results of other comparative investigations [5.76], the friction of an artificial hip joint with ceramic-to-ceramic articulation depends linearly on the load (Fig. 5.61), leading to the very low coefficient of friction μ = 0.035 ± 0.005.

Within the scattering range the friction was found to be influenced neither by the head diameter nor by the lubricant; distilled water, oil, Ringer's solution, and fresh synovial fluid were included in the tests. The sphericity of the head and the

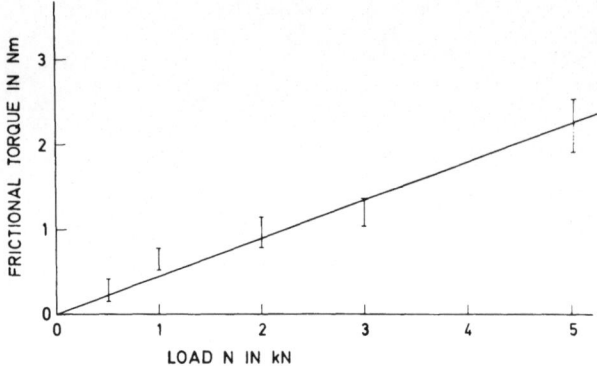

Fig. 5.61. Frictional torque vs. normal load; ceramic-to-ceramic articulation; lubricant:
Ringer's solution [5.76]

acetabulum component was found to be the only factor affecting the friction. As
soon as the deviation of the sphericity exceeds 18 μm a steep increase in friction
occurs (as shown in Fig. 5.62), obviously due to a decrease of the contact area
leading to an increase in the specific load and thus to a collapse of the lubri-
cating film. But this result is only of academic interest, for the sphericity of
ceramic hip joint components is normally found to be within 5 μm. In contrast to
this result, in the case of the ceramic-to-UHMWPE articulation the friction is also consistent above 18 μm, showing the same dependence on the sphericity as the
ceramic-to-ceramic articulation in the range below.

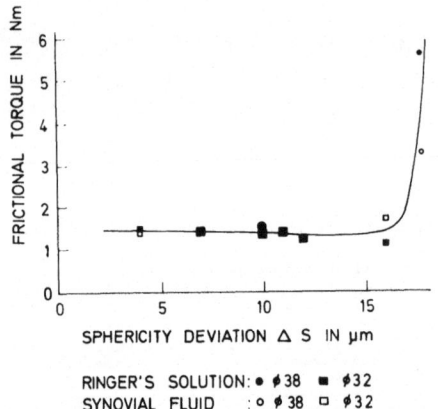

RINGER'S SOLUTION: ● ⌀38 ■ ⌀32
SYNOVIAL FLUID : ○ ⌀38 □ ⌀32

Fig. 5.62. Frictional torque vs. sphericity deviation; ceramic-to-ceramic articulation;
lubricant: Ringer's solution and synovial fluid [5.76]

270

As already outlined in section 5.1, the favorable tribology of the alumina surface depends on the fact that polar molecules are chemisorbed with high bond strength, forming surface layers which cause a reduction in wear of some magnitudes and a decrease in friction (Fig. 5.21).

5.3.1.4 Total Hip Prosthetic Systems

After making sure that high-purity alumina ceramics meet all requirements with respect to tribology and mechanical strength, particularly fatigue strength, systems of ceramic-metal hip prostheses first suggested by P. Boutin [5.77] were developed. Those components of the prostheses which are exposed to tensile and flexural stress are made of metal. Those components exposed to friction and wear are made of ceramics. The acetabulum component can be either ceramic or UHMWPE [5.75].

The first prosthesis to be described here consists of a metal stem, a ceramic femoral head, and a UHMWPE socket for implantation with bone cement (Fig. 5.63). The head diameter is 32 mm. The only difference between this design and the previously known hip prosthesis is the replacement of the metallic head by a ceramic head. Everything else is the same, including the operation technique.

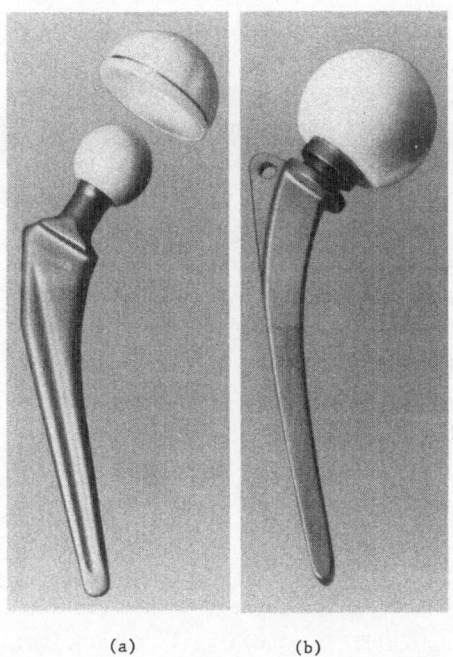

      (a)   (b)

Fig. 5.63. Ceramic-metal-UHMWPE prostheses; (a) after M.E. Müller; (b) after B.G. Weber (courtesy of Sulzer)

Figure 5.64 shows two designs with ceramic-to-ceramic articulation. The upper metal stem is determined for an implantation with bone cement; so are the two ceramic sockets at the left above this stem. The complete set of prosthetic devices includes three stems of different length. The metal stem at the bottom of Fig. 5.64 is prepared for an implantation without bone cement, as are the three ceramic sockets at the right. Both stems and sockets have self-locking ribs and screws for bone ingrowth [5.78,5.79].

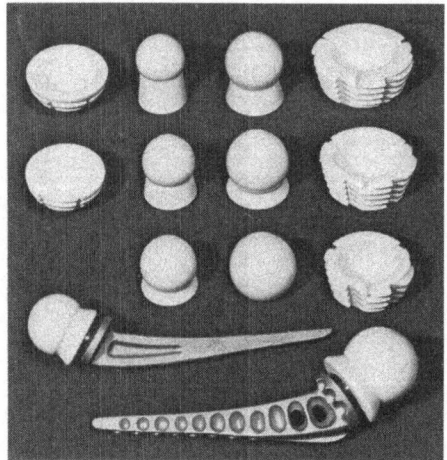

Fig. 5.64. Ceramic-metal-prostheses, after H. Mittelmeier (courtesy of Osteo)

The use of bone cement is normally the reason for the restriction of an artificial hip joint to older patients. The implantation without bone cement provides the first possibility of also using these prostheses for younger patients.

Figure 5.64 shows only one of four stems with different lengths. The ceramic heads of this program have diameters of 32 and 38 mm and three different neck lengths in order to obtain optimal muscle tension. All femoral heads fit all stems.

Figures 5.65 and 5.66 describe ceramic-metal prosthetic systems after Boutin [5.80] and Griss [5.81], having ceramic-to-ceramic articulation with a metal stem to be implanted with bone cement, some of the acetabulum components being also suitable for an implantation without bone cement.

5.3.1.5 Design

One of the most important features of these systems is the connection between stem and head. It has to be simple in application and entirely stable against rotation; any motion of metal on ceramics will cause metallic wear on the ceramic surface due to the great difference in hardness between these two materials.

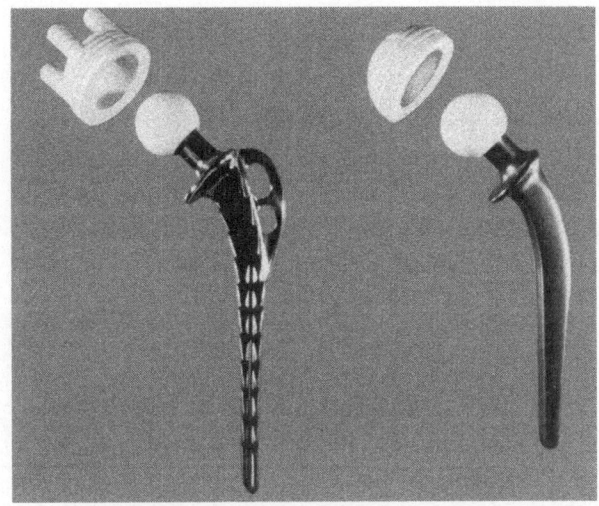

Fig. 5.65. Ceramic-metal prostheses, after P. Boutin (courtesy of Ceraver)

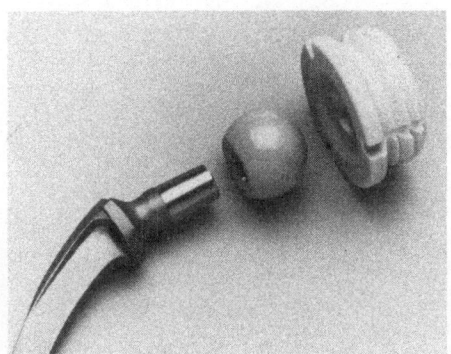

Fig. 5.66 Ceramic-metal prosthesis after P. Griss (courtesy of Friedrichsfeld)

Among the several possibilities of connection techniques already proven in technical application fields, such as adhesive bond, brazing, or mechanical fixing, only
the latter could be taken into consideration. Adhesive compounds degrade in the
body by resorption. This leads to a relative motion between metal and ceramics and
produces metal debris after a short period of time. Brazing is not suitable, either,
because most of the components of brazing alloys are not biocompatible. Therefore
mechanical clamping was the only possibility which remained for the connection between metal stem and ceramic head. It was decided to use a tapering cone. Since the
rotation stability as well as the wedge stress increase with decreasing angle, optimizing tests had to be carried out. According to these tests an angle of about 6°
was found to be a useful compromise [5.83]. Meanwhile this conical fixation has
proved to be the only reliable way of connecting the ceramic head to the metal

stem which meets both the mechanical and the biological requirements of a hip prosthesis.

Figure 5.67 gives a cross-section of the tapering cone which is used in the prostheses described in Figs. 5.63 and 5.64. This connection was found by simulator and animal tests to be absolutely stable against rotation. Since it leads to a high stress inside the ceramic head extensive mechanical tests were necessary in order to obtain the limits for static and dynamic loads. The tests were carried out with a six ton pulsator [5.83].

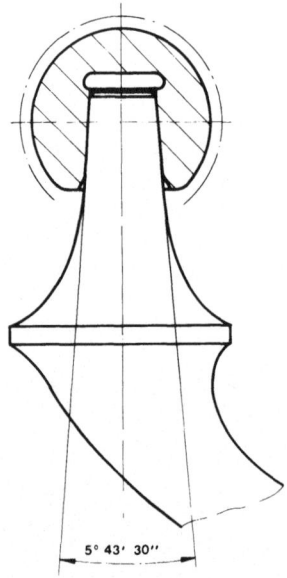

5° 43' 30''

Fig. 5.67. Conical fixation [5.83]

The objective of these experiments was to find out the static and dynamic load-carrying capacity of the whole prosthetic system consisting of metal stem, and a ceramic head and ceramic socket, both with 32 mm diameter (Fig. 5.68). The components were placed in a vertical position and the load was applied alternating between 6 and 50 kN. The tests were carried out with a frequency of 10 cycles/s in Ringer's solution in order to find out if there is any influence of body fluids on strength. After 20 million cycles, no sign of damage was observed among all systems tested [5.83]. The same results were obtained when the ceramic socket was replaced by a socket of UHMWPE or when the ceramic head was placed on a hardened steel plate or on a copper ring. These tests show that the ceramic components of the prostheses described in Figs. 5.63 and 5.64 can be loaded with 50 times the body weight for an arbitrarily long period of time. This seems to be a very high safety factor compared to loads occurring in vivo. No dynamic fatigue was observed under the test conditions described above.

Fig. 5.68. Tested system [5.83]

The same holds for static fatigue. According to an aging experiment with pre-stressed hip joint components in Ringer's solution and in sheep tissue, no time-dependent strength reduction was observed during a period of one year. The load-bearing capacities before and after the test were the same [5.83a].

5.3.1.6 Clinical Results

After a long period of extensive animal tests [5.71,5.84] the clinical application begun. With the prostheses described in Figs. 5.63 and 5.64 a total of about 60 000 cases could be evaluated up to 1982. This figure includes about 20 000 cases with ceramic-ceramic articulation implanted without bone cement [5.85]. The clinical cases described here, which constitute by far the majority of all implanted hip prostheses, have turned out successfully with respect to the ceramic components, given a correct application of the implantation technique. The ceramic components used here have surpassed by far all requirements of the respective standards (Table 5.8).

5.3.2 Dental Implants

Dental implants differ from orthopedic implants by the fact that one part of the implant is outside the tissue in the oral cavity. In this environment they are exposed to fluids with very different pH values, ranging from acidic to basic. Therefore, metallic dental implants cannot be prevented from being attacked chemically. This holds even for the most resistant metals having passivating layers, for the protection is given only if the layers are intact. Considering the rough implantation procedure of a dental implant, which is hammered into the compact bone in many cases, however, it is hard to believe that the passivating layer is not damaged. This is why the development of metallic dental implants has come to a halt. There was no way of further improving the biocompatibility, because all metals, even the most passive ones, such as cobalt base alloys, titanium, and tantalum release ions, and ion release is the first step of a chemical reaction.

In order to protect the surrounding tissue from contact with the metal surface, attempts were made to cover the metallic dental implants with alumina ceramic layers by means of plasma spraying or by similar covering techniques. However, since the layers are porous, the protection is insufficient. Additionally, there is a low bond strength between the ceramic layer and the metal surface; therefore, the layers chip off when the metal is deformed. This does not hold, however, for so-called dental ceramics being applied, not to implants, but to dental restorations such as crowns and bridges. In this case, metal structures are covered by a dense layer of a porcelain-like ceramic material having a good bond strength.

High-purity alumina ceramics, however, have proved successful for dental implants. They offer a high corrosion resistance leading to an excellent biocompatibility confirmed by numerous animal tests, as well as a strength comparable to that of metals [5.86]. Another important advantage of high-purity alumina ceramics is that, in contrast to metals, no calculus, plaque, or concrements can be deposited on the surface of ceramic dental implants, thus eliminating a main factor creating inflammations [5.87].

As early as in 1965 screw-type dental implants and diverging pins made of alumina ceramics were successfully tested clinically [5.88,5.89], followed by a step-shaped so-called instant implant (Fig. 5.71), an anchor-shaped extension implant (Fig. 5.69), and pin implants (Fig. 5.70). Obviously the extension implant seems to be a particularly favorable combination of material and design, because about one year after the implantation a so-called "lamina dura" appears in the X-ray image, evidence that the implant is functioning perfectly [5.92].

Whereas in the cases described above dental implants were used as anchors for artificial teeth or posts for dental bridges or dental prostheses, a new method of tooth replantation by means of a partial ceramic root has been developed in order to preserve teeth having defective roots [5.92]. The procedure is carried out by removing the tooth very carefully, cutting off a part of the root, drilling a hole into the residual root as well as into the alveolus, inserting a well fitting pin of high-purity alumina ceramics into the hole of the residual root, and replanting the tooth (Fig. 5.72).

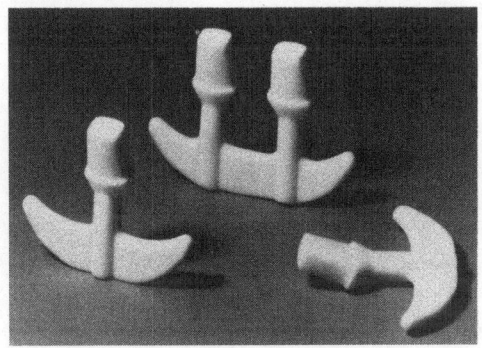

Fig. 5.69. Anchor-shaped extension implants made of high-purity alumina ceramics [5.86]

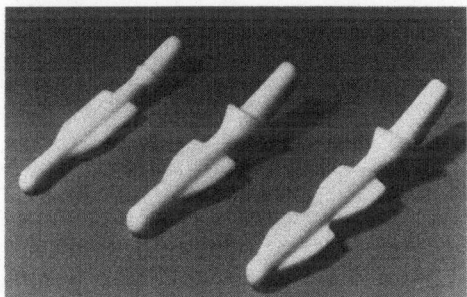

Fig. 5.70. Pin implants made of high-purity alumina ceramics [5.90]

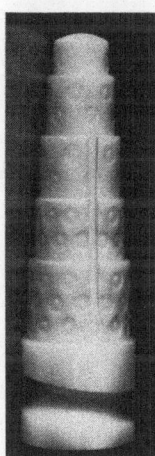

Fig. 5.71. Step-shaped instant dental implant made of high-purity alumina ceramics [5.91] (courtesy of Friedrichsfeld)

As of 1980, more than 2000 clinical cases were observed without complications during a period of about five years and demonstrate the successful application of high-purity alumina dental implants belonging to the different types and systems described here.

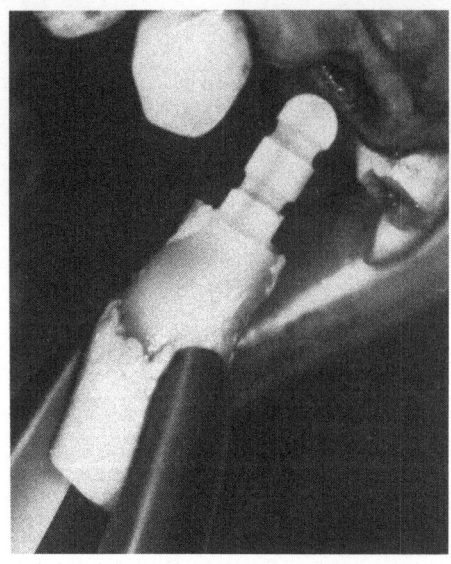

Fig. 5.72. Tooth replantation by means of a partial
ceramic root [5.93]

5.3.3 Maxillary Reconstructions

Among the possibilities of maxillary reconstructions by means of high alumina ceramics, the implantation of orbital floors after fractures of the midface [5.94]
seems to become an important and effective operation method in order to eliminate
defective vision (Fig. 5.73).

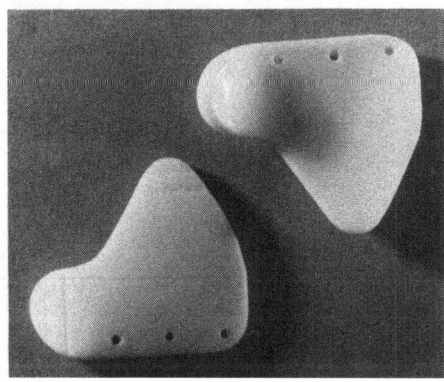

Fig. 5.73. Orbital floors for maxillary reconstructions
made of high-purity alumina ceramics

The upper side of the implant is polished in order to have a smooth surface for
the eyeball. Because there is almost no functional load, the implant can be designed with very thin walls. According to the results of about 100 clinical cases, a
fixation of the ceramic orbital floors by the surrounding tissue can be observed
even after a few weeks after the operation.

5.4 Armor Applications

The first successful applications of high alumina ceramics for armor purposes
date back to the early nineteensixties. Due to their extreme theoretical strength,
in the order of 100 000 MN/m^2 and their low specific weight, hard materials such
as oxide ceramics or carbides, particularly high alumina ceramics and boron carbide are potential materials for light-weight effective armor facing.

All projectiles that have to be stopped by the armoring material are characterized
by an extremely high velocity, up to 1000 m/s for bullets and 10 000 m/s for
shaped charges, particularly for the tip of the particle beam. Therefore, all interactions between the projectile and the armor facing material, i.e., deformation,
occur in the same range of high velocity. The rate of crack propagation in alumina
is about 3000 m/s, considerably lower than that of shock wave propagation, roughly
10 000 m/s. Thus, brittle fracture is too slow to accompany shock wave propagation,
and the resistance of the ceramic increases to the theoretical strength.

According to experimental data, there are considerable differences in the armoring
mechanism between metal and alumina ceramics. Whereas the bullet-absorbing effect
of metallic materials is based on plastic deformation, the mechanism taking place
in brittle materials can be described as a conversion of the kinetic energy of the
bullet into fracture energy absorbed by the ceramic material. For this reason it
is necessary for the ceramic to have a high hardness and compressive strength as
well as a high Young's modulus, which is the prerequisite for a high velocity of
sound.

When the projectile penetrates the ceramic surface, its tip becomes blunted due to
the extreme hardness of the ceramic material, leading to an increase in effective
cross-section of the bullet and accordingly to a reduction of the compressive
stress acting on the ceramic. A compressive shock wave is created at the point of
impact and travels through the material, leaving behind star-shaped radial cracks
and circumferential cracks (Fig. 5.74). These are described and interpreted in detail by [5.95] and [5.96]. The shock wave is reflected at the back surface and
travels back as a tensile shock wave.

An important prerequisite for the functioning of the mechanism described here is
a strong framing of the ceramic material on all sides, including a backing of a
material having a high fracture toughness. High-strength steel makes suitable
backup plates, but precipitation-hardened aluminum alloys or fiber reinforced
plastics appear to be superior because of the lower weights.

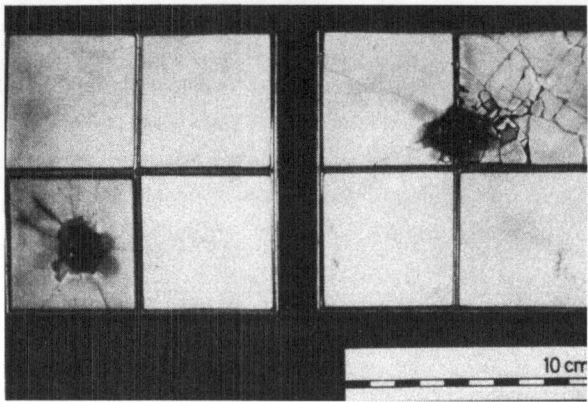

Fig. 5.74. High alumina ceramic armor plates being hit by 9mm bullets of armor piercing type

Figure 5.75 shows the assembly of a ceramic-metal combination plate consisting of a backup plate with frames made of an aluminum-magnesium-zinc alloy, and four high-purity alumina ceramic plates vulcanized to the metal by means of rubber foils. This process has turned out to be the only combination technique between ceramic and metal which is resistant to shock waves. All cements or adhesives, however, even the most ductile ones, would behave in a brittle fashion at the moment the shock wave passes through the material. They would burst and the ceramic plates would come off.

Fig. 5.75. Assembling of a ceramic-metal combination plate for bullet-proof life vests

The combination plates shown in Fig.5.75 are used in bullet-proof life vests. They have to be light-weight and effective. With a wall thickness of 2.5 mm for the ceramic plates and 4 mm for the aluminum backup plates, they withstand all bullets shot by pistol or revolver, including weapons of .357 Magnum type with a steel-core. They offer two considerable advantages compared to armor plates made of steel:

(1) Weight reduction.

(2) Higher energy absorption.

Due to the lower specific weight of alumina ceramics, as well as of aluminum, the weight reduction compared to steel plates is between 28% and 43%, depending on the special application and the types of bullets the plates have to withstand.

As indicated earlier, during the impact of the bullet a transformation of its kinetic energy into fracture energy absorbed by the ceramic material takes place. This leads to a significant reduction of the residual energy, which the wearer of the vest has to absorb. Lower residual energy means less risk of being injured. In order to demonstrate this a comparative animal test was carried out. A pig protected with a ceramic-metal combination plate at one side and with a steel plate at the other side was fired at by a gun with bullets which were withstood by both kinds of plates. After this the pig was killed. When inspecting the tissue behind the plates it was found that the volume of the hematom behind the steel plate was larger by thirty times than the hematom behind the ceramic-metal combination plate. This result was confirmed by a human test. The person tested felt no pain behind the ceramic plate in contrast to the part of the tissue behind the steel plate where an extended hematom developed.

Today the main applications of ceramic armors, particularly in the form of the combination plates described here, are restricted to bullet-proof life vests for policemen [5.97]. However, these plates can also be used for protection against larger bullet calibers if the wall thickness of the ceramic and of the metal is adapted to the kinetic energy of the bullet. In this way, vehicles can also be armored with high alumina ceramics. But this project is still under development, at least in Europe.

In the United States, however, ceramic armors have gained in significance even in the military range, particularly in those cases where weight reduction is of strategic importance, for example aircraft. Above all, components of tactical combat helicopters, such as pilot seats or base plates, are preferably armored with combination structures of ceramics and fiber reinforced plastics.

Another potential field of application of interest for ceramic armors could be explored with regard to the protection of armored vehicles and tanks from shaped charges. Armor facing materials show quite different interactions if subjected to shaped charges, in contrast to bullets. Due to the extreme velocities of the particle beam of a shaped charge, which is in the order of 10 000 m/s, armor facing materials showing brittle fracture, such as high alumina ceramics, turn out to be superior compared to such materials as metals, whose protective mechanism operates by plastic deformation [5.98]. However, this very special application is still in the first experimental stage. Due to the fact that all information about these applications is handled confidentially, particularly in the military range, most of the literature is classified and therefore not available.

5.5 Other Applications

In this section, a series of thermal and chemical applications of high alumina ceramics will be described. They are, however, of substancially minor importance compared to the applications of the foregoing sections.

Porous alumina ceramics such as standard refractories, will be omitted. As already indicated in the introduction, this book is restricted to dense materials.

Due to their outstanding resistance to elevated temperatures as well as to corrosion, high alumina ceramics were considered long ago for thermal applications as super refractories and for chemical applications such as laboratory and metallurgical equipment [5.99,5.100]. (Fig. 5.76a/b).

In contrast to the requirements in the field of mechanical engineering, the application described here require neither a high-strength nor a high-abrasion resistance. Therefore, there is no particular necessity for a very high density and a small grain size; only a high purity is desirable in order to obtain good temperature endurance and chemical stability.

Tubes for high-temperature furnaces are one of the most frequent applications among the fields described here. In the horizontal position they can be exposed to a temperature of up to $1250^{\circ}C$ without sag. When rotated about the axis the temperature can be raised to $1400^{\circ}C$ [5.101]. High alumina protective tubes for thermocouples and pyrometers as well as crucibles, trays, dishes, and boats for high temperature uses can generally be exposed to temperatures up to $1800^{\circ}C$.

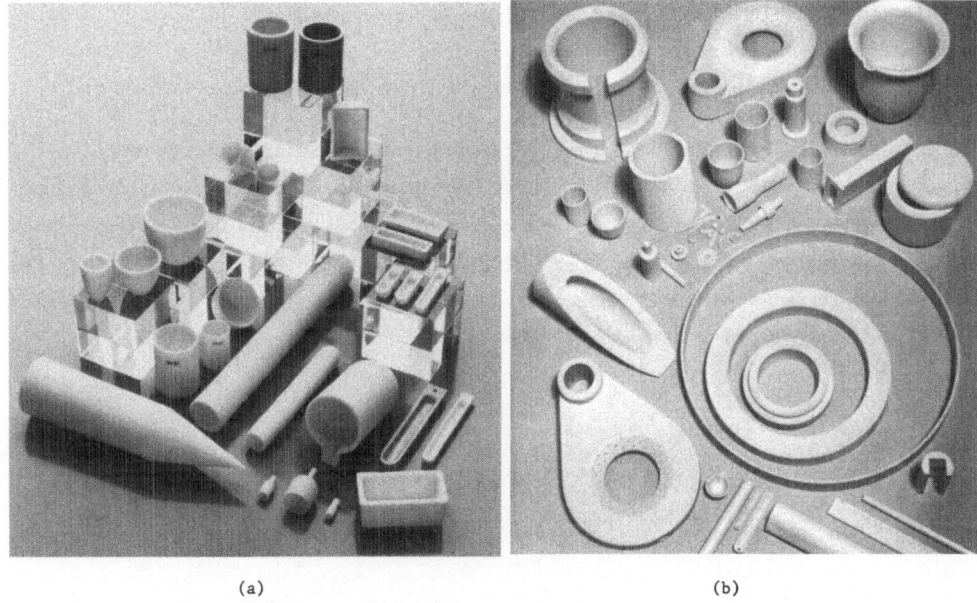

(a)   (b)

Fig. 5.76. Laboratory and metallurgical equipment made of high alumina ceramics;
(a: courtesy of Friedrichsfeld) (b: courtesy of Haldenwanger)

Besides the thermal and chemical applications described here, high alumina ceramics are used for grinding equipment such as grinding balls, grinding cases, mor-
tars and pestles.

With small additions of foreign oxides a coloration effect can be achieved leading
to an entirely new application: colored mar-proof watch housings. Polished discs
of synthetic sapphire, which are resistant to scratches, have long been used as
watch glasses. Therefore, it was obviously necessary to manufacture the housings
as well out of a mar-resistant material.

Microbearings for watches or for sensitive measuring instruments have been tentatively made of high alumina ceramics in order to replace sapphires and rubies.
But for economic reasons these efforts did not succeed. Therefore, microbearings
were and are now made out of synthetic sapphire and ruby, and are the main application field of these alumina single crystals. The manufacture of synthetic sapphire and ruby is usually carried out by means of the Verneuil process or the
Czochralski process by melting the raw material and growing the crystals out of
the melt. Due to their pronounced anisotropic nature the crystals have to be cut
and machined according to their crystallographic axes to obtain microbearings
with uniform wear and friction behavior.

Besides this, rubies are frequently used as laser crystals [5.102,5.103]. This application, however, requires an extreme purity of the raw material in order to obtain the specified properties by doping the crystal with tracer elements.

The important and extended application field of ceramic catalyst supports for emission control of motor vehicles is not included here. Materials for such purposes have to be as porous as possible to offer a high surface area. On the other hand, most of them are not made of alumina ceramics but of cordierite, which has a considerably lower thermal expansion and thus a higher thermal shock resistance compared to high alumina ceramics [5.104 to 5.108].

References

1.1   Kingery, W.D.; Bowen, H.K.; Uhlmann, D.R.: Introduction to ceramics. New York: John Wiley and Sons 1976

1.2   Evans, A.G.; Langdon, T.G.: Structural ceramics. Progress in Material Science 21(1976)171-441

1.3   D.P. 220394 from 1907

1.4   D.P. 560575 from 1929

1.5   Dawihl, W.: Untersuchungen im Gebiet hoher und höchster Temperaturen. Tonindustrie-Z. 58(1934)3-25

1.6   D.P. 715926 from 1936

1.7   Ryshkewitch, E.: Oxydkeramik der Einstoffsysteme vom Standpunkt der Physikalischen Chemie. Berlin, Göttingen, Heidelberg: Springer 1948

1.8   Ryshkewitch, E.: Oxide ceramics - physical chemistry and technology. New York and London: Academic Press 1960

1.9   Gitzen, W.H.: Alumina as a ceramic material. Columbus, Ohio: The American Ceramic Society 1970

1.10 Davidge, R.W.: Mechanical behaviour of ceramics. Cambridge (England): Cambridge University Press 1979

2.1   Gitzen, W.H. (Ed.): Alumina as a ceramic material. Columbus: The American Ceramic Society 1970

2.2   Winchell, H.: Navigation in crystallography. Bull.Geol.Soc.Amer. 57(1946)295-308

2.3   Kronberg, M.L.: Plastic deformation of single crystals of sapphire: basal slip and twinning. Acta Met. 5(1957)507-524

2.4   Kingery, W.D.; Bowen, H.K.; Uhlmann, D.R.: Introduction to ceramics, 2nd ed. New York: John Wiley and Sons 1976

2.5   Swanson, H.E.; Cook, M.L.; Isaacs, T.; Evans, E.H.: Standard X-Ray Powder Diffraction Patterns. Nat.Bur.Stand.Circ. 9(1960)539

2.6   Lynch, C.T. (Ed.): Handbook of materials science, Vol. I CRC Press 1975, pp. 357-363

2.7   Jan, J.P.; Steinemann, S.; Dinichert, P.: Density and lattice parameters of ruby. J.Phys.Chem.Sol. 12(1960)349-350

2.8   Rasmussen, J.J.; Kingery, W.D.: Effect of dopants on the defect structure of single-crystal aluminum oxide. J.Amer.Ceram.Soc. 53(1970)436-440

2.9   Rossi, L.R.; Lawrence, W.G.: Elastic properties of oxide solid solutions: the system $Al_2O_3-Cr_2O_3$. J.Amer.Ceram.Soc. 53(1970)604-608

2.10   Phillips, D.S.; Mitchell, T.E.; Heuer, A.H.: Precipitation in star sapphire, III. Chemical effects accompanying precipitation. Phil.Mag.A 42(1980)417-432

2.11   Goldsmith, A.; Waterman, T.E.; Hirschhorn, H.J., (Eds): Handbook of thermophysical properties of solid materials, Vol.III: Ceramics. New York: The MacMillan Company 1966

2.12   Weast, R.C.: Handbook of chemistry and physics, 54th ed. CRC Press 1974

2.13   Diamond, J.J.; Schneider, S.J.: Apparent temperatures measured at melting points of some metal oxides in a solar furnace. J.Amer.Ceram.Soc. 43(1960) 1-3

2.14   Davidge, R.W.: Mechanical behaviour of ceramics. London, New York, Melbourne: Cambridge University Press 1979

2.15   Norton, F.H.; Kingery, W.D.; Economos, G.; Humenik, Jr., M.: Study of metal ceramic interactions at elevated temperatures. USAEC-Report NYO-3144, 1953

2.16   Rhee, S.K.: Critical surface energies of Al_2O_3 and graphite. J.Amer.Ceram. Soc. 55(1972)300-303

2.17   Kingery, W.D.: Surface tension of some liquid oxides and their temperature coefficients. J.Amer.Ceram.Soc. 42(1959)6-10

2.18   Rasmussen, J.J.; Nelson, R.P.: Surface tension and density of molten Al_2O_3. J.Amer.Ceram.Soc. 54(1971)398-401

2.19   Rasmussen, J.J.: Surface tension, density, and volume change on melting of Al_2O_3 systems, Cr_2O_3 and Sm_2O_3. J.Amer.Ceram.Soc. 55(1972)326-327

2.20   Kingery, W.D.: Metal-ceramic interactions: IV, Absolute measurements of metal-ceramic interfacial energies. J.Amer.Ceram.Soc. 37(1954)42-45

2.21   Engel, T.K.: The heat capacities of Al_2O_3, UO_2 and PuO_2 from 300 to 1100 K. J.Nucl.Mater. 31(1969)211-214

2.22   Coble, R.L.; Kingery, W.D.: Effect of porosity on physical properties of sintered alumina. J.Amer.Ceram.Soc. 39(1956)377-385

2.23   Wachtman, Jr., J.B.; Scuderi, T.G.; Cleek, G.W.: Linear thermal expansion of aluminum oxide and thorium oxide from 100 to 1100 K.J.Amer.Ceram.Soc. 45(1962)319-323

2.24   Nielsen, T.H.; Leipold, M.H.: Thermal expansion in air of ceramic oxides to 2200°C.K.J.Amer.Ceram.Soc.46(1963)381-387

2.25   Kittel, C.: Introduction to solid state physics. New York: John Wiley and Sons 1968

2.26   Hoch, M; Vernardakis, T.: Specific heat and thermal expansion of solids at high temperatures. Scripta Met. 9(1975)1131-1133

2.27   Charvat, F,R.; Kingery, W.D.: Thermal conductivity: XIII, Effect of microstructure on conductivity of single-phase ceramics. J.Amer.Ceram.Soc. 40 (1957)306-312

2.28   Nishijima, T.; Kawada, T.; Ishihata, A.: Thermal conductivity of sintered UO_2 and Al_2O_3 at high temperatures. J.Amer.Ceram.Soc. 48(1965)31-34

2.29 Fitzer, E.; Weisenburger, S.: Cooperative measurements of heat transport properties of tungsten, alumina, and polycrystalline graphite up to 2700 K. In Reisbig, R.L. (Ed): Advances in thermal conductivity. Rolla (USA): 1974, pp.42-46

2.30 Hou, L.D.; Tiku, S.K.; Wang, H.A.; Kröger, F.A.: Conductivity and creep in acceptor-dominated polycrystalline Al_2O_3. J.Mater.Sci. 14(1979)1877-1889

2.31 Mitchell, T.E.; Hobbs, L.W.; Heuer, A.H.; Castaing, J.; Cadoz, J.; Philibert, J.: Interaction between point defects and dislocations in oxides. Acta Met. 27(1979)1677-1691

2.32 Mohapatra, S.K.; Tiku, S.K.; Kröger, F.A.: The defect structure of unintentionally doped α-Al_2O_3 crystals. J.Amer.Ceram.Soc. 62(1979)50-57

2.33 Oishi, Y.; Kingery, W.D.: Self diffusion of oxygen in single-crystal and polycrystalline aluminum oxide. J.Chem.Phys. 33(1960)480-486

2.34 Kröger, F.A.; Vink, V.J.: Relations between the concentrations of imperfections in crystalline solids. In Seitz, F.; Turnbull, D. (Eds.): Solid state physics, Vol.3.New York: Academic Press 1956, pp.307-435

2.35 Mohapatra, S.K.; Kröger, F.A.: The dominant type of atomic disorder in α-Al_2O_3 J.Amer.Ceram.Soc. 61(1978)106-109

2.36 Dienes, G.J.; Welch, D.O.; Fischer, C.R.; Hatcher, R.D.; Lazareth, O.; Samberg, M.: Shell-model calculations of some point-defect properties in α-Al_2O_3. Phys. Rev.B 11 (1975)3060-3070

2.37 Pappis, J.; Kingery, W.D.: Electrical properties of single-crystal and polycrystalline alumina at high temperatures. J.Amer.Ceram.Soc. 44(1961) 459-464

2.38 Brook, R.J.; Yee, J.; Kröger, F.A.: Electrochemical cells and electrical conduction in pure and doped Al_2O_3. J.Amer.Ceram.Soc. 54(1971)444-451

2.39 Brook, R.J.: Effect of TiO_2 on the initial sintering of Al_2O_3. J.Amer. Ceram.Soc. 55(1972)114-115

2.40 Yee, J.; Kröger, F.A.: Measurements of electromotive force in Al_2O_3 - pitfalls and results. J.Amer.Ceram.Soc. 56(1973)189-191

2.41 Hollenberg, G.W.; Gordon, R.S.: Effect of oxygen partial pressure on the creep of polycrystalline Al_2O_3 doped with Cr, Te, or Ti. J.Amer.Ceram.Soc. 56(1973)140-147

2.42 Raja Rao, W.; Cutler, I.B.: Effect of iron oxide on the sintering kinetics of Al_2O_3. J.Amer.Ceram.Soc. 56(1973)588-593

2.43 Rajo Rao, W.; Cutler, I.B.: Activated sintering of alumina by quenching of point defects. Mater.Sci.Res. 6(1973)253-260

2.44 Kitazawa, K.; Coble, R.L.: Electrical conduction in single-crystal and polycrystalline Al_2O_3 at high temperatures. J.Amer.Ceram.Soc. 57(1974) 245-250

2.45 Dutt, B.V.; Hurrell, J.P.; Kröger, F.A.: High-temperature defect structure of cobalt-doped α-alumina. J.Amer.Ceram.Soc. 58(1975)420-427

2.46 Dutt, B.V.; Kröger, F.A.: High-temperature defect structure of iron-doped α-alumina. J.Amer.Ceram.Soc. 58(1975)474-476

2.47 Frederikse, H.P.R.; Hosler, W.R.: High temperature electrical conductivity of aluminum oxide. In Cooper, A.R.; Heuer, A.H. (Eds.): Mass transport phenomena in ceramics. Materials Science Research, Vol.9. New York and London: Plenum Press 1975, pp.233-252

2.48   Mohapatra, S.K.; Kröger, F.A.: Defect structure of α-Al$_2$O$_3$ doped with magnesium. J.Amer.Ceram.Soc.60(1977)141-148

2.49   Mohapatra, S.K.; Kröger, F.A.: Defect structure of α-Al$_2$O$_3$ doped with titanium. J.Amer.Ceram.Soc. 60(1977)381-387

2.50   Yen, C.F.; Coble, R.L.: Defect centers in gamma-irradiated single-crystal Al$_2$O$_3$. J.Amer.Ceram.Soc. 62(1979)89-94

2.51   Harmer, H.; Roberts, E.W.; Brook, R.J.: Rapid sintering of pure and doped α-Al$_2$O$_3$. Trans.J.Brit.Ceram.Soc. 78(1979)22-25

2.52   Cox, R.T.: Chemical reactions and mass transport processes in donor and acceptor doped Al$_2$O$_3$ crystals. J.Phys.(Paris) 34(1973)suppl.C9,333-335

2.53   Freer, R.: Self-diffusion and impurity diffusion in oxides. J.Mater.Sci. 15 (1980)803-824

2.54   Hollenberg, G.W.; Gordon, R.S.: Origin of anomalously high activation energies in sintering and creep of impure refractory oxides. J.Amer.Ceram.Soc. 56(1973)109-110

2.55   Kitazawa, K.; Coble, R.L.: Chemical diffusion in polycrystalline Al$_2$O$_3$ as determined from electrical conductivity measurements. J.Amer.Ceram.Soc. 57 (1974)250-253

2.56   Coble, R.L.: Sintering crystalline solids: II, Experimental test of diffusion models in powder compacts. J.Appl.Phys. 32(1961)793-799

2.57   Bagley, R.D.; Cutler, I.B.; Johnson, D.L.: Effect of TiO$_2$ on the initial sintering of Al$_2$O$_3$. J.Amer.Ceram.Soc. 53(1970)136-141

2.58   Raja Rao, W.; Cutler, I.B.: Initial sintering and surface diffusion in Al$_2$O$_3$. J.Amer.Ceram.Soc. 55(1972)170-171

2.59   Cannon, R.M.; Coble, R.L.: Review of diffusional creep of Al$_2$O$_3$. In Bradt, R.C.; Tressler, R.E. (Eds.): Plastic deformation of ceramic materials. New York: Plenum Press 1975,pp.61-100

2.60   Paladino, A.E.; Kingery, W.D.: Aluminum ion diffusion in aluminum oxide. J.Chem.Phys. 37(1962)957-962

2.61   Wang, H.A.; Kröger, F.A.: Chemical diffusion in polycrystalline Al$_2$O$_3$. J. Amer.Ceram.Soc. 63(1980)613-619

2.62   Johnson, D.L.; Cutler, I.B.: Diffusion sintering: I, Initial stage sintering models and their application to shrinkage of powder compacts. J.Amer.Ceram. Soc. 46(1963)541-545

2.63   Johnson, D.L.: New method of obtaining volume, grain boundary, and surface diffusion coefficients from sintering data. J.Appl.Phys. 40(1969)192-200

2.64   Johnson, D.L.: A general model for the intermediate stage of sintering. J.Amer.Ceram.Soc. 53(1970)574-577

2.65   Gordon, R.S.: Mass transport in the diffusional creep of ionic solids. J. Amer.Ceram.Soc. 56(1973)147-152

2.66   Gordon, R.S.: Ambipolar diffusion and its application to diffusion creep. In Cooper, A.R.; Heuer, A.H. (Eds.): Mass transport phenomena in ceramics. Materials Science Research, Vol.9. New York and London: Plenum Press 1975,pp.445-464

2.67   Evans, A.G.; Langdon, T.G.: Structural ceramics. Progress in Materials Science, Vol.21. London: Pergamon 1976,pp.171-441

2.68   Harrop, P.J.: Self-diffusion in simple oxides (bibliography). J.Mater.Sci. 3(1968)206-222

2.69 Reed, D.J.; Wuensch, B.J.: Ion probe measurement of oxygen self-diffusion in single-crystal Al_2O_3. J.Amer.Ceram.Soc. 63(1980)88-92

2.70 Folweiler, R.C.: Creep behavior of pore-free polycrystalline aluminum oxide. J.Appl.Phys. 32(1961)773-778

2.71 Warshaw, S.I.; Norton, F.H.: Deformation behavior of polycrystalline aluminum oxide. J.Amer.Ceram.Soc. 45(1962)479-486

2.72 Coble, R.L.; Guerard, Y.H.: Creep of polycrystalline aluminum oxide. J.Amer. Ceram.Soc. 46(1963)353-354

2.73 Dawihl, W.; Klingler, E.: Über das Kriechverhalten von Sinterkörpern aus Al_2O_3 im Temperaturbereich von $1200^{\circ}C$ unter dem Einfluss von Druckspannungen. Ber.Dt.Keram.Ges. 42(1965)270-274

2.74 Hewson, C.W.; Kingery, W.D.: Effect of MgO and $MgTiO_3$ doping on diffusion-controlled creep of polycrystalline aluminum oxide. J.Amer.Ceram.Soc. 50 (1967) 218-219

2.75 Mocellin, A.; Kingery, W.D.: Creep deformation in MgO-saturated large-grain-size Al_2O_3. J.Amer.Ceram.Soc. 54(1971)339-341

2.76 Cannon, R.M.; Rhodes, W.H; Heuer, A.H.: Plastic deformation of fine-grained alumina (Al_2O_3): I, Interface controlled diffusional creep. J.Amer.Ceram. Soc. 63(1980)46-53

2.77 Engelhardt, G.; Thümmler, F.: Kriechuntersuchungen unter 4-Punkt-Biegebeanspruchung bei hohen Temperaturen, II. Messungen an polykristallinem Aluminiumoxid. Ber.Dt.Keram.Ges. 47(1970)571-577

2.78 Coble, R.L.: Initial sintering of alumina and hematite. J.Amer.Ceram.Soc. 41(1958)55-62

2.79 Heuer, A.H.; Cannon, R.M.; Tighe, N.J.: Plastic deformation in fine-grain ceramics. In: Burke, J.J.; Reed, N.L.; Weiss, V. (Eds.): Ultrafine-grain ceramics. New York: Syracuse University Press 1970,pp.339-365

2.80 Jones, T.P.; Coble, R.L.; Mogab, C.J.: Defect diffusion in single-crystal aluminum oxide. J.Amer.Ceram.Soc. 52(1969)331-334

2.81 Roy, S.K.; Coble, R.L.: Solubilities of magnesia, titania, and magnesium titanate in aluminum oxide. J.Amer.Ceram.Soc. 51(1968)1-6

2.82 Mistler, R.E.; Coble, R.L.: Rate-determining species in diffusion-controlled processes in Al_2O_3. J.Amer.Ceram.Soc. 54(1971)60-61

2.83 Passmore, E.M.; Vasilos, T.: Creep of dense, pure, fine-grained aluminum oxide. J.Amer.Ceram.Soc. 49(1966)166-168

2.84 Langdon, T.G.; Mohamed, F.A.: The incorporation of ambipolar diffusion in deformation mechanism maps for ceramics. J.Mater.Sci. 13(1978)473-482

2.85 Lessing, P.A.; Gordon, R.S.: Creep of polycrystalline alumina pure and doped with transition metal impurities. J.Mater.Sci. 12(1977)2291-2302

2.86 El-Aiat, M.M.; Hou, L.D.; Tiku, S.K.; Wang, H.A.; Kröger, F.A.: High-temperature conductivity and creep of polycrystalline Al_2O_3 doped with Fe and/or Ti. J.Amer.Ceram.Soc. 64(1981)174-182

2.87 Carter, C.B.; Kohlstedt, D.L.; Sass, S.L.: Electron diffraction and microscopy studies of the structure of grain boundaries in Al_2O_3. J.Amer.Ceram. Soc. 63(1980)623-627

2.88 Gupta, T.K.: Instability of cylindrical voids in alumina. J.Amer.Ceram.Soc. 61(1978)191-195

2.89 Robertson, W.M.; Chang, R.; in Kriegel, W.W.; Palmour III, H. (Eds.): Role of grain boundaries and surfaces in ceramics. Materials Science Research, Vol.3. New York: Plenum Press 1966,pp.49-60

2.90 Shackelford, J.F.; Scott, W.D.: Relative energies of ($\bar{1}$100) tilt boundaries in aluminum oxide. J.Amer.Ceram.Soc. 51(1968)688-692

2.91 Robertson, W.M.; Ekstrom, F.E.; in Gray, T.J.; Frechette, V.D. (Eds.): Kinetics and reactions in ionic systems. Materials Science Research, Vol.4. New York: Plenum Press 1969, pp.271-281

2.92 Prochazka, S.; Coble, R.L.: Surface diffusion in the initial sintering of alumina: I, II,and III. Phys.Sintering 2(1970)No.1,1-18,No.2,1-14, and No.2,15-34

2.93 Yen, C.F.; Coble, R.L.: Spheroidization of tubular voids in Al_2O_3 crystals at high temperatures. J.Amer.Ceram.Soc. 55(1972)507-509

2.94 Moriyoshi, Y.; Komatsu, W.: Kinetics of initial combined sintering. Yogyo Kyokai Shi 81(1973)102-107

2.95 Maruyama, T.; Komatsu, W.: Surface diffusion of single-crystal Al_2O_3 by scratch-smoothing method. J.Amer.Ceram.Soc. 58(1975)338-339

2.96 Wilks, R.S.: Neutron-induced damage in BeO, Al_2O_3 and MgO (a review). J.Mater.Sci. 26(1968)137-159

2.97 Hauth, W.E.; Stoddard, S.D.: State of the art-alumina ceramics for energy applications. Amer.Ceram.Soc.Bull. 57(1978)181-185

2.98 Bobleter, O.; Fiedler, W.; Grass, F.: Diffusion of the fission product Pm-147 in aluminum oxide. ATKE 10(1965)261-263

2.99 Fiedler, W.; Bobleter, O.: Diffusion of thorium 229 and uranium 233 in α-Al_2O_3. ATKE 12(1967)357-360

2.100 Fiedler, W.; Grass, F.: Diffusion of fission products in aluminum oxide. ATKE 11(1967)420-424

2.101 Matzke, H.: Rare-gas mobility in some anisotropic ceramic oxides: Al_2O_3, Cr_2O_3, Fe_2O_3, TiO_2, U_3O_8. J.Mater.Sci. 2(1967)444-456

2.102 Frischat, G.H.: Kationentransport in Aluminiumoxid. Ber.Dt.Keram.Ges. 48 (1971)441-447

2.103 Hirota, K.; Komatsu, W.: Concurrent measurement of volume, grain-boundary, and surface diffusion coefficients in the system NiO-Al_2O_3. J.Amer.Ceram. Soc. 60(1977)105-107

2.104 Fowler, J.D.; Chandra, D.; Elleman, T.S.; Payne, A.W.; Verghese, K.: Tritium diffusion in Al_2O_3 and BeO. J.Amer.Ceram.Soc. 60(1977)155-161

2.105 Roberts, R.M.; Elleman, T.S.; Palmour III, H.; Verghese, K.: Hydrogen permeability of sintered aluminum oxide. J.Amer.Ceram.Soc. 62(1979)495-499

2.106 Peters, D.W.; Feinstein, L.; Peltzer, C.: On the high-temperature conductivity of alumina. J.Chem.Phys. 42(1965)2345-2346

2.107 Moulson, A.J.; Popper, P.: Problems associated with the measurement of volume resistivity of insulating ceramics at high temperatures. Proc.Brit.Ceram. Soc. 10(1968)41-50

2.108 Özkan, O.T.; Moulson, A.J.: The electrical conductivity of single-crystal and polycrystalline aluminium oxide. Brit.J.Appl.Phys. 3(1970)983-987

2.109 Lackey, W.J.: Effect of temperature on electrical conductivity and transport mechanisms in sapphire. In Kriegel, W.W.; Palmour III, H. (Eds.): Ceramics in severe environments. Materials Science Research, Vol.5 New York and London: Plenum Press 1971,pp.489-502

2.110 Dilger, H.: Untersuchungen der mechanischen und elektrischen Eigenschaften von Al_2O_3 mit ZnO- und NiO- Zusätzen bei hohen Temperaturen, II. Elektrische Eigenschaften. Ber.Dt.Keram.Ges. 51(1974)123-126

2.111 Kingery, W.D.; Meiling, G.E.: Transference number measurements for aluminum oxide. J.Appl.Phys. 32(1961)556

2.112 Matsumara, T.: The elctrical properties of alumina at high temperatures. Can.J.Phys. 44(1966)1685-1698

2.113 Kemp, J.L.; Moulson, A.J.: The effect of iron additions on the electrical properties of alumina ceramics. Proc.Brit.Ceram.Soc. 18(1970)53-64

2.114 Hennicke, H.W.; Stuhrhahn, H.H.: Untersuchungen von Transportvorgängen im System Al_2O_3-Cr_2O_3. Ber.Dt.Keram.Ges. 48(1971)394-400

2.115 Tiku, S.K.; Kröger, F.A.: Energy levels of donor and acceptor dopants and electron and hole mobilities in α-Al_2O_3. J.Amer.Ceram.Soc. 63(1980)31-32

2.116 Vernetti, R.A.; Cook, R.L.: Effect of metal oxide additions on the high-temperature electrical conductivity of alumina. J.Amer.Ceram.Soc. 49(1966) 194-199

2.117 Cahoon, H.P.; Christensen, C.J.: Sintering and grain growth of α-alumina. J.Amer.Ceram.Soc. 39(1956)337-344

2.118 Coble, R.L.: Diffusion sintering in the solid state. In Kingery, W.D. (Ed.): Kinetics of high temperature processes. New York: John Wiley and Sons 1959, pp.147-163

2.119 Coble, R.L.: Sintering alumina: effect of atmospheres. J.Amer.Ceram.Soc. 45 (1962)123-127

2.120 Coble, R.L.; Burke, J.E.: Sintering in ceramics. In Burke, J.E. (Ed.): Progress in Ceramics Science, Vol.3. Oxford, London, New York and Paris: Pergamon Press 1963,pp.197-251

2.121 Johnson, D.L.; Cutler, I.B.: Diffusion sintering: II, Initial sintering kinetics of alumina. J.Amer.Ceram.Soc. 46(1963)545-550

2.122 Keski, J.R.; Cutler, I.B.: Effect of manganese oxide on sintering of alumina. J.Amer.Ceram.Soc. 48(1965)653-654

2.123 Wilcox, P.D.; Cutler, I.B.: Strength of partly sintered alumina compacts. J.Amer.Ceram.Soc. 49(1966)249-252

2.124 Keski, J.R.; Cutler, I.B.: Initial sintering of Mn_xO-Al_2O_3. J.Amer.Ceram.Soc. 51(1968)440-444

2.125 Young, W.S.; Cutler, I.B.: Initial sintering with constant rates of heating. J.Amer.Ceram.Soc. 53(1970)659-663

2.126 Kingery, W.D.; Berg, M.: Study of initial stages of sintering solids by viscous flow, evaporation-condensation, and self-diffusion. J.Appl.Phys. 26 (1955)1205-1212

2.127 Ashby, M.F.: A first report on sintering diagrams. Acta Met. 22(1974)275-289

2.128 Johnson, D.L.; Cutler, I.B.: The use of phase diagrams in the sintering of ceramics and metals. In Alper, A.M. (Ed.): Phase diagrams, Vol.II. New York and London: Academic Press 1970,pp.265-291

2.129 Kuczynski, G.C.; Abernethy, L.; Allan, J.: Sintering mechanisms of aluminum oxide. In Kingery, W.D. (Ed.): Kinetics of high temperature processes. New York: John Wiley and Sons 1959,pp.163-172

2.130 Greskovich, C.; Lay, K.W.: Grain growth in very porous Al_2O_3 compacts. J. Amer.Ceram.Soc. 55(1972)142-146

2.131 Wong, B.; Pask, J.A.: Models for kinetics of solid state sintering. J.Amer. Ceram.Soc. 62(1979)138-141

2.132 Wilson, T.L.; Shewmon, P.G.: Role of interfacial diffusion in the sintering of copper. Trans.AIME 236(1966)48-58

2.133 McAllister, P.V.;;Cutler, I.B.: Thermal grooving of MgO and Al_2O_3. J.Amer. Ceram.Soc. 55(1972)351-354

2.134 Coble, R.L.: Sintering crystalline solids. I. Intermediate and final stage diffusion models. J.Appl.Phys. 32(1961)787-792

2.135 Bruch, C.A.: Sintering kinetics for the high density alumina process. Amer. Ceram.Soc.Bull. 41(1962)799-806

2.136 Vink, H.J.: Scientific understanding of the manufacuture and properties of solid oxidic industrial materials. In Seltzer, M.S.; Jaffee, R.I. (Eds.): Defects and transport in oxides. New York and London: Plenum Press 1974,pp. 127-138

2.137 Peelen, J.G.J.: Influence of MgO on the evolution of the microstructure of alumina. In Kuczynski, G.C. (Ed.): Sintering and Catalysis. New York and London: Plenum Press 1975,pp.443-453

2.138 Johnson, W.C.; Coble, R.L.: A test of second-phase and impurity-segregation models for MgO-enhanced densification of sintered alumina. J.Amer.Ceram. Soc. 61(1978)110-114

2.139 Warman, M.O.; Budworth, D.W.: Criteria for the selection of additives to enable the sintering of alumina to proceed to theoretical density. Trans. Brit.Ceram.Soc. 66(1967)253-264

2.140 Rossi, G.; Burke, J.E.: Influence of additives on the microstructure of sintered Al_2O_3. J.Amer.Ceram.Soc. 56(1973)654-659

2.141 Brook, R.J.: Controlled grain growth. In Wang, F.F.Y. (Ed.): Ceramic fabrication processes. Treatise on Materials Science and Technology, Vol. 9. New York, San Francisco and London: Academic Press 1976,pp.331-364

2.142 Jorgensen, P.J.; Westbrook, J.H.: Role of solute segregation at grain boundaries during final-stage sintering of alumina. J.Amer.Ceram.Soc. 47(1964) 332-338

2.143 Jorgensen, P.J.: Modification of sintering kinetics by solute segregation in Al_2O_3. J.Amer.Ceram.Soc. 48(1965)207-210

2.144 Marcus, H.L.; Fine, M.E.: Grain-boundary segregation in MgO-doped Al_2O_3. J.Amer.Ceram.Soc. 55(1972)568-570

2.145 Johnson, W.C.; Stein, D.F: Additive and impurity distributions at grain boundaries in sintered alumina. J.Amer.Ceram.Soc. 58(1975)485-488

2.146 Johnson, W.C.: Mg distribution at grain boundaries in sintered alumina containing $MgAl_2O_4$ precipitates. J.Amer.Ceram.Soc. 61(1978)234-237

2.147 Clarke, D.R.: Grain-boundary segregation in an MgO-doped Al_2O_3. J.Amer.Ceram.Soc. 63(1980)339-341

2.148 Viechnicki, D.; Schmid, F.; McCauley, J.W.: Liquidus-solidus determinations in the system $MgAl_2O_4-Al_2O_3$. J.Amer.Ceram.Soc. 57(1974)47-48

2.149 Heuer, A.H.: The role of MgO in the sintering of alumina. J.Amer.Ceram.Soc. 62(1979)317-318

2.150 Harmer, M.P.; Brook, R.J.: The effect of MgO additions on the kinetics of hot pressing in Al_2O_3. J.Mater.Sci. 15(1980)3017-3024

2.151 Roy, S.K.; Coble, R.L.: Solubility of hydrogen in porous polycrystalline aluminum oxide. J.Amer.Ceram.Soc. 50(1967)435-436

2.152 Mocellin, A.; Kingery, W.D.: Microstructural changes during heat treatment of sintered Al_2O_3: J. Amer.Ceram.Soc. 56(1973)309-314

2.153 Monahan, R.D.; Halloran, J.W.: Single-crystal boundary migration in hot-pressed aluminum oxide. J.Amer.Ceram.Soc. 62(1979)564-567

2.154 Evans, P.F.: The activation energy for grain growth in alumina. In Kriegel, W.W.; Palmour III, H. (Eds.): The role of grain boundaries and surfaces in ceramics. Materials Science Research, Vol. 3. New York and London: Plenum Press 1966,pp.345-353

2.155 McHugh, C.O.; Whalen, T.J.; Humenik, Jr., M.: Dispersion-strengthened aluminum oxide. J.Amer.Ceram.Soc. 49(1966)486-491

2.156 Wang, H.A.; Kröger, F.A.: Pore formation during oxidative annealing of Al_2O_3:Fe and slowing of grain growth by precipitates and pores. J.Mater.Sci. 15(1980)1978-1986

2.157 Gupta, T.K.: Possible correlation between density and grain size during sintering. J.Amer.Ceram.Soc. 55(1972)276-277

2.158 Simpson, L.A.; Wasylyshin, A.: Fracture energy of Al_2O_3 containing Mo fibers J.Amer.Ceram.Soc. 54(1971)56-57

2.159 Rankin, D.T.; Siglich, J.J.; Petrak, D.R.; Ruh, R.: Hot-pressing and mechanical properties of Al_2O_3 with a Mo-dispersed phase. J.Amer.Ceram.Soc. 54 (1971)277-281

2.160 Steiner, C.J.P.; Spriggs, R.M.; Hasselman, D.P.H.: Synergetic pressure sintering of Al_2O_3. J.Amer.Ceram.Soc. 55(1972)115

2.161 Vasilos, T.; Spriggs, R.M.: Pressure sintering of ceramics. In Burke, J.E. (Ed.): Progress in Ceramic Science, Vol.4 Oxford, London, Edinburgh, New York, Toronto, Paris and Braunschweig: Pergamon Press 1966, pp.95-132

2.162 Coble, R.L.: Diffusion models for hot-pressing with surface energy and pressure effects as driving forces. J.Appl.Phys. 41(1970)4798-4807

2.163 Coble, R.L.; Ellis, J.S.: Hot-pressing alumina - mechanisms of material transport. J.Amer.Ceram.Soc. 46(1963)438-441

2.164 Kronberg, M.L.: Dynamical flow properties of single crystals of sapphire, I. J.Amer.Ceram.Soc. 45(1962)274-279

2.165 Hamano, Y.; Kinoshita, M.: Studies and applications of hot-pressing. In: Saito, S.; Somiya, S.; Kotera, Y.; Pask, J.A.; Batha, H.D. (Eds.): Equilibria and kinetics in modern ceramic processing. Tokyo Institute of Technology: Japan 1972,pp.28-33

2.166 Vasilos, T.; Spriggs, R.M.: Pressure sintering: mechanisms and microstructures for alumina and magnesia. J.Amer.Ceram.Soc. 46(1963)493-496

2.167 Rossi, R.C.; Fulrath, R.M.: Final stage densification in vacuum hot-pressing of alumina. J.Amer.Ceram.Soc. 48(1965)558-564

2.168 Rossi, R.C.; Buch, J.D.; Fulrath, R.M.: Intermediate-stage densification in vacuum hot-pressing of alumina. J.Amer.Ceram.Soc. 53(1970)629-633

2.169 Nabarro, F.R.N.: Deformation of crystals by the motion of single ions. Rep. Conf.Strength Sol. 1948,pp.75-90

2.170 Herring, C.: Diffusional viscosity of a polycrystalline solid. J.Appl.Phys. 21(1950)437-445

2.171 Johnson, W.C.: Grain boundary segregation in ceramics. Met.Trans.A 8(1977) 1413-1422

2.172 Taylor, R.I.; Coad, J.P.; Brook, R.J.: Grain boundary segregation in Al_2O_3. J.Amer.Ceram.Soc. 57(1974)539-540

2.173 Taylor, R.I.; Coad, J.P.; Hughes, A.E.: Grain boundary segregation in MgO-doped Al$_2$O$_3$. J.Amer.Ceram.Soc. 59(1976)374-375

2.174 Dufek, G.; Vendl, A.; Wruss, W.; Kieffer, R.: Bestimmungen der Verteilung von Sinterzusätzen in gesinterter Tonerde durch Mikrosondenanalyse. Ber. Dt.Keram.Ges. 53(1976)336-338

2.175 Franken, P.E.C.; Gehring, A.P.: Grain boundary analysis of MgO-doped Al$_2$O$_3$. J.Mater.Sci. 16(1981)384-388

2.176 Funkenbusch, A.W.; Smith, D.W.: Influence of calcium on the fracture strength of polycrystalline alumina. Met.Trans.A 6(1975)2299-2301

2.177 Jupp, R.S.; Stein, D.F.; Smith, D.W.: Observations on the effect of calcium segregation on the fracture behaviour of polycrystalline alumina. J. Mater.Sci. 15(1980)96-102

2.178 Sinharoy, S.; Levenson, L.L.; Pallard, W.V.; Day, D.E.: Surface segregation of calcium in dense alumina exposed to steam and steam-CO. Amer.Ceram.Soc. Bull. 57(1978)231-233

2.179 Aust, K.T.; Hanneman, R.E.; Niessen, P.; Westbrook, J.H.: Solute induced hardening near grain boundaries in zone refined metals. Acta Met. 16(1968) 291-302

2.180 Tong, S.S.C.; Williams, J.P.: Chemical analysis of grain boundary impurities in polycrystalline ceramic materials by spark source mass spectrometry. J. Amer.Ceram.Soc. 53(1970)58-59

2.181 Johnson, W.C.; Stein, D.F.; Rice, R.W.: Analysis of grain boundary impurities and fluoride additives in hot-pressed oxides by Auger electron spectroscopy. J.Amer.Ceram.Soc. 57(1974)342-344

2.182 Bender, B.; Williams, D.B.; Notis, M.R.: Investigation of grain-boundary segregation in ceramic oxides by analytical scanning transmission electron microscopy. J.Amer.Ceram.Soc. 63(1980)542-546

2.183 Nanni, P.; Stoddart, C.T.H.; Hondros, E.D.: Grain boundary segregation and sintering in alumina. Mater.Chem. 1(1976)297-320

3.1 Kingery, W.D.; Bowen, H.K.; Uhlmann, D.R.: Introduction to ceramics, 2nd ed., New York: John Wiley and Sons 1976

3.2 Evans, A.G.; Langdon, T.G.: Structural ceramics. Progress in Materials Science 21(1976)171-441

3.3 Rice, R.W.: Microstructure dependence of mechanical behavior of ceramics. In MacCrone, R.K. (Ed.): Treatise on materials science and technology, Vol.11: Properties and microstructure. New York, San Francisco, London: Academic Press 1977, pp.199-381

3.4 Davidge, R.W.: Mechanical behaviour of ceramics. Cambridge, London, New York, and Melbourne: Cambridge University Press 1979

3.5 Wachtman Jr., J.B.; Tefft, W.E.; Lam Jr., D.G.; Stinchfield, R.P.: Elastic constants of synthetic single crystal corundum at room temperature. J.Res. Nat.Bur.Stds. 64A(1960)213-228

3.6 Wachtman Jr., J.B.; Tefft, W.E.; Lam Jr., D.G.: Young's modulus of single crystal corundum from 77 to 850 K. In Kriegel, W.W.; Palmour III, H. (Eds.): Mechanical properties of engineering ceramics. New York and London: Interscience Publishers 1961, pp.221-223

3.7   Tefft, W.E.: Elastic constants of synthetic single crystal corundum. J.Res. Nat.Bur.Stds. 70A(1966)277-280

3.8   Coble, R.L.; Kingery, W.D.: Effect of porosity on physical properties of sintered alumina. J.Amer.Ceram.Soc. 39(1956)377-385

3.9   Crandall, W.B.; Chung, D.H.; Gray, T.J.: The mechanical properties of ultra-fine hot-pressed alumina. In Kriegel, W.W.; Palmour III, H. (Eds.): Mechanical properties of engineering ceramics. New York and London: Interscience Publishers 1961, pp.349-376

3.10   Spriggs, R.M.; Mitchell, J.B.; Vasilos, T.: Mechanical properties of pure, dense aluminum oxide as a function of temperature and grain size. J.Amer. Ceram.Soc. 47(1964)323-327

3.11   Chung, D.H.; Simmons, G.: Pressure and temperature dependence of the isotropic elastic moduli of polycrystalline alumina. J.Appl.Phys. 39(1968) 5316-5326

3.12   Soga, N.; Anderson, O.L.: High-temperature elastic properties of polycrystalline MgO and Al_2O_3. J.Amer.Ceram.Soc. 49(1966)355-359

3.13   Bailey, J.E.; Hill, N.A.: The effect of porosity and microstructure on the mechanical properties of ceramics. Proc.Brit.Ceram.Soc. 15(1970)15-35

3.14   Knudsen, F.P.: Effect of porosity on Young's modulus of alumina. J.Amer. Ceram.Soc. 45(1962)94-95

3.15   Wachtman, J.B.: Elastic deformation of ceramics and other refractory materials. In Wachtman, J.B. (Ed.): Mechanical and thermal properties of ceramics, NBS Special Publication 303. Washington: Nat. Bur. Stands. 1969, pp.303-385

3.16   Mackenzie, J.K.: The elastic constants of a solid containing spherical holes. Proc.Phys.Soc. 63B(1950)2-11

3.17   Hashin, Z.: Elasticity of ceramic systems. In Fulrath, R.M.; Pask, J.A. (Eds.): (Eds.): Ceramic microstructures. New York: John Wiley and Sons 1968, pp.313

3.18   Spriggs, R.M.: Expression for effect of porosity on elastic modulus of polycrystalline refractory materials, particularly aluminum oxide. J.Amer. Ceram.Soc. 44(1961)628-629

3.19   Soga, N.; Anderson, O.L.: Simplified method for calculating elastic moduli of ceramic powders from compressibility and Debye temperature data. J.Amer.Ceram.Soc. 49(1966)318-322

3.20   Kelly, A.: The strength of ceramics. Proc.Brit.Ceram.Soc. 15(1970)1-14

3.21   Davidge, R.W.; Evans, A.G.: The strength of ceramics. Mater.Sci.Engng. 6 (1970)281-298

3.22   Davidge, R.W.: Effects of microstructure on the mechanical properties of ceramics. In Bradt, R.C.; Hasselman, D.P.H.; Lange, F.F. (Eds.): Fracture mechanics of ceramics, Vol.2. New York and London: Plenum Press 1974, pp.447-468

3.23   Lawn, B.R.; Wilshaw, T.R.: Fracture of brittle solids. Cambridge, London, New York, and Melbourne: Cambridge University Press 1975

3.24   Griffith, A.A.: The phenomena of rupture and flow in solids. Phil.Trans. Roy.Soc. A221(1920)163-198

3.25   Wiederhorn, S.M.: Fracture of sapphire. J.Amer.Ceram.Soc. 52(1969)485-491

3.26   Wiederhorn, S.M.; Hockey, B.J.; Roberts, D.E.: Effect of temperature on the fracture of sapphire. Phil.Mag. 28(1973)783-796

3.27 Iwasa, M.; Ueno, T.; Bradt, R.C.: Fracture toughness of quartz and sapphire single crystals at room temperature. J.Soc.Mater.Sci.Japan 30(1981)1001-1004

3.28 Anderson, N.C.: Basal plane cleavage cracking of synthetic sapphire arc lamp envelopes. J.Amer.Ceram.Soc. 62(1979)108-109

3.29 Rice, R.W.: Comments on "Surface-finish effects on strength-vs-grain-size relations in polycrystalline Al_2O_3". J.Amer.Ceram.Soc. 58(1975)154-155

3.30 Rice, R.W.; Freiman, S.W.: Grain-size dependence of fracture energy in ceramics: II, A model for noncubic materials. J.Amer.Ceram.Soc. 64(1981) 350-354

3.31 Claussen, N.; Mussler, B.; Swain, M.V.: Grain-size dependence of fracture energy in ceramics. J.Amer.Ceram.Soc. 65(1982)C14-16

3.32 Kingery, W.D.: Metal-ceramic interactions: IV, Absolute measurements of metal-ceramic interfacial energies. J.Amer.Ceram.Soc. 37(1954)42-45

3.33 Evans, A.G.; Tappin, G.: Effects of microstructure on the stress to propagate inherent flaws. Proc.Brit.Ceram.Soc. 20(1972)275-297

3.34 Simpson, L.A.: Effect of microstructure on measurements of fracture energy of Al_2O_3. J.Amer.Ceram.Soc. 56(1973)7-11

3.35 Pabst, R.F.: Determination of K_{Ic} factors with diamond saw cuts in ceramic materials. In Bradt, R.C.; Hasselman, D.P.H.; Lange, F.F. (Eds.): Fracture mechanics of ceramics, Vol.2. New York and London: Plenum Press 1974, pp.555-565

3.36 Claussen, N.; Pabst, R.; Lahmann, C.P.: Influence of microstructure of Al_2O_3 and ZrO_2 on K_{Ic}. Proc.Brit.Ceram.Soc. 25(1975)139-149

3.37 Evans, A.G.: Fracture mechanics determinations. In Bradt, R.C.; Hasselman, D.P.H.; Lange, F.F. (Eds.): Fracture mechanics of ceramics, Vol.1. New York and London: Plenum Press 1974, pp.17-48

3.38 Nakayama, J.: Direct measurement of fracture energies of brittle heterogeneous materials. J.Amer.Ceram.Soc. 48(1965)583-587

3.39 Dalgleish, B.J.; Pratt, P.L.; Sandford, J.: The fracture toughness-grain size relationship in polycrystalline alumina. Science of Ceramics 8(1976) 225-238

3.40 Simpson, L.A.: Discrepancy arising from measurements of grain-size dependence of fracture energy of Al_2O_3. J.Amer.Ceram.Soc. 56(1973)610-611

3.41 Veldkamp, J.D.B.; Hattu, N.: On the fracture toughness of brittle materials, Philips J.Res. 34(1979)1-25

3.42 Bansal, G.; Duckworth, W.: Comments on "Sub-critical crack extension and crack resistance in polycrystalline alumina". J.Mater.Sci. 13(1978)215-216

3.43 Dalgleish, B.J.; Fakhr, A.; Pratt, P.L.; Rawlings, R.D.: Fracture toughness of alumina. Ber.Dtsch.Keram.Ges. 55(1978)511-514

3.44 Davidge, R.W.; Tappin, G.: The effective surface energy of brittle materials. J.Mater.Sci. 3(1968)165-173

3.45 Funkenbusch, A.W.; Smith, D.W.: Influence of calcium on the fracture strength of polycrystalline alumina. Met.Trans. 6A(1975)2299-2301

3.46 Hübner, H.: Mechanical behavior of polycrystalline Al_2O_3/Cr_2O_3 alloys. Science of Ceramics 10(1979)529-536

3.47 Hübner, H.; Jillek, W.: Sub-critical crack extension and crack resistance in polycrystalline alumina. J.Mater.Sci. 12(1977)117-125

3.48   Jupp, R.S.; Stein, D.F.; Smith, D.W.: Observations on the effect of calcium
       segregation on the fracture behavior of polycrystalline alumina. J.Mater.Sci.
       15(1980)96-102

3.49   Mai, Y.W.: Thermal-stress resistance and fracture toughness of two tool
       ceramics. J.Mater.Sci. 11(1976)1430-1438

3.50   Munz, D.; Bubsey, R.T.; Shannon Jr., J.L.: Fracture toughness determination
       of Al_2O_3 using four-point-bend specimens with straight-through and chevron
       notches. J.Amer.Ceram.Soc. 63(1980)300-305

3.51   Simpson, L.A.: Use of notched-beam test for evaluation of fracture energies
       of ceramics. J.Amer.Ceram.Soc. 57(1974)151-154

3.52   de With, G.; Hattu, N.: The influence of CaO-doping on the fracture tough-
       ness of hot-pressed Al_2O_3. J.Mater.Sci. 16(1981)841-844

3.53   Bertolotti, R.L.: Fracture toughness of polycrystalline Al_2O_3. J.Amer.Ceram.
       Soc. 56(1973)107

3.54   Simpson, L.A.; Ritchie, I.G.; Lloyd, D.J.: Cause of discrepancy resulting
       from testing methods in the relation of grain size and fracture energy in
       Al_2O_3. J.Amer.Ceram.Soc. 58(1975)537-538

3.55   Barker, L.M.: Short rod and K_{Ic} measurements of Al_2O_3. In Bradt, R.C.;
       Hasselman, D.P.H.; Lange, F.F. (Eds.): Fracture mechanics of ceramics,
       Vol.3. New York and London: Plenum Press 1978, pp.483-494

3.56   Devezas, T.C.: Über das thermisch aktivierte Bruchverhalten von Aluminium-
       oxidkeramik. PhD-Thesis University of Erlangen-Nürnberg, 1981

3.57   Freiman, S.W.; McKinney, K.R.; Smith, H.L.: Slow crack growth in poly-
       crystalline ceramics. In Bradt, R.C.; Hasselman, D.P.H.; Lange, F.F. (Eds.):
       Fracture mechanics of ceramics, Vol.2 New York and London: Plenum Press
       1974, pp.659-676

3.58   Gutshall, P.L.; Gross, G.E.: Observations and mechanisms of fracture in
       polycrystalline alumina. Engng.Fracture Mech. 1(1969)463-471

3.59   Rice, R.W.; Freiman, S.W.; Becher, P.F.: Grain-size dependence of fracture
       energy in ceramics: I, Experiments. J.Amer.Ceram.Soc. 64(1981)345-350

3.60   Swanson, G.D.: Fracture energies of ceramics. J.Amer.Ceram.Soc. 55(1972)
       48-49

3.61   Swanson, G.D.; Gross, G.E.: Factor analysis of fracture-toughness test
       parameters for Al_2O_3. J.Amer.Ceram.Soc. 54(1971)382-384

3.62   Rice, R.W.; Freiman, S.W.; Pohanka, R.C.; Mecholsky Jr., J.J.; Wu, C.C.:
       Microstructural dependence of fracture mechanics parameters in ceramics.
       In Bradt, R.C.; Hasselman, D.P.H.; Lange, F.F. (Eds.): Fracture mechanics
       of ceramics, Vol.4 New York and London: Plenum Press 1978, pp.849-876

3.63   Gesing, A.; Bradt, R.C.: A microcracking model for the effect of grain
       size on slow crack growth in polycrystalline Al_2O_3. In Bradt, R.C.;
       Evans, A.G.; Hasselman, D.P.H.; Lange, F.F. (Eds.): Fracture mechanics
       of ceramics, Vol.5. New York and London: Plenum Press 1983, pp.569-590

3.64   Tattersall, H.G.; Tappin, G.: The work of fracture and its measurement in
       metals, ceramics and other materials. J.Mater.Sci. 1(1966)296-301

3.65   Binns, D.B.; Popper, P.: Mechanical properties of some commercial alumina
       ceramics. Proc.Brit.Ceram.Soc. 6(1966)71-82

3.66   Coppola, J.A.; Bradt, R.C.: Effects of porosity on fracture of Al_2O_3.
       J.Amer.Ceram.Soc. 56(1973)392-393

3.67 Kleinlein, F.W.; Hübner, H.: The evaluation of crack resistance and crack velocity from controlled fracture experiments of ceramic bend specimens. In Taplin, D.M.R. (Ed.): Fracture 1977, Vol. 3, Waterloo (Canada): University of Waterloo Press 1977, pp.883-891

3.68 Rice, R.W.: Strength/grain-size effects in ceramics. Proc.Brit.Ceram.Soc. 20(1972)205-257

3.69 Grabner, L.: Spectroscopic technique for the measurement of residual stress in sintered Al_2O_3. J.Appl.Phys. 49(1978)580-583

3.70 Hoagland, R.G.; Hahn, G.T.; Rosenfield, A.R.: Influence of microstructure on fracture propagation in rock. Rock Mechanics 5(1973)77-106

3.71 Noone, M.J.; Mehan, R.L.: Observation of crack propagation in polycrystalline ceramics and its relationship to acoustic emission. In Bradt, R.C.; Hasselman, D.P.H.; Lange, F.F. (Eds.): Fracture mechanics of ceramics, Vol.1, New York and London: Plenum Press 1974, pp.201-229

3.72 Wu, C.C.; Freiman, S.W.; Rice, R.W.; Mecholsky, J.J.: Microstructural aspects of crack propagation in ceramics. J.Mater.Sci. 13(1978)2659-2670

3.73 Evans, A.G.: Microfracture from thermal expansion anisotropy - I. Single phase systems. Acta Met. 26(1978)1845-1853

3.74 Evans, A.G.; Linzer, M.; Russell, L.R.: Acoustic emission and crack propagation in polycrystalline alumina. Mater.Sci.Engng. 15(1974)253-261

3.75 Hoagland, R.G.; Embury, J.D.: A treatment of inelastic deformation around a crack tip due to microcracking. J.Amer.Ceram.Soc. 63(1980)404-410

3.76 Kesler, C.E.; Naus, D.J.; Lott, J.L.: Mechanical behavior of materials, Vol.VI, Concrete and cement paste, glass and ceramics. The Society of Materials Science (Japan) 1972, pp.113-124

3.77 Buresch, F.E.: A structure sensitive K_{Ic} value and its dependence on grain size distribution, density and microcrack interaction. In Bradt, R.C.; Hasselman, D.P.H.; Lange, F.F. (Eds.): Fracture mechanics of ceramics, Vol.4. New York and London: Plenum Press 1978, pp.835-847

3.78 Kreher, W.; Pompe, W.: Increased fracture toughness of ceramics by energy-dissipative mechanisms. J.Mater.Sci. 16(1981)694-706

3.79 Pabst, R.F.; Steeb, J.; Claussen, N.: Microcracking in a process zone and its relation to continuum fracture mechanics. In Bradt, R.C.; Hasselman, D.P.H.; Lange, F.F. (Eds.): Fracture mechanics of ceramics, Vol.4. New York and London: Plenum Press 1978, pp.821-833

3.80 Davidge, R.W.; Tappin, G.: The effects of temperature and environment on the strength of two polycrystalline aluminas. Proc.Brit.Ceram.Soc. 15(1970)47-60

3.81 Kobayashi, A.S.; Emery, A.F.; Gorum, A.E.; Basu, T.: Fracture toughness of alumina at room temperature to $1600^{\circ}C$. Proc. of ICM3, Vol.3. Cambridge (England), 1979, pp.3-9

3.82 Rice, R.W.: Machining of ceramics. In Burke, J.J.; Gorum, A.E.; Katz, R.N. (Eds.): Ceramics for high performance applications. Chestnut Hill (U.S.A.): Brook Hill Publishing Co. 1974, pp. 287-343

3.83 Kirchner, H.P.; Gruver, R.M.; Sotter, W.A.: Characteristics of flaws at fracture origins and fracture stress-flaw size relations in various ceramics. Mater.Sci.Engng. 22(1976)147-157

3.84 Tressler, R.E.; Langensiepen, R.A.; Bradt, R.C.: Surface-finish effects on strength-vs-grain-size relations in polycrystalline Al_2O_3. J.Amer. Ceram.Soc. 57(1974)226-227

3.85 Rice, R.W.: Fractographic identification of strength-controlling flaws and microstructures. In Bradt, R.C.; Hasselman, D.P.H.; Lange, F.F. (Eds.): Fracture mechanics of ceramics, Vol.1. New York and London: Plenum Press 1974, pp.323-345

3.86 Meredith, H.; Pratt, P.L.: The observed fracture stress and measured values of K_{Ic} in commercial polycrystalline alumina. Special Ceramics 6, The British Ceramic Research Association 1975, pp.107-122

3.87 Kirchner, H.P.; Gruver, R.M.: A fractographic criterion for subcritical crack-growth boundaries in hot-pressed alumina. J.Mater.Sci. 14(1979) 2110-2118

3.88 Kirchner, H.P.; Gruver, R.M.: Fractographic criteria for subcritical crack growth boundaries in 96% Al_2O_3. J.Amer.Ceram.Soc. 63(1980)169-174

3.89 Tressler, R.E.; Bradt, R.C.: Surface-finish effects on strength-vs-grain size relations in polycrystalline Al_2O_3 - reply. J.Amer.Ceram.Soc. 58 (1975)155

3.90 Stofel, E.; Conrad H.: Fracture and twinning in sapphire (alpha-Al_2O_3 crystals). Trans. AIME 227(1963)1053-1060

3.91 Bayer, P.D.; Cooper, R.E.: Size-strength effects in sapphire and silicon nitride whiskers at 20°C. J.Mater.Sci. 2(1967)233-237

3.92 Bayer, P.D.; Cooper, R.E.: Fracture mechanics in sapphire whiskers. In Pratt, P.L. (Ed.): Fracture 1969. London: Chapman and Hall Ltd. 1969, p. 372 ff

3.93 Pollock, J.T.A.; Hurley, G.F.: Dependence of room temperature fracture strength on strain-rate in sapphire. J.Mater.Sci. 8(1973)1595-1602

3.94 Evans, A.G.; Wiederhorn, S.M.; Hockey, B.J.: Comments on Dependence of room temperature fracture strength on strain-rate in sapphire. J.Mater. Sci. 9(1974)1367-1370

3.95 Heuer, A.H.: Deformation twinning in corundum. Phil.Mag. 13(1966)379-393

3.96 Lankford, J.: Compressive strength and microplasticity in polycrystalline alumina. J.Mater.Sci. 12(1977)791-796

3.97 Lankford, J.: Tensile failure in unflawed polycrystalline Al_2O_3. J.Mater. Sci. 13(1978)351-357

3.98 Duckworth, W.: Discussion of a Ryskewitch paper. J.Amer.Ceram.Soc. 36 (1953)68

3.99 Passmore, E.M.; Spriggs, R.M.; Vasilos, T.: Strength-grain size-porosity relations in alumina. J.Amer.Ceram.Soc. 48(1965)1-7

3.100 Steele, B.R.; Rigby, F.; Hesketh, M.C.: Investigations on the modulus of rupture of sintered alumina bodies. Proc.Brit.Ceram.Soc. 6(1966)83-94

3.101 Spriggs, R.M.; Vasilos, T.: Effect of grain size on transverse bend strength of alumina and magnesia. J.Amer.Ceram.Soc. 46(1963)224-228

3.102 Meredith, H.; Newey, C.W.A.; Pratt, P.L.: The influence of texture on some mechanical properties of debased polycrystalline alumina. Proc. Brit.Ceram.Soc. 20(1972)299-316

3.103 Davidge, R.W.: The texture of special ceramics with particular reference to mechanical properties. Proc.Brit.Ceram.Soc. 20(1972)364-378

3.104 Heuer, A.H.; Cannon, R.M.; Tighe, N.J.: Plastic deformation in fine-grain ceramics. In Burke, J.J.; Reed, N.L.; Weiss, V. (Eds.): Ultrafine-grain ceramics. New York: Syracuse University Press 1970, pp.339-365

3.105 Passmore, E.M.; Moschetti, A.; Vasilos, T.: Brittle-ductile transition in polycrystalline aluminum oxide. Phil.Mag. 13(1966)1157-1162

3.106 Congleton, J.; Petch, J.A.: Dislocation movement in the brittle fracture of alumina. Acta Met. 14(1966)1179-1182

3.107 Petch, N.J.: Metallographic aspects of fracture. In Liebowitz, H. (Ed.): Fracture, Vol.1. New York and London: Academic Press 1968, pp.351-393

3.108 Gandhi, C.; Ashby, M.F.: Fracture-mechanism maps for materials which cleave: F.C.C., B.C.C. and H.C.P. metals and ceramics. Acta Met. 27(1979) 1565-1602

3.109 Evans, P.R.V.; In: Weil, N.A. (Ed.): Studies of the brittle behavior of ceramic materials. Armour Res. Foundation of Illinois Inst. of Tech.Rep. ASD-TR-61-628, Part II. Cited in [3.3]

3.110 Dawihl, W.; Dörre, E.: Über Festigkeits- und Verformungseigenschaften von Sinterkörpern aus Al$_2$O$_3$ in Abhängigkeit von Zusammensetzung und Gefüge. Ber.Dt.Keram.Ges. 41(1964)85-96

3.111 Rice,R.W.: The compressive strength of ceramics. In Kriegel,W.W., Palmour III, H.(Eds.): Ceramics in severe environments. Materials Science Research, Vol.5. New York and London: Plenum Press 1971,pp.195-229

3.112 Lankford,J.: Temperature-strain rate dependence of compressive strength and damage mechanisms in aluminium oxide. J.Mater.Sci. 16(1981)1567-1578

3.113 Lankford,J.: Compressive microfracture and indentation damage in Al$_2$O$_3$. In Bradt,R.C.; Hasselman,D.P.H.; Lange,F.F. (Eds): Fracture mechanics of ceramics, Vol.3. New York and London: Plenum Press 1978, pp.245-255

3.114 Lankford,J.; Davidson,D.L.: The effect of compressive strength on the mechanical performance of strong ceramics. Proc. of ICM 3, Vol.3., Cambridge(England) 1979, pp.35-43

3.115 Carniglia,S.C.: Petch relation of single-phase oxide ceramics. J.Amer.Ceram. Soc. 48(1965)580-583

3.116 Carniglia,S.C.: Reexamination of experimental strength-vs-grain-size data for ceramics. J.Amer.Ceram.Soc. 55(1972)243-249

3.117 Davidge,R.W.: Combination of fracture mechanics, probability, and micromechanical models of crack growth in ceramic systems. Metal Sci. Aug.-Sept. 1980, pp.459-462

3.118 Rice,R.W.: Comment on "Fracture stress as related to flaw and fracture mirror sizes". J.Amer.Ceram.Soc. 61(1978)466-467

3.119 Wiederhorn,S.M.: Mechanical and thermal properties of ceramics. Washington: NBS Special Publication No 303, 1969

3.120 McKinney,K.R.; Herbert,C.M.: Effect of surface finish on structural ceramic failure. J.Amer.Ceram.Soc. 53(1970)513-516

3.121 Sedlacek,R.D.; Halden,F.A.; Jorgensen,P.J.: The science of ceramic machining and surface finishing. Washington: NBS Special Publication No 348, 1972

3.122 Rhodes,W.H.; Cannon,R.M.: Microstructure studies of polycrystalline refractory compounds. U.S.Naval Air System Report N00019-73-C-0376, 1974

3.123 Singh,J.P.; Virkar,A.V.; Shetty,D.K.; Gordon,R.S.: Strength-grain size relations in polycrystalline ceramics. J.Amer.Ceram.Soc. 62(1979)179-183

3.124 Evans,A.G.: A dimensional analysis of the grain-size dependence of strength. J.Amer.Ceram.Soc. 63(1980)115-116

3.125 Kirchner,H.P.; Ragosta,J.M.: Crack growth from surface flaws in larger grains in alumina. J.Amer.Ceram.Soc. 63(1980)490-495

3.126 Virkar,A.V.; Shetty,D.K.; Evans,A.G.: Grain-size dependence of strength. J.Amer.Ceram.Soc. 64(1981)C56-57

3.127 Pearson,S.: Delayed fracture of sintered alumina. Proc.Phys.Soc.(London) 69 (444B)(1956)1293-1296

3.128 Williams,J.: Stress endurance of sintered alumina. Trans.Brit.Ceram.Soc. 55 (1956)287-312

3.129 Wiederhorn,S.M.: Influence of water vapor on crack propagation in soda-lime glass. J.Amer.Ceram.Soc.50(1967)407-414

3.130 Evans,A.G.: A method for evaluating the time-dependent failure characteristics of brittle materials - and its application to polycrystalline alumina. J.Mater.Sci. 7(1972)1137-1146

3.131 Zhurkov,S.N.: Kinetic concept of the strength of solids. Int.J.Fracture Mech. 1(1965)311-323

3.132 Evans,A.G.: Slow crack growth in brittle materials under dynamic loading conditions. Int.J.Fracture 10(1974)251-259

3.133 Evans,A.G.; Wiederhorn,S.M.: Proof-testing of ceramic materials - an analytical basis for failure prediction. Int.J.Fracture 10(1974)379-392

3.134 Evans,A.G.; Johnson,H.: The fracture stress and its dependence on slow crack growth. J.Mater.Sci.10(1975)214-222

3.135 Davidge,R.W.; McLaren,J.R.; Tappin,G.: Strength-probability-time (SPT) relationships in ceramics. J.Mater.Sci. 8(1973)1699-1705

3.136 Wiederhorn,S.M.: Subcritical crack growth in ceramics. In Bradt,R.C.; Hasselman,D.P.H.; Lange,F.F.(Eds.): Fracture mechanics of ceramics, Vol.2. New York and London: Plenum Press 1974,pp.613-646

3.137 Ritter,J.E.,Jr.; Meisel,J.A.: Strength and failure prediction of glass and ceramics. J.Amer.Ceram.Soc.59(1976)478-481

3.138 Ritter,J.E.,Jr.: Engineering design and fatigue failure of brittle materials. In Bradt,R.C.; Hasselman,D.P.H.; Lange,F.F.(Eds.): Fracture mechanics of ceramics, Vol.4. New York and London: Plenum Press 1978,pp.667-686

3.139 Evans,A.G.; Russell,R.L.; Richerson,D.W.: Slow crack growth in ceramic materials at elevated temperatures. Met.Trans.A 6(1975)707-716

3.140 Ritter,J.E.,Jr.; Wulf,S.A.: Evaluation of proof testing to assure against delayed failure. Amer.Ceram.Soc.Bull.57(1978)186-190

3.141 Evans,A.G.; Wiederhorn,S.M.: Crack propagation and failure prediction in silicon nitride at elevated temperatures. J.Mater.Sci. 9(1974)270-278

3.142 Wiederhorn,S.M.: Moisture assisted crack growth in ceramics. Int.J.Fracture Mech.4(1968)171-177

3.143 Bansal,K.G.; Duckworth,W.H.: Effects of moisture-assisted slow crack growth on ceramic strength. J.Mater.Sci. 13(1978)239-242

3.144 Kotchick,D.M.; Tressler,R.E.: Surface damage and environmental effects on the strain-rate sensitivity of the strength of sapphire and silicon carbide filaments. J.Mater.Sci. 10(1975)608-612

3.145 McLaren,J.R.; Davidge,R.W.: The combined influence of stress, time, and temperature on the strength of polycrystalline alumina. Proc.Brit.Ceram.Soc. 25 (1975)151-167

3.146 Rockar,E.M.; Pletka,B.J.: Fracture mechanics of alumina in a simulated biological environment. In Bradt,R.C.; Hasselman,D.P.H.; Lange,F.F. (Eds): Fracture mechanics of ceramics, Vol.4. New York and London: Plenum Press 1978,pp.725-735

3.147 Jakus,K.; Service,T.; Ritter Jr.,J.E.: High-temperature fatigue behavior of polycrystalline alumina. J.Amer.Ceram.Soc. 63(1980)4-7

3.148 Xavier,C.; Hübner,H.W.: The time-dependent strength degradation of pure alumina. Part I: Determination of the parameters of subcritical crack growth. Ceramica 28(1982)161-176

3.149 Dawihl,W.; Klingler,E.: Sinterkörper auf Aluminiumoxidgrundlage und die Abhängigkeit ihrer Zeitstandsfestigkeit von grenzflächenaktiven Stoffen. Ber. Dtsch.Keram.Ges. 43(1966)473-476

3.150 Kirchner,H.P.; Walker,R.E.: Delayed fracture of alumina ceramics with compressive surface layers. Mater.Sci.Engng. 8(1971)301-309

3.151 Krohn,D.A.; Hasselman,D.P.H.: Static and cyclic fatigue behavior of a polycrystalline alumina. J.Amer.Ceram.Soc. 55(1972)208-211

3.152 Guiu,F.: Cyclic fatigue of polycrystalline alumina in direct push-pull. J. Mater.Sci. 13(1978)1357-1361

3.153 Ritter Jr.,J.E.; Humenik,J.N.: Static and dynamic fatigue of polycrystalline alumina. J.Mater.Sci. 14(1979)626-632

3.154 Freiman,S.W.; Mulville,D.R.; Mast,P.W.: Crack propagation studies in brittle materials. J.Mater.Sci. 8(1973)1527-1533

3.155 Ferber,M.K.; Brown,S.D.: Subcritical crack growth in dense alumina exposed to physiological media. J.Amer.Ceram.Soc. 63(1980)424-429

3.156 Xavier,C.; Hübner,H.W.: Proof testing of alumina to assure against premature failure. Science of Ceramics 11(1981)495-502

3.157 Frakes,J.T.; Brown,S.D.; Kenner,G.H.: Delayed failure and aging of porous alumina in water and physiological media. Amer.Ceram.Soc.Bull. 53(1974)183-187

3.158 Dailly,D.F.; Hastings,G.W.; Lach,S.: Stress corrosion of debased alumina. Proc.Brit.Ceram.Soc. 31(1981)191-200

3.158a Dörre,E.; Dawihl,W.; Krohn,U.; Altmeyer,G.; Semlitsch,M.: Do ceramic components of hip joints maintain their strength in human bodies? In Vincenzini,P. (Ed.): Ceramic in surgery. Amsterdam: Elsevier Scient. Publ. Co. 1983,pp. 61-72

3.159 Dawihl,W.; Dörre,E.; Altmeyer,G.: Über die Druck- und Biegeschwellfestigkeit von Sinterkörpern auf Aluminiumoxidgrundlage. Ber.Dtsch.Keram.Ges. 42(1965) 243-247

3.160 Dawihl,W.; Klingler,E.: Über die Temperaturabhängigkeit der Schwingfestigkeit von Sinterkörpern auf Aluminiumoxidgrundlage. Ber.Dtsch.Keram.Ges. 42(1965) 311-313

3.161 Sarkar,B.K.; Glinn,T.G.T.: Fatigue behavior of high-Al_2O_3 ceramics. Trans. Brit.Ceram.Soc. 69(1970)199-203

3.162 Evans,A.G.; Fuller,R.M.: Crack propagation in ceramic materials under cyclic loading conditions. Met.Trans. 5.(1974)27-33

3.163 Wiederhorn,S.M.; Fuller Jr.,E.R.; Thomson,R.: Micromechanisms of crack growth in ceramics and glasses in corrosive environments. Metal Sci. August-September 1980,450-458

3.164 Pollet,J.C.; Burns,S.J.: Thermally activated crack propagation theory. Int. J.Fracture 13(1977)667-679

3.165 Thomson,R.: Theory of chemically assisted fracture, part 1. General reaction rate theory and thermodynamics. J.Mater.Sci. 15(1980)1014-1026

3.166 Stevens,R.N.; Dutton,R.: The propagation of Griffith cracks at high temperatures by mass transport processes. Mater.Sci.Engng. 8(1971)220-234

3.167 Thomson,R.; Hsieh,C.; Rana,V.: Lattice trapping of fracture cracks. J.Appl. Phys. 42(1971)3154-3160

3.168 Lawn,B.R.: An atomistic model of kinetic crack growth in brittle solids. J. Mater.Sci. 10(1975)469-480

3.169 Fuller Jr.,E.R.; Thomson,R.M.: Lattice theory of fracture. In Bradt,R.C.; Hasselman,D.P.H.; Lange,F.F.(Eds.): Fracture mechanics of ceramics,Vol.4. New York and London: Plenum Press 1978,pp.507-548

3.170 Fuller Jr.,E.R.; Lawn,B.R.; Thomson,R.M.: Atomic modelling of chemical interactions at crack tips. Acta Met. 28(1980)1407-1414

3.171 Gilman,J.J.; Tong,H.C.: Quantum tunneling as an elementary fracture process. J.Appl.Phys. 42(1971)3479-3486

3.172 Clarke,F.J.P.; Tattersall,H.G.; Tappin,G.: Toughness of ceramics and their work of fracture. Proc.Brit.Ceram.Soc. 6(1966)163-172

3.173 Davidge,R.W.; Tappin,G.: Thermal shock and fracture in ceramics. Trans.Brit. Ceram.Soc. 66(1967)405-422

3.174 Hasselman,D.P.H.: Unified theory of thermal shock fracture initiation and crack propagation in brittle ceramics. J.Amer.Ceram.Soc. 52(1969)600-604

3.175 Hasselman,D.P.H.: Thermal stress resistance parameters for brittle refractory ceramics: a compendium.Ceram.Bull. 49(1970)1033-1037

3.176 Evans,A.G.: Thermal fracture in ceramic materials. Proc.Brit.Ceram.Soc. 25 (1975)217-237

3.177 Ainsworth,J.H.; Moore,R.E.: Fracture behavior of thermally shocked aluminum oxide. J.Amer.Ceram.Soc. 52(1969)628-629

3.178 Hasselman,D.P.H.: Strength behavior of polycrystalline alumina subjected to thermal shock. J.Amer.Ceram.Soc. 53(1970)490-495

3.179 Gupta,T.K.: Strength degradation and crack propagation in thermally shocked Al_2O_3. J.Amer.Ceram.Soc. 55(1972)249-253

3.180 Krohn,D.A.; Larson,D.R.; Hasselman,D.P.H.: Comparison of thermal-stress re-sistance of polycrystalline Al_2O_3 and BeO. J.Amer.Ceram.Soc. 56(1973)490-491

3.181 Bertsch,B.E.; Larson,D.R.; Hasselman,D.P.H.: Effect of crack density on strength loss of polycrystalline Al_2O_3 subjected to severe thermal shock. J.Amer.Ceram.Soc. 57(1974)235-236

3.182 Smith,R.D.; Anderson,H.U.; Moore,R.E.: Influence of induced porosity on the thermal shock characteristics of Al_2O_3. Amer.Ceram.Soc.Bull. 55(1976)979-982

3.183 Larson,D.R.; Coppola,J.A.; Hasselman,D.P.H.; Bradt,R.C.: Fracture toughness and spalling behavior of high-Al_2O_3 refractories. J.Amer.Ceram.Soc. 57(1974) 417-421

3.184 Sarkar,B.K.; Glinn,T.G.J.: Impact fatigue of an alumina ceramic. J.Mater.Sci. 4(1969)951-954

3.185 Huffine,C.L.; Berger,C.M.: Impact fatigue of polycrystalline alumina. Amer. Ceram.Soc.Bull. 56(1977)201-203

3.186 Bertolotti,R.L.: Strength and absorbed energy in instrumented impact tests of polycrystalline Al_2O_3. J.Amer.Ceram.Soc. 57(1974)300-302

3.187 Lankford,J.; Davidson,D.L.: The crack-initiation threshold in ceramic materi-als subject to elastic/plastic indentation. J.Mater.Sci. 14(1979)1662-1668

3.188 Evans,A.G.: High-temperature slow crack growth in ceramic materials. In Burke J.J.; Gorum,A.E.; Katz,R.N. (Eds.): Ceramics for high performance applica-tions. Chestnut Hill (U.S.A.): Brook Hill Publishing Co.,1974,pp.373-396

3.189 Lange,F.F.; Radford,K.C.: Healing of surface cracks in polycrystalline Al_2O_3. J.Amer.Ceram.Soc.53(1970)420-421

3.190 Dawihl,W.; Altmeyer,G.: Mikrorißbildung, Spannungsintensitätsfaktor und Aus-heilvorgänge bei Sinterkörpern aus Aluminiumoxid. Ber.Dt.Keram.Ges. 51(1974) 69-72

3.191 Gupta,T.K.: Crack healing and strengthening of thermally shocked alumina. J.Amer.Ceram.Soc. 59(1976)259-262

3.192 Evans,A.G.; Charles,E.A.: Strength recovery by diffusive crack healing. Acta Met. 25(1977)919-927

3.193 Lino,U.R.A.; Hübner,H.W.: Effect of surface condition on the strength of alu-minum oxide. Science of Ceramics 12(1983)607-612

3.194 Conrad,H.: Mechanical behavior of sapphire. J.Amer.Ceram.Soc. 48(1965)195-201

3.195 Heuer,A.H.: Plastic deformation in polycrystalline alumina. Proc.Brit.Ceram. Soc. 15(1970)173-183

3.196 Snow,J.D.; Heuer,A.H.: Slip systems in Al_2O_3. J.Amer.Ceram.Soc. 56(1973)153-157

3.197 Chin,G.Y.: Slip and twinning systems in ceramic oxides. In Bradt,R.C.; Tress-ler,R.E.(Eds): Deformation of ceramic materials. New York and London: Plenum Press 1975,pp.25-59

3.198 Mitchell,T.E.: Application of transmission electron microscopy to the study of deformation in ceramic oxides. J.Amer.Ceram.Soc. 62(1979)254-267

3.199 Becher,P.F.: Deformation substructure in polycrystalline alumina. J.Mater. Sci. 6(1971)275-280

3.200 Cutter,I.A.; McPherson,R.: Plastic deformation of Al_2O_3 during abrasion. J. Amer.Ceram.Soc. 56(1973)266-269

3.201 Becher,P.F.: Abrasive surface deformation in sapphire. J.Amer.Ceram.Soc. 59 (1976)143-145

3.202 Wachtman Jr.,J.B.; Maxwell,L.H.: Plastic deformation of ceramic oxide single crystals. J.Amer.Ceram.Soc. 37(1954)291-299

3.203 Wachtman Jr.,J.B.; Maxwell,L.H.: Plastic deformation of ceramic oxide single crystals,II. J.Amer.Ceram.Soc. 40(1957)377-385

3.204 Kronberg,M.L.: Plastic deformation of single crystals of sapphire: basal slip and twinning. Acta Met. 5(1957)507-524

3.205 Chang,R.: Creep of Al_2O_3 single crystal. J.Appl.Phys.31(1960)484-487

3.206 Kronberg,M.L.: Dynamical flow properties of single crystals of sapphire, I. J.Amer.Ceram.Soc. 45(1962)274-279

3.207 Conrad,H.; Stone,G.; Janowski,K.: Yielding and flow of sapphire (alpha-Al_2O_3 crystals) in tension and compression. Trans.AIME 233(1965)889-897

3.208 Scheuplein,R.; Gibbs,P.: Surface structure in corundum: I, Etching of dis-locations. J.Amer.Ceram.Soc. 43(1960)458-467

3.209 Gooch,D.J.; Groves,G.W.: Prismatic slip in sapphire. J.Amer.Ceram.Soc. 55 (1972)105

3.210 Gooch,D.J.; Groves,G.W.: Non-basal slip in sapphire. Phil.Mag. 28(1973)623-637

3.211 Bilde-Sørensen,J.B.; Thölen,A.R.; Gooch,D.J.; Groves,D.W.: Structure of the <01$\bar{1}$0> dislocation in sapphire. Phil.Mag. 33(1976)877-899

3.212 Kotchick,D.M.; Tressler,R.E.: Deformation behavior of sapphire via the pris-matic slip system. J.Amer.Ceram.Soc. 63(1980)429-434

3.213 Gulden,T.D.: Direct observation of nonbasal dislocations in sintered alumina. J.Amer.Ceram.Soc. 50(1967)472-475

3.214 Bayer,P.D.; Cooper,R.E.: A new slip system in sapphire. J.Mater.Sci. 2(1967) 301-302

3.215 Heuer,A.H.; Firestone,R.F.; Snow,J.D.; Tullis,J.: Nonbasal slip in alumina at high temperatures and pressures. In Kriegel,W.W.; Palmour III,H.(Eds.): Materials in severe environments. Materials Science Research, Vol.5.New York and London: Plenum Press 1971,pp.331-340

3.216 Cadoz,J.; Pellissier,B.: Influence of three-fold symmetry on pyramidal slip of alumina single crystals. Scripta Met. 10(1976)597-600

3.217 Hockey,B.J.: Plastic deformation of aluminum oxide by indentation and abrasion. J.Amer.Ceram.Soc. 54(1971)223-231

3.218 Gooch,D.J.; Groves,G.W.: The creep of sapphire filaments with orientations close to the c-axis. J.Mater.Sci. 8(1973)1238-1246

3.219 Michael,D.J.; Tressler,R.E.: Deformation dynamics of pore-free Ti^{4+}-doped and pure c-axis sapphire crystals. J.Mater.Sci. 9(1974)1781-1788

3.220 Tressler,R.E.; Barber,D.J.: Yielding and flow of c-axis sapphire filaments. J.Amer.Ceram.Soc. 57(1974)13-19

3.221 Hockey,B.J.: Pyramidal slip on $(11\bar{2}3)<\bar{1}100>$ and basal twinning in Al_2O_3. In Bradt,R.C.; Tressler,R.E.(Eds.): Deformation of ceramic materials. New York and London: Plenum Press 1975,pp.167-179

3.222 Champion,J.A.; Clemence,M.A.: Etch pits in flux-grown corundum. J.Mater.Sci.2 (1967)153-159

3.223 Engelhardt,G.; Kalb,S.: Ätzverfahren zum Anätzen von Versetzungen in poly-kristallinem Aluminiumoxid. Ber.Dt.Keram.Ges. 51(1974)231-233

3.224 Firestone,R.F.; Heuer,A.H.: Creep deformation of 0° sapphire. J.Amer.Ceram. Soc. 59(1976)24-29

3.225 Barber,D.J.; Tighe,N.J.: Electron microscopy and diffraction of synthetic corundum crystals. Phil.Mag. 14(1966)531-544

3.226 Caslavsky,J.L.; Gazzara,C.P.; Middleton,R.M.: The study of basal dislocations in sapphire. Phil.Mag. 25(1972)35-44

3.227 Mitchell,T.E.; Pletka,B.J.; Phillips,D.S.; Heuer,A.H.: Climb dissociation of dislocations in sapphire $(\alpha-Al_2O_3)$. Phil.Mag. 34(1976)441-451

3.228 May,C.A.; Shah,J.S.: Dislocation reactions and cavitation studies in melt-grown sapphire. J.Mater.Sci.4(1969)179-188

3.229 Firestone,R.F.; Heuer,A.H.: Yield point of sapphire. J.Amer.Ceram.Soc. 56 (1973)136-139

3.230 Castaing,J.; Cadoz,J.; Kirby,S.H.: Prismatic slip of Al_2O_3 single crystals below 1000^oC in compression under hydrostatic pressure. J.Amer.Ceram.Soc. 64 (1981)504-511

3.231 Radford,K.C.; Pratt,P.L.: Mechanical properties of impurity-doped alumina single crystals. Proc.Brit.Ceram.Soc. 15(1970)185-202

3.232 Bertolotti,R.L.; Scott,W.D.: Compressive creep of Al_2O_3 single crystals. J. Amer.Ceram.Soc. 54(1971)286-291

3.233 Tressler,R.E.; Michael,D.J.: Dynamics of flow of c-axis sapphire. In Bradt, R.C; Tressler,R.E.(Eds): Deformation of ceramic materials. New York and London: Plenum Press 1975,pp.195-215

3.234 Weertman,J.: Steady-state creep through dislocation climb. J.Appl.Phys. 28 (1957)362-364

3.235 Nabarro,F.R.N.: Steady-state diffusional creep. Phil.Mag. 16(1967)231-237

3.236 Pletka,B.J.; Heuer,A.H.; Mitchell,T.E.: Work-hardening in sapphire (α-Al$_2$O$_3$). Acta Met. 25(1977)25-33

3.237 Pletka,B.J.; Mitchell,T.E.; Heuer,A.H.: Dislocation structures in sapphire deformed by basal slip. J.Amer.Ceram.Soc. 57(1974)388-393

3.238 Pletka,B.J.; Mitchell,T.E.; Heuer,A.H.: Strengthening mechanisms in sapphire. In Bradt,R.C.; Tressler,R.E.(Eds.): Deformation of ceramic materials. New York and London: Plenum Press 1975,pp.181-194

3.239 Mitchell,T.E.; Hobbs,L.W.; Heuer,A.H.; Castaing,J.; Cadoz,J.; Philibert,J.: Interaction between point defects and dislocations in oxides. Acta Met. 27 (1979)1677-1691

3.240 Busovne Jr.,B.J.; Kotchick,D.M.; Tressler,R.E.: Deformation history effects on the precipitation hardening behavior of Ti^{4+}-doped sapphire. Phil.Mag. A39(1979)265-276

3.241 Conrad,H.; Janowski,K.; Stofel,E.: Additional observations on twinning in sapphire (α-Al$_2$O$_3$ crystals) during compression. Trans.AIME 233(1965)255-256

3.242 Becher, P.F.; Palmour III, H.: High-temperature deformation of alumina double bicrystals. J.Amer.Ceram.Soc. 53(1970)119-123

3.243 Achutaramayya,G.; Scott,W.D.: Interfacial energies of coherent twin boundaries in alumina. Acta Met. 23(1975)1469-1472

3.244 Scott,W.D.; Orr,K.K.: Rhombohedral twinning in alumina. J.Amer.Ceram.Soc. 66 (1983)27-32

3.245 Becher,P.F.: Deformation behavior of alumina at elevated temperatures. In Kriegel,W.W.; Palmour III,H.(Eds.): Ceramics in severe environments. Materials Science Research, Vol.5. New York and London: Plenum Press 1971,pp. 315-329

3.246 Scott,W.D.: Rhombohedral twinning in aluminum oxide. In Bradt,R.C.; Tressler, R.E.(Eds.): Deformation of ceramic materials. New York and London: Plenum Press 1975,pp.151-166

3.247 Westbrook,J.H.; Jorgensen,P.J.: Indentation creep of solids. Trans.AIME 233 (1965)425-428

3.248 Koester,R.D.; Moak,D.P.: Hot hardness of selected borides, oxides, and carbides to 1900°C. J.Amer.Ceram.Soc. 50(1967)290-296

3.249 Bradt,R.C.: Cr$_2$O$_3$ solid solution hardening of Al$_2$O$_3$. J.Amer.Ceram.Soc. 50 (1967)54-55

3.250 Belon,L.; Forestier,H.; Bigot,Y.: The hardness of some solid solutions of alumina. In Popper,P.(Ed.): Special Ceramics,Vol.4. Manchester: The British Ceramic Research Association 1968,pp.203-211

3.251 Ghate,B.B.; Smith,W.C.; Kim,C.H.; Hasselman,D.P.H.; Kane,G.E.: Effect of chromia alloying on machining performance of alumina ceramic cutting tools. Amer.Ceram.Soc.Bull. 54(1975)210-215

3.252 Shinozaki,K.; Ishikura,Y.; Uematsu,K.; Mitsutani,N.; Kato,M.: Vickers micro-hardness of solid solutions in the system Cr_2O_3-Al_2O_3. J.Mater.Sci. 15(1980) 1314-1316

3.253 Kennedy,C.R.; Bradt,R.C.: Softening of Al_2O_3 by solid solution of $MgO \cdot TiO_2$. J.Amer.Ceram.Soc. 56(1973)608

3.254 Skrovanek,S.D.; Bradt,R.C.: Microhardness of a fine-grain-size Al_2O_3. J.Amer. Ceram.Soc. 62(1979)215-216

3.255 Richerson,D.W.: Modern ceramic engineering. New York and Basel: Marcel Dek-ker, Inc. 1982

3.256 Evans,A.G.: Abrasive wear in ceramics: An assessment. In Hockey,B.J.; Rice, R.W.: The science of ceramic machining and surface finishing II. Washington: NBS Special Publication 562,1979,pp.1-14

3.257 Rice,R.W.; Mecholsky Jr.,J.J.; Becher,P.F.: The effect of grinding direction on flaw character and strength of single crystal and polycrystalline ceramics. J.Mater.Sci. 16(1981)853-862

3.258 Swain,M.V.: Microcracking associated with the scratching of brittle solids. In Bradt,R.C.; Hasselman,D.P.H.; Lange,F.F. (Eds.): Fracture mechanics of ceramics, Vol.3. New York and London: Plenum Press 1978,pp.257-272

3.259 Steijn,R.P.: On the wear of sapphire. J.Appl.Phys. 32(1961)1951-1958

3.260 Hockey,B.J.; Lawn,B.R.: Electron microscopy of microcracking about indenta-tions in aluminum oxide and silicon carbide. J.Mater.Sci. 10(1975)1275-1284

3.261 Rice,R.W.; Speronello,B.K.: Effect of microstructure on rate of machining of ceramics. J.Amer.Ceram.Soc. 59(1976)330-333

3.262 Lange,F.F.; James,M.R.; Green,D.J.: Determination of residual surface stresses caused by grinding in polycrystalline Al_2O_3. J.Amer.Ceram.Soc. 66 (1983)C16-17

3.263 Dawihl,W.; Klingler,E.; Dörre,E.: Zusammenhänge zwischen Verschleißwider-stand und Eigenschaften von Sinterkörpern auf Al_2O_3-Grundlage. Ber.Dtsch. Keram.Ges. 46(1969)409-415

3.264 Westwood,A.R.C.; MacMillan,N.H.; Kalyoncu,R.S.: Environment-sensitive hard-ness and machinability of Al_2O_3. J.Amer.Ceram.Soc. 56(1973)258-262

3.265 Gruver,R.M.; Kirchner,H.P.: Effect of environment on penetration of surface damage and remaining strength. J.Amer.Ceram.Soc. 57(1974)220-223

3.266 Swain,M.V.; Latamision,R.M.; Westwood,A.R.C.: Further studies on environ-ment-sensitive hardness and machinability of Al_2O_3. J.Amer.Ceram.Soc. 58 (1975)372-376

3.267 Rice,R.W.: The effect of grinding direction on the strength of ceramics. In Schneider,S.J.; Rice,R.W. (Eds.): The science of ceramic machining and surface finishing. Washington: NBS Special Publication 348,1972,pp.365-376

3.268 Groves, G.W.: The plasticity of ceramics. Proc.Brit.Ceram.Soc. 15(1970) 103-112

3.269 Langdon, T.G.; Cropper, D.R.; Pask, J.A.: Creep mechanisms in ceramic materials at elevated temperatures. In Kriegel, W.W.; Palmour III, H.(Eds.): Ceramics in severe environments. Materials Science Research Vol. 5. New York and London: Plenum Press 1971, pp. 297-313

3.270 Terwilliger, G.R.; Radford, K.C.: High temperature deformation of ceramics: I, Background. Amer.Ceram.Soc.Bull. 53(1974)172-179, and Radford, K.C.; Terwilliger, G.R.: High temperature deformation of ceramics: II, Specific behavior. Amer.Ceram.Soc.Bull 53(1974)465-472

3.271 Dokko, P.C.; Pask, J.A.: Plastic deformation of ceramic materials. Mater. Sci.Engng. 25(1976)77-86

3.272 Cannon, W.R.; Langdon, T.G.: Creep of ceramics, part 1: Mechanical characteristics. J.Mater.Sci 18(1983)1-50

3.273 Cannon, R.M.; Coble, R.L.: Review of diffusional creep of Al_2O_3. In Bradt, R.C.; Tressler, R.E. (Eds.): Deformation of ceramic materials. New York and London: Plenum Press 1975, pp. 61-100

3.274 Nabarro, F.R.N.: Deformation of crystals by the motion of single ions. Rep. Conf. Strength Solids, Physical Society, London 1948, pp. 75-90

3.275 Herring, C.: Diffusional viscosity of a polycrystalline solid. J.Appl. Phys. 21(1950)437-445

3.276 Coble, R.L.: A model for boundary diffusion controlled creep in polycrystalline materials. J.Appl.Phys. 34(1963)1679-1682

3.277 Gordon, R.S.: Mass transport in the diffusional creep of ionic solids. J.Amer.Ceram.Soc. 56(1973)147-152

3.278 Gordon, R.S.: Ambipolar diffusion and its application to diffusion creep. In Cooper, A.R.; Heuer, A.H. (Eds.): Mass transport phenomena in ceramics. Materials Science Research, Vol.9. New York and London: Plenum Press 1975, pp. 445-464

3.279 Cannon, W.R.; Sherby, O.D.: Third-power stress dependence in creep of polycrystalline nonmetals. J.Amer.Ceram.Soc. 56(1973)157-160

3.280 Weertman, J.: Dislocation climb theory of steady state creep. Trans ASM 61 (1968)681-694

3.281 Weertman, J.: Steady-state creep of crystals. J.Appl.Phys. 28(1957)1185-1189

3.282 Barrett, C.R.; Nix, W.D.: A model for steady state creep based on the motion of jogged screw dislocations. Acta.Met. 13(1965)1247-1258

3.283 Evans, H.E.; Knowles, G.: Dislocation creep in non-metallic materials. Acta. Met. 26(1978)141-145

3.284 Langdon, T.G.; Mohamed, F.A.: The incorparation of ambipolar diffusion in deformation mechanism maps for ceramics. J.Mater.Sci. 13(1978)473-482

3.285 Cannon, R.M.; Rhodes, W.H.; Heuer, A.H.: Plastic deformation of fine-grained alumina (Al_2O_3): I, Interface-controlled diffusional creep. J.Amer.Ceram.Soc. 63(1980)46-53

3.286 Ikuma, Y.; Gordon, R.S.: Effect of doping simultaneously with iron and titanium on the diffusional creep of polycrystalline Al_2O_3. J.Amer.Ceram.Soc. 66(1983)139-147

3.287 Ashby, M.F.: On interface-reaction control of Nabarro-Herring creep and sintering. Scripta Met. 3(1969)837-842

3.288 Gifkins, R.C.: Diffusional creep mechanisms. J.Amer.Ceram.Soc. 51(1968) 69-72

3.289 Ikuma, Y.; Gordon, R.S.: Role of interfacial defect creation-annihilation processes at grain boundaries on the diffusional creep of polycrystalline alumina. In Pask, J.A.; Evans, A.G. (Eds.): Surfaces and interfaces in ceramic and ceramic-metal systems. Materials Science Research, Vol. 14. New York and London: Plenum Press 1981, pp. 283-294

3.290 Ashby, M.F.; Verrall, R.A.: Diffusional-accommodated flow and superplasticity. Acta Met. 21(1973)149-163

3.291 Warshaw, S.I.; Norton, F.H.: Deformation behavior of polycrystalline aluminum oxide. J.Amer.Ceram.Soc. 45(1962)479-486

3.292 Chang, R.: High temperature creep and anelastic phenomena in polycrystalline refractory oxides. J.Nucl.Mater. 2(1959)174-181

3.293 Folweiler, R.C.: Creep behavior of pore-free polycrystalline aluminum oxide. J.Appl.Phys. 32(1961)773-778

3.294 Dawihl, W.; Klingler, E.: Über das Kriechverhalten von Sinterkörpern aus Al_2O_3 im Temperaturbereich von 1200°C unter dem Einfluß von Druckspannungen. Ber.Dtsch.Keram.Ges. 42(1965)270-274

3.295 Passmore, E.M.; Vasilos, T.: Creep of dense, pure, fine-grained aluminum oxide. J.Amer.Ceram.Soc. 49(1966)166-168

3.296 Sugita, T.; Pask, J.A.: Creep of doped polycrystalline Al_2O_3. J.Amer.Ceram. Soc. 53(1970)609-613

3.297 Engelhardt, G.; Thümmler, F.: Kriechuntersuchungen unter 4-Punkt-Biegebeanspruchung bei hohen Temperaturen, II. Messungen an polykristallinem Aluminiumoxid. Ber.Dtsch.Keram.Ges. 47(1970)571-577

3.298 Mocellin, A.; Kingery, W.D.: Creep deformation in MgO-saturated large-grain-size Al_2O_3. J.Amer.Ceram.Soc. 54(1971)339-341

3.299 Crosby, A.; Evans, P.E.: Creep in pure and two-phase nickel-doped alumina. J.Mater.Sci. 8(1973)1573-1580

3.300 Davies, C.K.L.: Creep and fracture of polycrystalline alumina deformed in tension. In Harris, J.E.; Sykes, E.C. (Eds): Physical metallurgy of reactor fuel elements. London: Metals Society 1975, pp. 99-106

3.301 Cannon, W.R.; Sherby, O.D.: Creep behavior and grain-boundary sliding in polycrystalline Al_2O_3. J.Amer.Ceram.Soc. 60(1977)44-47

3.302 Lessing, P.A.; Gordon, R.S.: Creep of polycrystalline alumina, pure and doped with transition metal impurities. J.Mater.Sci. 12(1977)2291-2302

3.303 Hou, L.D.; Tiku, S.K.; Wang, H.A.; Kröger, F.A.: Conductivity and creep in acceptor-dominated polycrystalline Al_2O_3. J.Mater.Sci. 14(1979)1877-89

3.304 Porter, J.R.; Blumenthal, W.; Evans, A.G.: Creep fracture in ceramic polycrystals-I. Creep cavitation effects in polycrystalline alumina. Acta Met. 29(1981)1899-1906

3.305 El-Aiat, M.M.; Hou, L.D.; Tiku, S.K.; Wang, H.A.; Kröger, F.A.: High-temperature conductivity and creep of polycrystalline Al_2O_3 doped with Fe and/or Ti. J.Amer.Ceram.Soc. 64(1981)174-182

3.306 Carry, C.; Mocellin A.: Superplastic forming of alumina. Proc.Brit.Ceram. Soc. 33(1983)101-115

3.307 Coble, R.L.; Guerard, Y.H.: Creep of polycrystalline aluminum oxide. J.Amer. Ceram.Soc. 46(1963)353-354

3.308 Fryer, G.M.; Roberts, J.P.: Tensile creep of porous polycrystalline alumina. Proc.Brit.Ceram.Soc. 6(1966)225-232

3.309 Hewson, C.W.; Kingery, W.D.: Effect of MgO and MgTiO$_3$ doping on diffusion-controlled creep of polycrystalline aluminum oxide. J.Amer.Ceram.Soc. 50 (1967)218-219

3.310 Spriggs, R.M.; Vasilos, T.: Functional relation between creep rate and porosity for polycrystalline ceramics. J.Amer.Ceram.Soc. 47(1964)47-48

3.311 Heuer, A.H.; Tighe, N.J.; Cannon, R.M.: Plastic deformation of fine-grained alumina (Al$_2$O$_3$): II, Basal slip and nonaccommodated grain-boundary sliding. J.Amer.Ceram.Soc. 63(1980)53-58

3.312 Frost, H.J.; Ashby, M.F.: Deformation-mechanism maps, The plasticity and creep of metals and ceramics. Oxford: Pergamon Press 1982

3.313 Mohamed, F.A.; Langdon, T.G.: Deformation mechanism maps based on grain size. Met.Trans. 5(1974)2339-2345

3.314 Langdon, T.G.; Mohamed, F.A.: A new type of deformation mechanism map for high-temperature creep. Mater.Sci.Engng. 32(1978)103-112

3.315 Hollenberg, G.W.; Gordon, R.S.: Effect of oxygen partial pressure on the creep of polycrystalline Al$_2$O$_3$ doped with Cr, Fe, or Ti. J.Amer.Ceram.Soc. 56(1973)140-147

3.316 Lessing, P.A.; Gordon, R.S.: Impurity and grain size effects on the creep of polycrystalline magnesia and alumina. In Bradt, R.C.; Tressler, R.E. (Eds.): Deformation of ceramic materials. New York and London: Plenum Press 1975, pp. 271-296

3.317 Raja Rao, W.; Cutler, I.B.: Effect of iron oxide on the sintering kinetics of Al$_2$O$_3$. J.Amer.Ceram.Soc. 56(1973)588-593

3.318 Paladino, A.E.; Kingery, W.D.: Aluminum ion diffusion in aluminum oxide. J.Chem.Phys. 37(1962)957-962

3.319 Coble, R.L.: Sintering alumina: effect of atmospheres. J.Amer.Ceram.Soc. 45 (1962)123-127

3.320 Evans, A.G.; Rana, A.: High temperature failure mechanisms in ceramics. Acta Met. 28(1980)129-141

3.321 Evans, A.G.; Blumenthal, W.: High temperature failure in ceramics. In Bradt, R.C.; Evans, A.G.; Hasselman, D.P.H.; Lange, F.F.: Fracture mechanics of ceramics, Vol.6. New York and London: Plenum Press 1983, pp.423-448

3.322 Miller, D.A.; Langdon, T.G.: Creep fracture maps for 316 stainless steel. Met.Trans. 10A(1979)1635-1641

3.323 Vasilos, T.; Passmore, E.M.: Effect of microstructures on deformation of ceramics. In Fulrath, R.M.; Pask, J.A. (Eds.): Ceramic microstructures. New York: John Wiley and Sons 1968, pp.406-430

3.324 Hull, D.; Rimmer, D.E.: The growth of grain-boundary voids under stress. Phil.Mag. 4(1959)673-687

3.325 Raj, R.; Ashby, M.F.: Intergranular fracture at elevated temperature. Acta Met. 23(1975)653-666

3.326 Hsueh, C.H.; Evans, A.G.: Creep fracture in ceramic polycrystals-II. Effects of inhomogeneity on creep rupture. Acta Met. 29(1981)1907-1917

3.327 Page, R.A.; Lankford, J.: Characterization of creep cavitation in sintered alumina by small-angle neutron scattering. J.Amer.Ceram.Soc. 66(1983)C146-148

3.328 Raj, R.; Baik, S.: Creep crack propagation by cavitation. Metal Sci. 14 (1980)385-394

3.329 Chuang, T.: A diffusive crack growth model for creep fracture. J.Amer.Ceram. Soc. 65(1982)93-103

3.330 Tree, Y.; Venkateswaran, A.; Hasselman, D.P.H.: Observations on the fracture and deformation behavior during annealing of residually stressed Al_2O_3. J. Mater.Sci. 18(1983)2135-2148

3.331 Lange, F.F.: The interaction of a crack front with a second-phase dispersion. Phil.Mag. 22(1970)983-992

3.332 Evans, A.G.: The strength of brittle materials containing second phase dispersions. Phil.Mag. 26(1972)1327-1344

3.333 McHugh, C.O.; Whalen, T.J.; Humenik Jr., M.: Dispersion-strengthened aluminum oxide. J.Amer.Ceram.Soc. 49(1966)486-491

3.334 Rankin, D.T.: Stiglich, J.J.; Petrak, D.R.; Ruh, R.: Hot-pressing and mechanical properties of Al_2O_3 with a Mo-dispersed phase. J.Amer.Ceram.Soc. 54(1971)277-281

3.335 Simpson, L.A.; Wasylyshin, A.: Fracture energy of Al_2O_3 containing Mo fibers. J.Amer.Ceram.Soc. 54(1971)56-57

3.336 Lloyd, D.J.; Tangri, K.: Acoustic emission from Al_2O_3-Mo fibre composites. J.Mater.Sci. 9(1974)482-486

3.337 Claussen, N.: Hot-pressed eutetics of oxides and metal fibers. J.Amer.Ceram.Soc. 56(1973)442

3.338 Rasmussen, J.J.; Stringfellow, G.B.; Cutler, I.B.; Brown, S.D.: Effect of impurities on the strength of polycrystalline magnesia and alumina. J.Amer. Ceram.Soc. 48(1965)146-150

3.339 Wahi, R.P.; Hübner, H.: Bruchverhalten von Aluminiumoxid mit Zusätzen von Bariumoxid. Ber.Dt.Keram.Ges. 53(1976)423-427

3.340 Wahi, R.P.; Hübner, H.; Ilschner, B.: Fracture behavior of two-phase ceramic alloys based on aluminum oxide. Proceedings of 2nd International Conference on Materials (ICM II), Boston (U.S.A.) 1976, pp. 1183-1187

3.341 Grellner, W.: Elastische Kenngrößen und Bruchverhalten von Werkstoffen des Systems Al_2O_3+TiC bei erhöhten Temperaturen. Ber.Dt.Keram.Ges. 55(1978) 484-487

3.342 Grellner, W.; Hübner, H.; Ilschner, B.; Kleinlein, F.W.: On high temperature strength of a two-phase Al_2O_3-base material. Science of Ceramics 10 (1980)513-519

3.343 Wahi, R.P.; Ilschner, B.: Fracture behavior of composites based on Al_2O_3-TiC. J.Mater.Sci. 15(1980)875-885

3.344 Kirchner, H.P.: Strengthening of ceramics. Treatments, tests, and design applications. New York and Basel: Marcel Dekker, Inc. 1979

3.345 Gruver, R.M.; Kirchner, H.P.: Residual stress and flexural strength of thermally conditioned 96 % alumina rods. J.Amer.Ceram.Soc. 51(1968)232-233

3.346 Kirchner, H.P.; Gruver, R.M.; Walker, R.E.: Strengthening alumina by glazing and quenching. Amer.Ceram.Soc.Bull. 47(1968)798-802

3.347 Marshall, D.B.; Lawn, B.R.; Kirchner, H.P.; Gruver, R.M.: Contact-induced strength degradation of thermally treated Al_2O_3. J.Amer.Ceram.Soc. 61(1978) 271-272

3.348 Kirchner, H.P.; Gruver, R.M.: The elevated temperature flexural strength and impact resistance of alumina ceramics strengthened by quenching. Mater. Sci.Engng. 13(1974)63-69

3.349 Kirchner, H.P.; Gruver, R.M.; Walker, R.E.: Strengthening of hot-pressed Al_2O_3 by quenching. J.Amer.Ceram.Soc. 56(1973)17-21

3.350 Buessem, W.R.; Gruver, R.M.: Computation of residual stresses in quenched Al_2O_3. J.Amer.Ceram.Soc. 55(1972)101-104

3.351 Kirchner, H.P.; Gruver, R.M.: Fracture mirrors in alumina ceramics. Phil. Mag. 27(1973)1433-1446

3.352 Stolz, R.; Varner, J.R.: Festigkeit thermisch gehärteter Aluminiumoxid-Keramiken. Ber.Dt.Keram.Ges. 54(1977)396-398

3.353 Kirchner, H.P.; Gruver, R.M.; Walker, R.E.: Chemical strengthening of polycrystalline alumina. J.Amer.Ceram.Soc. 51(1968)251-255

3.354 Frasier, J.T.; Jones, J.T.; Raghavan, K.S.; McGee, T.D.; Bell, H.: Chemical strengthening of Al_2O_3. Amer.Ceram.Soc.Bull. 50(1971)541-544

3.355 Rahman, S.F.; Jones, J.T.: Strengthening of sapphire rods by BaO. J.Amer. Ceram.Soc. 56(1973)602

3.356 Platts, D.R.; Kirchner, H.P.; Gruver, R.M.; Walker, R.E.: Strengthening of glazed alumina by ion exchange. J.Amer.Ceram.Soc. 53(1970)281

3.357 Garvie, R.C.; Hannink, R.H.; Pascoe, R.T.: Ceramic steel? Nature 258(1975) 703-704

3.358 Claussen, N.: Fracture toughness of Al_2O_3 with an unstabilized ZrO_2 dispersed phase. J.Amer.Ceram.Soc. 59(1976)49-51

3.359 Porter, D.L.; Heuer, A.H.: Mechanisms of toughening partially stabilized zirconia (PSZ). J.Amer.Ceram.Soc. 60(1977)183-184

3.360 Claussen, N.; Cox, R.L.; Wallace, J.S.: Slow growth of microcracks: evidence for one type of ZrO_2 toughening. J.Amer.Ceram.Soc. 65(1982)C190-191

3.361 Heuer, A.H.: Alloy design in partially stabilized zirconia. In Heuer, A.H.; Hobbs, L.W. (Eds.): Science and technology of zirconia. Advances in Ceramics, Vol. 3. Columbus (U.S.A.): The American Ceramic Society 1981, pp. 98-115

3.362 Lange, F.F.: Transformation toughening: thermodynamic approach to phase retention and toughening. In Bradt, R.C.; Evans, A.G.; Hasselman, D.P.H.; Lange, F.F. (Eds.): Fracture mechanics of ceramics, Vol. 6. New York and London: Plenum Press 1983, pp. 255-274

3.363 Evans, A.G.; Heuer, A.H.: Transformation toughening in ceramics: martensitic transformations in crack-tip stress fields. J.Amer.Ceram.Soc. 63(1980) 241-248

3.364 Claussen, N.; Rühle, M.: Design of transformation-toughened ceramics. In Heuer, A.H.; Hobbs, L.W. (Eds.): Science and technology of zirconia. Advances in Ceramics, Vol. 3. Columbus (U.S.A.): The American Ceramic Society 1981, pp. 137-163

3.365 McMeeking, R.M.; Evans, A.G.: Mechanics of transformation toughening in brittle materials. J.Amer.Ceram.Soc. 65(1982)242-246

3.366 Dworak, U.; Olapinski, H.; Thamerus, G.: Mechanical strengthening of alumina and zirconia ceramics through the introduction of secondary phases. Science of Ceramics 9(1977)543-550

3.367 Pilyankevich, A.N.; Claussen, N.: Toughening of BN by stress-induced phase transformation. Mater.Res.Bull. 13(1978)413-417

3.368 Claussen, N.: Mechanical properties of sintered and hot-pressed Si_3N_4-ZrO_2 composites. J.Amer.Ceram.Soc. 61(1978)94-95

3.369 Fagherazzi, G.; Enzo, S.; Gohardi, V.; Scarinci, G.: A structural study of metastable tetragonal zirconia in an Al_2O_3-ZrO_2-Na_2O glass ceramic system. J.Mater.Sci. 15(1980)2693-2700

3.370 Lange, F.F.; Davies, B.I.; Raleigh, D.O.: Transformation strengthening of ß"-Al_2O_3 with tetragonal ZrO_2. J.Amer.Ceram.Soc. 66(1983)C50-52

3.371 Prochazka, S.; Wallace, J.S.; Claussen, N.: Microstructure of sintered mullite-zirconia composites. J.Amer.Ceram.Soc. 66(1983)C125-127

3.372 Evans, A.G.; Burlingame, N.; Drory, M.; Kriven, W.M.: Martensitic transformations in zirconia - particle size effects and toughening. Acta Met. 29 (1981)447-456

3.373 Evans, A.G.; Marshall, D.B.; Burlingame, N.H.: Transformation toughening in ceramics. In Heuer, A.H.; Hobbs, L.W. (Eds.): Science and technology of zirconia. Advances in Ceramics, Vol. 3. Columbus (U.S.A.): The American Ceramic Society 1981, pp. 202-216

3.374 Claussen, N.; Steeb, J.; Pabst, R.F.: Effect of induced microcracking on the fracture toughness of ceramics. Amer.Ceram.Soc.Bull 56(1977)559-562

3.375 Claussen, N.; Jahn, J.: Umwandlungsverhalten von ZrO_2-Teilchen in einer keramischen Matrix. Ber.Dt.Keram.Ges. 11(1978)487-491

3.376 Claussen, N.: Stress-induced transformation of tetragonal ZrO_2 particles in ceramic matrices. J.Amer.Ceram.Soc. 61(1978)85-86

3.377 Kosmac, T.; Wallace, J.S.; Claussen, N.: Influence of MgO additions on the microstructure and mechanical properties of Al_2O_3-ZrO_2 composites. J.Amer. Ceram.Soc. 65(1982)C66-67

3.378 Becher, P.F.; Tennery, V.J.: Fracture toughness of Al_2O_3-ZrO_2 composites. In Bradt, R.C.; Evans, A.G.; Hasselman, D.P.H.; Lange, F.F. (Eds.): Fracture mechanics of ceramics, Vol. 6. New York and London: Plenum Press 1983, pp. 383-399

3.379 Green, D.J.: Critical microstructures for microcracking in Al_2O_3-ZrO_2 composites. J.Amer.Ceram.Soc. 65(1982)610-614

3.380 Heuer, A.H.; Claussen, N.; Kriven, W.M.; Rühle, M.: Stability of tetragonal particles in ceramic matrices. J.Amer.Ceram.Soc. 65(1982)642-650

4.1 D.P. 43977 from 1887

4.2 D.P. 65604 from 1892

4.3 Gitzen, W.H. (Ed.): Alumina as a ceramic material. Columbus, Ohio: The American Ceramic Society 1970

4.4 Fulda, N.; Ginsberg, H.: Tonerde und Aluminium. Berlin: Walter de Gruyter 1951

4.5 Kohl, H.: Sinterkorund, ein neues keramisches Material aus reinem Aluminiumoxyd. Ber.Dtsch.Keram.Ges. 13(1932)70-85

4.6 Rumpf, H.; Behrens, D.: Symposion Zerkleinern. Weinheim: Verlag Chemie 1962

4.7 Brunauer, S.; Emmett, P.H.; Teller, E.: Adsorption of gases in multimolecular layers. J.Amer.Ceram.Soc. 21(1938)309-319

4.8 Kingery, W.D. (Ed.): Ceramic fabrication processes. The Technology Press of Massachusetts Institute of Technology 1958, p 57

4.9 Suzuki, A.; Tanaka, T.: Measurement of flow properties of powders along an inclined plane. Ind.Eng.Chem.Fundam. 10(1971)84-91

4.10 Hennicke, H.W.; Neuenfeld, K.: Spritzgußtechnik als Formgebungsverfahren in der Keramik. Ber.Dtsch.Keram.Ges. 45(1968)469-473

4.11 Dawihl W.; Dörre, E.: Adsorption reactions on the surface of powders, sintered metals, and oxides. Powder Met.Int. 1(1969)1-5

4.12 McCreight, L.R.: A laboratory hydrostatic pressing chamber. Amer.Ceram.Soc. Bull. 30(1951)127-129

4.13 Wagner, H.E.; Harmon, C.G.: Hydrostatic pressing as fabrication technique. Amer.Ceram.Soc.Bull. 30(1951)341-344

4.14 Navias, L.: Extrusion of refractory oxide insulators for vacuum tubes. J. Amer.Ceram.Soc. 15(1932)234-251

4.15 D.P. 680250 from 1936

4.16 Dawihl, W.; Hinnüber, J.: Über den Aufbau der Hartmetallegierungen. Kolloid-Z.104(1943)233-236

4.17 Goetzel, C.G.: The physics of powder metallurgy. New York, London: McGraw-Hill Book Co. 1950,pp.256-276

4.18 Rice, R.W.: Effect of gaseous impurties on the hot-pressing and behavior of MgO, CaO and Al_2O_3. Proc.Brit.Ceram.Soc. 12(1969)99-123

4.19 US.P. 3,732056 from 1973

4.20 Rankin, D.T.; Stiglich, J.J.; Petrak, D.R.; Ruh, R.: Hot-pressing and mechanical properties of Al_2O_3 with an Mo-dispersed phase. J.Amer.Ceram.Soc. 54(1971)277-281

4.21 Dawihl, W.; Dörre, E.; Dworak, U.; Stannek, W.; Sturhahn, H.: Oxidkeramische Werkstoffe im allgemeinen Maschinenbau. Forschungskuratorium Maschinenbau, H. 37(1975)

4.22 Rossi, R.C.; Fulrath, R.H.: Final stage densification in vacuum hot-pressing of alumina. J.Amer.Ceram.Soc. 48(1965)558-564

4.23 Budworth, D.W.; Roberts, E.W.; Scott, W.D.: Joining of alumina components by hot-pressing. Trans.Brit.Ceram.Soc. 62(1963)949-954

4.24 Haerdtl, K.H.: Gas isostatic hot-pressing without molds. Amer.Ceram.Soc. Bull. 54(1975)201-205,207

4.25 Hedwall, J.A.: Sintering and reactivity of solids. Ceram.Age 65(1955)13-17

4.26 Coble, R.L.: Sintering Crystalline solids: I, Intermediate and final state diffusion models. J.Appl.Phys. 32(1961)787-792

4.27 Cahoon, H.P.; Christensen, C.J.: Sintering and grain growth of alpha-alumina. J.Amer.Ceram.Soc. 39(1956)337-344

4.28 Sinharoy, S.; Levenson, L.L.; Ballard, W.V.; Day, D.E.: Surface segregation of calcium in dense alumina exposed to steam and steam-CO. Amer.Ceram.Soc. Bull. 57(1978)231-233

4.29 Sinharoy, S.; Levenson, L.L.; Day, D.E.: Influence of calcium migration on the strength reduction of dense alumina exposed to steam. Amer.Ceram.Soc. Bull. 58(1979)464-466

4.30 Held, K.; Reckziegel, A.: Oxidkeramik. Ullmanns Enzyklopädie d. techn. Chemie 17. Weinheim: Verlag Chemie 1979,S.515-529

5.1 Dawihl, W.; Klingler, E.: Reines Aluminiumoxid als Sinterwerkstoff. VDI-Nachr. 9(1958)1-4

5.2 Fereira, L.E.; Briggs, D.D.; Barnhart, R.G.: Engineering with high-alumina ceramics. Metal Progr. 98(1970)78-82

5.3 Klingler, E.; Dörre, E.: Neue Anwendungsgebiete für reines Aluminiumoxid in der Elektrotechnik und Elektronik. Ber.Dtsch.Keram.Ges. 44(1967)498-505

5.4 Cappelletti, C.M.; Busse, C.A.; Dörre, E.: Metal-to-ceramic seals for thermionic converters. Proc. 2nd Int. Conf. on Thermionic El. Power Generation (1968)613-632

5.5 Heldt, K.; Haase, G.: Der elektrische Widerstand von reinem hochvakuumgesintertem Aluminiumoxyd. Z.angew.Phys. 6(1954)157-160

5.6 Pincus, A.G.: Mechanism of ceramic-to-metal adherence. Ceramic Age 70(1954) 16-32

5.7 Meyer, A.: Zum Haftmechanismus von Molybdän/Mangan-Metallisierungsschichten auf Korund-Keramik. Ber.Dtsch.Keram.Ges. 42(1965)405-415, 452-454

5.8 Klomp, J.T.; Botden, Th.P.J.: Sealing pure alumina ceramics to metals. Amer. Ceram.Soc.Bull. 49(1970)204-211

5.9 Arthur, M.E.; Fusell, L.E.: Effect of sintering conditions on Al_2O_3-MoMn bond strengths. Amer.Ceram.Soc.Bull. 50(1971)982-984

5.10 Dörre, E.; Cappelletti, C.M.: Ceramic-to-metal sealing processes for high operating temperatures. DVS-Ber. 32(1974)99-106

5.11 Dawihl, W.; Klingler, E.: Mechanische und thermische Eigenschaften von Schweißverbindungen des Al_2O_3 mit Metallen. Ber.Dtsch.Keram.Ges. 46(1969) 12-18

5.12 Klomp, J.T.: Solid-state bonding of metals to ceramics. Science of Ceramics 5(1969)501-522

5.13 Dörre, E.; Ziegele, W.: Keramik-Metall-Verbindungen auf Al_2O_3-Basis für hohe Betriebstemperaturen. Ber.Dtsch.Keram.Ges. 47(1970)622-626

5.14 Elßner, G.; Pabst, R.; Puhr-Westerheide, J.: Schichtverbundkombinationen aus hochschmelzenden Metallen und Oxiden. Z.Werkstofft. 5(1974)61-69

5.15 Blume, H.; Hinckeldey, A.; Keller, J.: Vergleich der Keramik- und Glasausführung der Scheibentriode 2 C 39. Siemens Rev. 25(1958)498-504

5.16 Heimke, G.: Oxidkeramische Gehäuse. Keram.Z. 24(1972)74-77

5.17 Syunry, T.S.: Design and development of ceramic-metal seals for vacuum and ultrahigh-vacuum work. Res.Ind. 22(1977)164-168

5.18 Droscha, H.: Particle accelerator vacuum chambers in aluminum oxide ceramics. Kerntechnik 12(1970)477-479

5.19 Coble, R.L.: Sintering alumina: Effect of atmospheres. J.Amer.Ceram.Soc. 45 (1962)123-127

5.20 Dörre, E.: Keramische Werkstoffe auf Aluminiumoxid-Basis und ihre Anwendungen. VDI-Ber. 174(1971)19-28

5.21 US.P. 3 311 482 from 1971

5.22 Peelen, J.G.J.: Transparent hot-pressed Alumina. Ceramurgia Int. 5(1979) 115-119

5.23 Hill, G.J.; Strudwick, P.; Banett, M.J.: Alumina substrates with very good surface finish. Proc.Brit.Ceram.Soc. 18(1970)263-270

5.24 Shanefield, D.J.; Mistler, R.E.: Manufacture of fine-grained alumina substrates for thin films. West. Elec. Eng. 15(1971)26-31

5.25 Class, W.: Supersmooth substrates. Res/Develop. 22(1971)36-39

5.26 Shanefield, D.J.; Mistler, R.E.: Fine grained alumina substrates: I, Manufacturing process. Amer.Ceram.Soc.Bull. 53(1974)416-420

5.27 Gardner, R.A.; Nufer, R.W.: Properties of multilayer ceramic green sheets. Solid State Technol. 17(1974)38-43

5.28 US.P. 3 999 004 from 1976

5.29 Dawihl, W.; Dörre, E.: Über Festigkeits- und Verformungseigenschaften von Sinterkörpern aus Aluminiumoxid in Abhängigkeit von Zusammensetzung und Gefüge. Ber.Dtsch.Keram.Ges. 41(1964)85-96

5.30 Dawihl, W.; Klingler, E.: Über das Kriechverhalten von Sinterkörpern aus Aluminiumoxid im Temperaturbereich von 1200° C unter dem Einfluß von Druckspannungen. Ber.Dtsch.Keram.Ges. 42(1965)270-274

5.31 Dawihl, W.; Klingler, E.: Der Korrosionswiderstand von Aluminiumoxideinkristallen und von gesinterten Werkstoffen auf Aluminiumoxidgrundlage gegen anorganische Säuren. Ber.Dtsch.Keram.Ges. 44(1967)1-4

5.32 Dawihl, W.; Klingler, E.: Verwendbarkeit von gesintertem Aluminiumoxid als verschleißbeständiger und warmfester Werkstoff. Z.Stahl u.Eisen 87(1967) 273-280

5.33 Dawihl, W.; Klingler, E.: Eigenschaftsvergleiche von Hartmetallen und Oxidwerkstoffen als Grundlage ihrer Anwendungsbereiche und der Erfahrungsstand im technischen Einsatz von Aluminiumoxidsinterwerkstoffen in verschiedenen Industriegebieten. Techn.Mitt. 58(1965)610-625

5.34 Dörre, E.; Lipp, A.; Rauschert, D.; Schüller, K.-H.; Vogel, H.: Verschleißbeanspruchte dichte keramische Werkstoffe im Maschinenbau. Ber.Dtsch.Keram. Ges. 50(1973)115-121

5.35 Dawihl, W.; Klingler, E.: Anwendungsbereiche von Aluminiumoxiderzeugnissen als verschleiß- und korrosionsfeste Werkstoffe und ihre Verschleißgleichungen. Schweiz.Arch.angew.Wiss.u.Techn. 33(1967)257-268

5.36 Dörre, E.: Oxidkeramische Werkstoffe - ihre Eigenschaften und Anwendungen unter besonderer Berücksichtigung des Verschleißverhaltens. VDI-Ber. 194 (1973)121-129

5.37 Dawihl, W.; Klingler, E.; Dörre, E.: Zusammenhänge zwischen Verschleißwiderstand und Eigenschaften von Sinterkörpern auf Al_2O_3-Grundlage. Ber.Dtsch. Keram.Ges. 46(1969)409-415

5.38 Dawihl, W.; Dörre, E.: Adsorption behaviour of high-density alumina ceramics exposed to fluids. Evaluation of Biomaterials. Chichester: John Wiley & Sons. 1980, pp.239-245

5.39 Wegener, W.; Jammers, H.-Ch.; Veltmann, E.: Falschdrahtverfahren und -maschinen. Textilind. 67(1967)439-446,520-530,619-627

5.40 Dworak, U.; Weinz, E.A.: Biegebruchfestigkeit und Oberflächen von Drallgebern für das Falschzwirnen. Melliand Text.Ber. 54(1973)578-583

5.41 Dörre, E.: Step cone performance better with high purity alumina. Wire J. 4 (1971)52-57

5.42 Stannek, W.: Design features of drawing equipment using ceramic oxide components. Wire 45(1978)933-935

5.43 Dawihl, W.; Dörre, E.; Dworak, U.; Stannek, W.; Sturhahn, H.: Oxidkeramische Werkstoffe im allgemeinen Maschinenbau. Forschungshefte Forschungskuratorium Maschinenbau e.V. Heft 37(1975)

5.44 Sturhahn, H; Dawihl, W.; Thamerus, G.: Anwendungsmöglichkeiten und Werkstoffeigenschaften von ZrO_2-Sintererzeugnissen. Ber.Dtsch.Keram.Ges. 52(1975)59-62

5.45 Taylor, G.I.: Drainage at a table roll and a foil. Pulp and Paper Mag. of Canada 59(1958)172-176

5.46 Fuchs, K.D.: Oxide ceramic hard covers on modern fourdriniers. Allg. Papier-Rdsch.(1970)14-16,149-150,250-251

5.47 Fuchs, K.D. Sieb, Belag und Füllstoffe - Bedeutung und Einfluß auf den Verschleißmechanismus in der Siebpartie. Wochenblatt f. Papierfabr. 103(1975) 348-352

5.48 Petruschke, K.: Erste Erfahrungen mit Foils in schnellaufenden Papiermaschinen. Wochenblatt f. Papierfabr. 94(1966)557-562

5.49 US.P. 3,711 171 from 1973

5.50 Mayer, E.: Axiale Gleitringdichtungen. Düsseldorf: VDI 1965

5.51 King, A.G.; Wheildon, W.M.: Ceramics in machining processes. New York, London: Academic Press 1966

5.52 Kim, C.H.; Koper, W.; Hasselman, D.P.H.; Kane, G.E.: Machining performance of sintered vs hot-pressed ceramic cutting tools. Amer.Ceram.Soc.Bull. 54 (1975)589-590

5.53 Dworak, U.; Olapinski, H.; Thamerus, G.: Festigkeitssteigerung von mehrphasigen keramischen Werkstoffen am Beispiel der Systeme ZrO_2-ZrO_2/Al_2O_3-ZrO_2/Al_2O_3-TiC. Ber.Dtsch.Keram.Ges. 55(1978)98-101

5.54 Whitney, E.D.: New advances in ceramic tooling in the USA. Powder Met.Int. 10(1978)18-21

5.55 Dawihl, W.; Dörre, E.; Dworak, U.: Application of ceramic cutting tools in machining steel and cast iron. Powder Met.Int. 3(1971)189-192

5.56 Dörre, E.: Stand der Entwicklung keramischer Schneidwerkzeuge. Techn.Mitt. 64(1971)7-9

5.57 Dawihl, W.; Klingler, E.: Standzeit und Wirtschaftlichkeit oxidkeramischer Drehwerkzeuge. VDI-Z. 100(1958)559-563

5.58 Dawihl, W.; Klingler, E.: Fortschritte beim Drehen mit Aluminiumoxidschneidplatten. Werkst. u. Betr. 92(1959)271-274

5.59   Dawihl, W.; Altmeyer, G.; Sutter, H.: Über Schnittemperatur- und Schnitt-
       kraftmessungen beim Drehen mit Hartmetall- und Aluminiumoxidwerkzeugen.
       Werkst. u. Betr. 98(1965)691-697

5.60   Dawihl, W.; Dörre, E.: Gleit- und Drehverschleiß von Aluminiumoxidsinter-
       körpern beim Gleiten an oder Zerspanen von Stahl. Arch. Eisenhüttenwesen
       41(1970)763-767

5.61   Lenz, E.: Der Einfluß der Schnittemperatur auf die Standzeit der kerami-
       schen Schneidstoffe. Maschinenmarkt 69(1963)1-12

5.62   Dworak, U.; Gomoll, V.: Wirtschaftlicher Einsatz von Schneidkeramik. ZwF
       71(1976)421-425

5.63   Gomoll, V.: Keramische Schneidstoffe - Stand der Technik und Ausblicke.
       VDI-Z. 122(1980)160-174

5.64   Dörre, E.; Dawihl, W.: Mechanische und tribologische Eigenschaften kera-
       mischer Endoprothesen. Biomed. Techn. 23(1978)305-310

5.65   Dörre, E.: Aluminiumoxidkeramik als Implantatwerkstoff. Med.-Orthop.Techn.
       96(1976)104-105

5.66   Zweymüller, K.: Knochen- und Gelenkersatz mit biokeramischen Endoprothe-
       sen. Wien: Facultas 1978

5.67   Dörre, E.; Beutler, H.; Geduldig, D.: Anforderungen an oxidkeramische
       Werkstoffe als Biomaterial für künstliche Gelenke. Arch. orthop. Unfall-
       Chir. 83(1975)269-278

5.68   Griss, P.; Krempien, B.; v. Andrian-Werburg, H.; Heimke, G.; Fleiner, R.:
       Experimentelle Untersuchung zur Gewebeverträglichkeit oxidkeramischer
       (Al$_2$O$_3$)-Abriebteilchen. Arch.orthop. Unfall-Chir. 76(1973)273-279

5.69   Harms, J.; Mäusle, E.: Biologische Verträglichkeitsuntersuchungen von
       Implantatwerkstoffen im Tierversuch. Med.-Orthop. Techn. 96(1976)103-104

5.70   Harms, J.; Mäusle, E.: Tissue reaction to ceramic implant materials. J.Bio-
       med. Mat.Res. 13(1979)67-87

5.71   Dörre, E.; Geduldig, D.; Happel, M.; Lade, R.; Prüssner, P.; Willert, H.-G.;
       Zichner, L.: Animal studies on bone ingrowth kinetics of ceramic material
       under dynamic stress. J. Biomed. Mat. Res. Symp. 7(1976)493-502

5.72   DIN 58835 and ISO 6474

5.73   Dawihl, W.; Altmeyer, G.; Dörre, E.: Statische und dynamische Dauerfestig-
       keit von Aluminiumoxid-Sinterkörpern. Z. Werkstofft. 8(1977)328-330

5.74   Ungethüm, M.: Tribologische Aspekte beim totalen Gelenkersatz der Hüfte.
       Z. Orthop. 112(1974)168-176

5.75   Semlitsch, M.; Lehmann, M.; Weber, H.; Dörre, E.; Willert, H.-G.: Neue Per-
       spektiven zu verlängerter Funktionsdauer künstlicher Hüftgelenke durch Werk-
       stoffkombinationen Polyäthylen-Aluminiumoxidkeramik-Metall. Med.-Orthop.
       Techn. 96(1976)152-157

5.76   Dawihl, W.; Mittelmeier, H.; Dörre, E.; Altmeyer, G.; Hanser, U.: Zur Tri-
       bologie von Hüftgelenk-Endoprothesen aus Aluminiumoxidkeramik. Med.-Orthop.
       Techn. 99(1979)114-118

5.77   Boutin, P.: Arthroplastic totale de la hanche par prothese en alumine frittée.
       Rev. Chir. Orthop. 58(1972)229-246

5.78   Mittelmeier, H.: Zementlose Verankerung von Endoprothesen nach dem Tragrip-
       penprinzip. Z. Orthop. 112(1974)27-30

5.79   Mittelmeier, H.: Selbsthaftende Keramik-Metall-Verbundprothesen. Med.-Orthop.
       Techn. 95(1975)152-159

5.80 Boutin, P.: Les prothéses totales de la hanche en alumine. L'ancrage direct sans ciment dans 50 cas. Rev. Chir. Orthop. 60(1974)233-235

5.81 Griss, P.; Heimke, G.; Krempien, B.; Silber, R.; Haehner, K.: Erste Erfahrungen mit der Keramik-Metall-Verbundprothese. Med.-Orthop. Techn. 96(1975)159-162

5.82 US.P. 4 012 795 from 1977

5.83a Dörre, E.; Dawihl, W.; Krohn, U.; Altmeyer, G.; Semlitsch, M.: Do ceramic components of hip joints maintain their strength in human bodies? In Vincenzini, P. (Ed): Ceramic in surgery. Amsterdam: Elsevier Scient. Publ. Co. 1983, pp. 61-72

5.83 Dörre, E.; Dawihl, W.; Altmeyer, G.: Dauerfestigkeit keramischer Hüftendoprothesen. Biomed. Techn. 22(1977)3-7

5.84 Geduldig, D.; Dörre, E.; Happel, M.; Lade, R.; Prüssner, P.; Willert, H.-G.; Zichner, L.: Welche Aussicht hat die Biokeramik als Implantatmaterial in der Orthopädie? Med.-Orthop. Techn. 95(1975)138-143

5.85 Mittelmeier, H.; Harms, J.: Die Anwendung von Keramik in der Gelenkersatz-Chirurgie. Med.-Orthop. Techn. 97(1977)55-57

5.86 Mutschelknauss, E.; Dörre, E.: Extensions-Implantate aus Aluminiumoxidkeramik. Quintessenz 28(1977)1-10

5.87 Brinkmann, E.: Das Keramik-Anker-Implantat nach E. Mutschelknauss. Zahnärztl. Prax. 29(1978)148-151

5.88 Sandhaus, S.: Nouveaux aspects de l'implantologie. L-implant CBS. Stuttgart: Medica 1969

5.89 Mutschelknauss, E.: Enossale Implantation von Porzellankörpern. Quintessenz 21(1970)19-22

5.90 Mutschelknauss, E.; Dörre, E.: Enossale Stiftimplantate aus Aluminiumoxidkeramik. Zahnärztl. Prax. 29(1978)362-369

5.91 Schulte, W.; Heimke, G.: Das Tübinger Sofortimplantat. Quintessenz 27(1976) 17-20

5.92 Brinkmann, E.: Enossale Implantate aus Aluminiumoxidkeramik. Zahnärztl. Prax. 31(1980)328-329

5.93 Kirschner, H.; Bolz, U.; Enomoto, S.; Hüttemann, R.W.; Meinel, W.; Sturm, J.: Eine neue Methode kombinierter autoalloplastischer Zahnreplantation mit partieller Al_2O_3-Keramikwurzel. Dtsch. Zahnärztl. Z. 33(1978)594-598

5.94 Frenkel, G.; Nowak, K.; Schulz-Freywald, G.; Bertram, J.J.; Gruh, W.; Dörre, E.: Untersuchungen mit nichtmetallischen Werkstoffen in der Zahn-, Mund- und Kieferchirurgie. Dtsch. Zahnärztl. Z. 32(1977)295-297

5.95 Frechette, V.D.; Cline, C.F.: Fractography of ballistically tested ceramics. Amer.Ceram.Soc.Bull. 49(1979)994-997

5.96 Long, W.D.; Brantley, W.A.: Fractographic analyses of ceramics and glass-ceramics subjected to ballistic impact. Amer.Ceram.Soc.Bull. 50(1971)662-665

5.97 Droscha, H.: Schußsichere Westen mit Keramik-Metall-Verbundplatten bieten mehr Schutz und sind leichter. Deutsche Polizei(1974)150-152

5.98 Trinks, W.; Geiger, W.; Kollmannsperger, H.: Grenzen der Schutzwirkung von Panzerwerkstoffen gegen Hohlladungen. Jahrb.d.Wehrtechn. 6(1971)46-50

5.99 D.P. 220394 from 1907

5.100 Dawihl, W.: Untersuchungen im Gebiete hoher und höchster Temperaturen. Ton-ind. Z. 58(1934)3-25

5.101 Ryshkewitch, E.: Oxide Ceramics - physical chemistry and technology. New York and London: Academic Press 1960

5.102 Mahan, A.I.; Bitterli, C.; Cannon, S.M.; Grant, D.G.: Ruby as a macroscopic fluorescing and laser material. J.Opt.Soc.Amer. 59(1969)49-59

5.103 Seifert, W.: Growth of ruby for lasers by an improved verneuil process. J. Cryst. Growth 12(1972)17-20

5.104 US.P. 3 726 811 from 1973

5.105 US.P. 3 799 796 from 1974

5.106 US.P. 3 804 781 from 1974

5.107 US.P. 3 856 708 from 1974

5.108 Brit.P. 1 493 909 from 1977

Subject Index

Diamond tools 194,209,213

Dielectric losses 220,225

Dielectric strength 218

Diffusion

    activation energy 34,35

    ambipolar 31,155,156

    coefficient 30

    extrinsic 26-29

    grain-boundary 34,36,165

    impurity 40,41

    intrinsic 19,23,27,29

    lattice 32-34,164

    surface 38

    welding 257

Diffusional crack growth 180

Dihedral angle 12

Dimensional

    deviations 212

    inspection 212

    stability 234,237,251,252

Discharge lamps 220,229-232

Dislocation reactions 136

Dislocations 135,136

Disorder

    extrinsic 27

    Frenkel 20-22,24,25,28,57

    intrinsic 20,27

    Schottky 20-22,24,25,56

Dispersion strengthening 181-183

Diverting step cone 241

Donor doping 47

Drawing step cone 241

Drying process 202

Dry pressing 193,197-199

Elastic anisotropy 75

Elastic constants 76-79

    porosity dependence 78,79,101

Electrical conductivity 41-46

    activation energy of 45

    pressure dependence 44-46

    temperature dependence 44

Electrical properties 218

Electronic applications 220

Electronic circuits 232

Electron tubes 224-226

Electrostatic charges 241

Energy absorption 281

Energy consumption 207

Equilibrium cavities 178,179

Extrusion 193,197,201

Fabrication 193-215

False-twist-spinning 238-240

Fatigue

    cyclic 122

    dynamic 110,112,116-118,267,274

    impact 130,268

    static 110,112,117,118,120,275

Feed rate 260,265

Feedthroughs 226

Felt 249

Felt suction boxes 249

Femoral head 271

Filler 245

Final density 197,198,205,213

Fine grinding 209

Fixation methods 246

Flank wear 258

Flat suction boxes 248

Flaws

    extrinsic 96,98,99,108

    intrinsic 97-99,108

Flow stress 138,169,170

Foils 246-248

Foreign oxides 206

Forming 197

    boards 247

    pressure 197,198

Fourdrinier section 245,246

Powder metallurgical process 255
Precompaction 193,208
Precompression 193
Preparation of powders 194
Pressure distribution 200
Pressure sintering 204
Principles of design 213
Production volume distribution 7
Projectile 279
Protective layer 235
Pump parts 249-253
Pumps 249,252
Purity control 212

Quality control 210,211

R curve 91
Recrystallization 205
Rectifier tubes 226
Requirements 217
Residual energy 281
Residual porosity 62,205,218
Rheological requirements 197
Ribbons 202
Ringer's solution 121,268-270
Rough grinding 209
Ruby 283

Sapphire 1,283
Seal discs 253
Seal ring packing 250
Seal rings 249-252
Segregation 69-73
   enrichment factors 72
Shaped charges 279,282
Sheet forming operation 245
Shock wave 279
   propagation 279
Shrinkage 51,199,201,208,213
   rate 208

Silicate additions 254
   binder 216
   phase 218,224
Silicates 217,254
Silicon
   carbide 2
   nitride 2
   rectifiers 227
Sintering 49-63,194,197,205-208
   activation energy 52
   atmosphere 62,207,229
   cycle 207
   final stage 53,62,68
   initial 50,51
   intermediate 53,68
   kinetics 53-55,57
   mechanisms 50,52,59,61,65
   temperature 207
   time 207
Skimming edge 245
Sleeve bearings 252
Sliding surfaces 250,252
Slip
   casting 193
   systems 134,135,137,139
Smoothness 209
Socket 271
Sodium vapor 231
Solid-state bonding 223
Solubility of gases 229
Solution hardening 142
Spark-plug insulators 223,224
Specific
   load 270
   surface 196
Sphericity 269,270
Spinel 221
Spray drier 197
Stacking sequence 10
Standards 266,267,275